The CRC Press Series on

DISCRETE MATHEMATICS

AND

ITS APPLICATIONS

Series Editor

Kenneth H. Rosen, Ph.D.

AT&T Bell Laboratories

Charles J. Colbourn and Jeffrey H. Dinitz,
The CRC Handbook of Combinatorial Designs

Steven Furino, Ying Miao, and Jianxing Yin,
Frames and Resolvable Designs: Uses, Constructions,
and Existence

Jacob E. Goodman and Joseph O'Rourke,
Handbook of Discrete and Computational Geometry

Charles C. Lindner and Christopher A. Rodgers
Design Theory

*Daryl D. Harms, Miroslav Kraetzl, Charles J. Colbourn,
and John S. Devitt,*
Network Reliability: Experiments with A Symbolic
Algebra Environment

*Alfred J. Menezes, Paul C. van Oorschot,
and Scott A. Vanstone,*
Handbook of Applied Cryptography

Richard A. Mollin, Quadratics

Richard A. Mollin, Fundamental Number Theory
with Applications

Douglas R. Stinson, Cryptography: Theory and Practice

COMBINATORIAL ALGORITHMS

Generation, Enumeration, and Search

Donald L. Kreher

Department of Mathematical Sciences
Michigan Technological University

Douglas R. Stinson

Department of Combinatorics and Optimization
University of Waterloo

CRC Press
Taylor & Francis Group
Boca Raton London New York

CRC Press is an imprint of the
Taylor & Francis Group, an **informa** business

CRC Press
Taylor & Francis Group
6000 Broken Sound Parkway NW, Suite 300
Boca Raton, FL 33487-2742

First issued in paperback 2019

ISBN-13: 978-0-8493-3988-2 (hbk)
ISBN-13: 978-0-367-40015-6 (pbk)
Library of Congress Card Number 98-41243

Library of Congress Cataloging-in-Publication Data

Kreher, Donald L.
    Combinatorial algorithms : generation, enumeration, and search /
Donald L. Kreher, Douglas R. Stinson.
        p. cm. — (CRC Press series on discrete mathematics and its
applications)
    Includes bibliographical references and index.
    ISBN 0-8493-3988-X (alk. paper)
    1. Combinatorial analysis. 2. Algorithms. I. Stinson, Douglas
R. (Douglas Robert), 1956-   . II. Title. III. Series.
QA164.K73 1998
511′.6—dc21   98-41243
                                                           CIP

Visit the Taylor & Francis Web site at
http://www.taylorandfrancis.com

and the CRC Press Web site at
http://www.crcpress.com

Preface

Our objective in writing this book was to produce a general, introductory textbook on the subject of combinatorial algorithms. Several textbooks on combinatorial algorithms were written in the 1970s, and are now out-of-date. More recent books on algorithms have either been general textbooks, or books on specialized topics, such as graph algorithms to name one example. We felt that a new textbook on combinatorial algorithms, that emphasizes the basic techniques of generation, enumeration and search, would be very timely.

We have both taught courses on this subject, to undergraduate and graduate students in mathematics and computer science, at Michigan Technological University and the University of Nebraska-Lincoln. We tried to design the book to be flexible enough to be useful in a wide variety of approaches to the subject.

We have provided a reasonable amount of mathematical background where it is needed, since an understanding of the algorithms is not possible without an understanding of the underlying mathematics. We give informal descriptions of the many algorithms in this book, along with more precise pseudo-code that can easily be converted to working programs. C implementations of all the algorithms are available for free downloading from the website

$$\texttt{http://www.math.mtu.edu/~kreher/cages.html}$$

There are also many examples in the book to illustrate the workings of the algorithms.

The book is organized into eight chapters. Chapter 1 provides some background and notation for fundamental concepts that are used throughout the book. Chapters 2 and 3 are concerned with the generation of elementary combinatorial objects such as subsets and permutations, to name two examples. Chapter 4 presents the important combinatorial search technique called backtracking. It includes a discussion of pruning methods, and the maximum clique problem is studied in detail. Chapter 5 gives an overview of the relatively new area of heuristic search algorithms, including hill-climbing, simulated annealing, tabu search and genetic algorithms. In Chapter 6, we study several basic algorithms for permutation groups, and how they are applied in solving certain combinatorial enumeration and search problems. Chapter 7 uses techniques from the previous chapter to develop algorithms for testing isomorphism of combinatorial objects. Finally, Chapter 8 discusses the technique of basis reduction, which is an important technique in solving certain combinatorial search problems.

There is probably more material in this book than can be covered in one semester. We hope that it is possible to base several different types of courses on this book. An introductory course suitable for undergraduate students could cover most of the material in Chapters 1–5. A second or graduate course could concentrate on the more advanced material in Chapters 6–8. We hope that, aside from its primary purpose as a textbook, researchers and practitioners in all areas of combinatorial computing will find this book useful as a source of algorithms for practical use.

We would like to thank the many people who provided encouragement while we wrote this book, pointed out typos and errors, and gave us useful suggestions on material to include and how various topics should be treated. In particular, we would like to convey our thanks to Mark Chateauneuf, Charlie Colbourn, Bill Kocay, François Margot, Wendy Myrvold, David Olson, Partic Östergård, Jamie Radcliffe, Stanisław Radziszowski and Mimi Williams.

Donald L. Kreher
Douglas R. Stinson

To our wives, Carol and Janet.

Contents

1

Structures and Algorithms

1.1   What are combinatorial algorithms?

In this book, we are primarily interested in the study of algorithms to investigate combinatorial structures. We will call such algorithms *combinatorial algorithms*, and informally classify them according to their desired purpose, as follows.

generation *Construct all the combinatorial structures of a particular type.*
   Examples of the combinatorial structures we might wish to generate include subsets, permutations, partitions, trees and Catalan families. A generation algorithm will list all the objects under consideration in a certain order, such as a lexicographic order. It may be desirable to predetermine the position of a given object in the generated list without generating the whole list. This leads to the discussion of *ranking*, which is studied in Chapters 2 and 3. The inverse operation of ranking is *unranking* and is also studied in these two chapters.

enumeration *Compute the number of different structures of a particular type.*
   Every generation algorithm is also an enumeration algorithm, since each object can be counted as it is generated. The converse is not true, however. It is often easier to enumerate the number of combinatorial structures of a particular type than it is to actually list them. For example, the number of k-subsets of an n-element set is

$$\binom{n}{k} = \frac{n!}{(n-k)!k!},$$

which is easily computed. On the other hand, listing all of the k-subsets is more difficult.

There are many situations when two objects are different representations of the "same" structure. This is formalized in the idea of isomorphism of structures. For example, if we permute the names of the vertices of a graph, the resulting two graphs are isomorphic. Enumeration of the number of non-isomorphic structures of a given type often involves algorithms for isomorphism testing, which is the main topic studied in Chapter 7. Algorithms for testing isomorphism depend on various group-theoretic algorithms, which are studied in Chapter 6.

search *Find at least one example of a structure of a particular type (if it exists).*
A typical example of a search problem is to find a clique of a specified size in a given graph. Generating algorithms can sometimes be used to search for a particular structure, but for many problems, this may not be an efficient approach. Often, it is easier to find one example of a structure than it is to enumerate or generate all the structures of a specified type.

A variation of a search problem is an optimization problem, where we want to find the optimal structure of a given type. Optimality will be defined for a particular structure according to some specified measure of "profit" or "cost". For example, the maximum clique problem requires finding a clique of largest size in a given graph. (The size of a clique is the number of vertices it contains.)

Many interesting and important search and optimization problems belong to the class of NP-hard problems, for which no efficient (i.e., polynomial-time) algorithms are known to exist. The maximum clique problem mentioned above falls into this class of problems. For NP-hard problems, we will often use algorithms based on the idea of backtracking, which is the topic of Chapter 4. An alternative approach is to try various types of heuristic algorithms. This topic is discussed in Chapters 5 and 8.

1.2 What are combinatorial structures?

The structures we study in this book are those that can be described as collections of k-element subsets, k-tuples, or permutations from a parent set. Given such a structure, we may be interested in all of the substructures contained within it of a particular or optimal type. On the other hand, we may wish to study all structures of a given form. We introduce some of the types of combinatorial structures in this section that will be used in later parts of the book.

1.2.1 Sets and lists

The basic building blocks of combinatorial structures are finite sets and lists. We review some basic terminology and notation now.

A (finite) *set* is a finite collection of objects called the *elements* of the set. We write the elements of a set in brace brackets. For example, if we write $X = \{1, 3, 7, 9\}$, then we mean that X is a set that contains the elements $1, 3, 7$ and 9.

A set is an unordered structure, so $\{1, 3, 7, 9\} = \{7, 1, 9, 3\}$, for example. Also, the elements of a set are distinct. We write $x \in X$ to indicate that x is an element of the set X.

The *cardinality* (or size) of a set X, denoted $|X|$, is the number of elements in X. For example, $|\{1, 3, 7, 9\}| = 4$. For a nonnegative integer k, a *k-set* is a set of cardinality k. The *empty set* is the (unique) set that contains no elements. It is a 0-set and is denoted by $\emptyset$.

If X and Y are sets, then we say that X is a *subset* of Y if every element of X is an element of Y. This is equivalent to the following condition:

$$x \in X \Rightarrow x \in Y.$$

If X is a subset of Y, then we write $X \subseteq Y$. A *k-subset* of Y is a subset of Y that has cardinality k.

A (finite) *list* is an ordered collection of objects which are called the *items* of the list. We write the items of a list in order between square brackets. For example, if we write $X = [1, 3, 1, 9]$, then we mean that X is the list that contains the items $1, 3, 1$ and 9 in that order. Since a set is an ordered structure, it follows that $[1, 3, 1, 9] \neq [1, 1, 9, 3]$, for example. Note that the items of a list need not be distinct.

The *length* of a list X is the number of items (not necessarily distinct) in X. For example, $[1, 3, 1, 9]$ is a list of length 4. For a nonnegative integer n, an *n-tuple* is a list of length n. The *empty list* is the (unique) list that contains no elements; it is written as [ ]. If X is a list of length n, then the items in X are denoted $X[0], X[1], \ldots, X[n-1]$, in order. We usually denote the first item in the list X as $X[0]$, as is done in the C programming language. Thus, if $X = [1, 3, 1, 9]$, then $X[0] = 1$, $X[1] = 3$, $X[2] = 1$ and $X[3] = 9$. However, in some situations, we may list the elements of X as $X[1], X[2], \ldots, X[n]$. An alternative notation for a list, that we will sometimes use, is to write the items in the list X in subscripted form, as $X_0, X_1, \ldots, X_{n-1}$.

The *Cartesian product* (or cross product) of the sets X and Y, denoted by $X \times Y$, is the set of all ordered pairs whose first item is in X and whose second item is in Y. Thus

$$X \times Y = \{[x, y] : x \in X \text{ and } y \in Y\}.$$

For example, if $X = \{1, 3, 7, 9\}$ and $Y = \{0, 2, 4\}$, then

$$\{1, 3, 7, 9\} \times \{0, 2\} = \{[1, 0], [1, 2], [3, 0], [3, 2], [7, 0], [7, 2], [9, 0], [9, 2]\}.$$

If X is a finite set of cardinality n, then a *permutation* of X is a list π of length n such that every element of X occurs exactly once in the list π. There are exactly $n! = n(n-1) \cdots 1$ permutations of an n-set. For a positive integer $k < n$, a *k-permutation* of X is a list π of length k such that every element of X occurs at most once in the list π. There are exactly

$$\frac{n!}{(n-k)!} = n(n-1) \cdots (n-k+1)$$

k-permutations of an n-set.

1.2.2   Graphs

We begin by defining the concept of a graph.

Definition 1.1:   A *graph* consists of a finite set $\mathcal{V}$ of *vertices* and a finite set $\mathcal{E}$ of *edges*, such that each edge is a two element subset of vertices. It is customary to write a graph as an ordered pair, $(\mathcal{V}, \mathcal{E})$.

A *complete graph* is a graph in which $\mathcal{E}$ consists of all two element subsets of $\mathcal{V}$. If $|\mathcal{V}| = n$, then the complete graph is denoted by K_n.

We will usually represent the vertices of a graph $(\mathcal{V}, \mathcal{E})$ by dots, and join two vertices x and y by a line whenever $\{x, y\} \in \mathcal{E}$. A vertex x is *incident* with an edge e if $x \in e$. The *degree* of a vertex $x \in \mathcal{V}$, denoted by $\deg(x)$, is the number of edges that are incident with the vertex x. A graph is *regular* of degree d if every vertex has degree d. In Example 1.1 we present a graph that is regular of degree three.

Example 1.1   *A graph*
Let $\mathcal{V} = \{0, 1, 2, 3, 4, 5, 6, 7\}$, and let $\mathcal{E} = \{\{0, 1\}, \{0, 2\}, \{2, 3\}, \{1, 3\}, \{0, 4\}, \{1, 5\}, \{2, 6\}, \{3, 7\}, \{4, 5\}, \{4, 6\}, \{6, 7\}, \{5, 7\}\}$. This graph is called the *cube* and can be represented by the diagram in Figure 1.1.   ▯

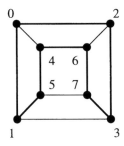

FIGURE 1.1
The cube and a Hamiltonian circuit.

One of the many interesting substructures that can occur in a graph is a *Hamiltonian circuit*. This is a closed path that passes through every vertex exactly once. The list $[0, 2, 3, 7, 6, 4, 5, 1]$ describes a Hamiltonian circuit in the graph in Figure 1.1. It is indicated by thick lines in the diagram. Note that different lists can represent the same Hamiltonian circuit. In fact, there are $2n$ such lists, where n is the number of vertices in the graph, since we can pick any of the n vertices as the "starting point" and then traverse the circuit in two possible directions.

A graph $(\mathcal{V}, \mathcal{E})$ is a *weighted graph* if there is a weight function $w : \mathcal{E} \to \mathbb{R}$ associated with it. The weight of a substructure, such as a Hamiltonian circuit, is defined to be the sum of the weights of its edges. Finding a smallest weight Hamiltonian circuit in a weighted graph is called the Traveling Salesman problem and is discussed in Chapter 4.

Another example of a substructure in a graph is a clique. A *clique* in a graph $\mathcal{G} = (\mathcal{V}, \mathcal{E})$ is a subset $S \subseteq \mathcal{V}$ such that $\{x, y\} \in \mathcal{E}$ for all $x, y \in S$, $x \neq y$. A clique that has the maximum size among all cliques of $\mathcal{G}$ is called a *maximum clique*. In Figure 1.1, every edge $\{x, y\}$ determines a maximum clique of the graph, since there are no cliques of size 3. Finding a maximum clique in a graph is called the Maximum Clique problem and is discussed in Chapter 4.

1.2.3 Set systems

We next define a generalization of a graph called a set system.

> **Definition 1.2:** A *set system* consists of a finite set $\mathcal{X}$ of *points* and a finite set $\mathcal{B}$ of *blocks*, such that each block is a subset of $\mathcal{X}$. We use the notation $(\mathcal{X}, \mathcal{B})$ to denote a set system.

Observe that a graph is just a set system in which every block has cardinality two. Another simple example of a set system is a partition of a set $\mathcal{X}$. A *partition* is a set system $(\mathcal{X}, \mathcal{B})$ in which $A \cap B = \emptyset$ for all $A, B \in \mathcal{B}$ with $A \neq B$, and $\cup_{A \in \mathcal{B}} A = \mathcal{X}$.

We now define another, more complicated, combinatorial structure, and then formulate it as a special type of set system.

> **Definition 1.3:** A *Latin square* of order n is an n by n array A, whose entries are chosen from an n-element set $\mathcal{Y}$, such that each symbol in $\mathcal{Y}$ occurs in exactly one cell in each row and in each column of A.

Example 1.2 *A Latin square of order four*

Let $\mathcal{Y} = \{1, 2, 3, 4\}$ and let

$$A = \begin{array}{|c|c|c|c|} \hline 1 & 2 & 3 & 4 \\ \hline 2 & 1 & 4 & 3 \\ \hline 3 & 4 & 1 & 2 \\ \hline 4 & 3 & 2 & 1 \\ \hline \end{array}.$$

◻

Suppose that A is a Latin square of order n on a symbol set $\mathcal{Y}$. Label the n rows of A with the n elements of $\mathcal{Y}$, and likewise label the n columns of A with the n elements of $\mathcal{Y}$. Now define a set system $(\mathcal{X}, \mathcal{B})$ as follows:

$$\mathcal{X} = \mathcal{Y} \times \{1, 2, 3\},$$

and

$$\mathcal{B} = \{\{(y_1, 1), (y_2, 2), (A[y_1, y_2], 3)\} : y_1, y_2 \in \mathcal{Y}\}.$$

If we start with the Latin square of order four that was presented in Example 1.2, then we obtain the following set system $(\mathcal{X}, \mathcal{B})$:

$$\mathcal{X} = \{1, 2, 3, 4\} \times \{1, 2, 3\}$$

$$\mathcal{B} = \left\{ \begin{array}{l} \{[1,1], [1,2], [1,3]\}, \{[1,1], [2,2], [2,3]\}, \\ \{[1,1], [3,2], [3,3]\}, \{[1,1], [4,2], [4,3]\}, \\ \{[2,1], [1,2], [2,3]\}, \{[2,1], [2,2], [1,3]\}, \\ \{[2,1], [3,2], [4,3]\}, \{[2,1], [4,2], [3,3]\}, \\ \{[3,1], [1,2], [3,3]\}, \{[3,1], [2,2], [4,3]\}, \\ \{[3,1], [3,2], [1,3]\}, \{[3,1], [4,2], [2,3]\}, \\ \{[4,1], [1,2], [4,3]\}, \{[4,1], [2,2], [3,3]\}, \\ \{[4,1], [3,2], [2,3]\}, \{[4,1], [4,2], [1,3]\} \end{array} \right\}$$

If we look at the blocks in this set system we see that every pair of the form $\{(y_1, i), (y_2, j)\}$, where $i \neq j$, occurs in exactly one block in $\mathcal{B}$. This motivates the following definition.

Definition 1.4:   A *transversal design* $TD(n)$ is a set system $(\mathcal{X}, \mathcal{B})$ for which there exists a partition $\{\mathcal{X}_1, \mathcal{X}_2, \mathcal{X}_3\}$ of $\mathcal{X}$ such that the following properties are satisfied:

1. $|\mathcal{X}| = 3n$, and $|\mathcal{X}_i| = n$ for $1 \leq i \leq 3$,
2. For every $B \in \mathcal{B}$ and for $1 \leq i \leq 3$, $|B \cap \mathcal{X}_i| = 1$.
3. For every $x \in \mathcal{X}_i$ and every $y \in \mathcal{X}_j$ with $i \neq j$, there exists a unique block $B \in \mathcal{B}$ such that $\{x, y\} \subseteq B$.

If we chose the partition $\{\mathcal{X}_1, \mathcal{X}_2, \mathcal{X}_3\}$ of $\mathcal{X} = \{1, 2, 3, 4\} \times \{1, 2, 3\}$ so that

$$\mathcal{X}_1 = \{[1,1], [2,1], [3,1], [4,1]\}$$
$$\mathcal{X}_2 = \{[1,2], [2,2], [3,2], [4,2]\},$$

and

$$\mathcal{X}_3 = \{[1,3], [2,3], [3,3], [4,3]\},$$

then it is easy to check that the set system that we constructed above from the Latin square of order four is a $TD(4)$. In general, we can construct a $TD(n)$ from a Latin square of order n, and vice versa.

1.3   What are combinatorial problems?

The combinatorial search problems we study are of various types, depending on the nature of the desired answer. To illustrate the basic terminology we use, we define four variants of the **Knapsack** problem.

Problem 1.1:   Knapsack (decision)

Instance: *profits* $p_0, p_1, p_2, \ldots, p_{n-1}$;

weights $w_0, w_1, w_2, \ldots, w_{n-1}$;

capacity M; and

target profit P

Question: does there exist an n-tuple $[x_0, \ldots, x_{n-1}] \in \{0, 1\}^n$ such that

$$\sum_{i=0}^{n-1} p_i x_i \geq P$$

and

$$\sum_{i=0}^{n-1} w_i x_i \leq M?$$

Problem 1.1 is an example of a *decision problem*, in which a question is to be answered "yes" or "no". All that is required for an algorithm to "solve" a decision problem is for the algorithm to provide the correct answer ("yes" or "no") for any problem instance.

Problem 1.2:   Knapsack (search)

Instance: *profits* $p_0, p_1, p_2, \ldots, p_{n-1}$;

weights $w_0, w_1, w_2, \ldots, w_{n-1}$;

capacity M; and

target profit P

Find: an n-tuple $[x_0, \ldots, x_{n-1}] \in \{0, 1\}^n$ such that

$$\sum_{i=0}^{n-1} p_i x_i \geq P$$

and

$$\sum_{i=0}^{n-1} w_i x_i \leq M$$

if such an n-tuple exists.

Problem 1.2 is an example of a *search problem*. It is exactly the same as the corresponding decision problem, except that we are asked to find an n-tuple $[x_0, \ldots, x_{n-1}]$ in the case where the answer to the search problem is "yes".

Problem 1.3:   Knapsack (optimal value)

Instance: *profits* $p_0, p_1, p_2, \ldots, p_{n-1}$;

                 weights $w_0, w_1, w_2, \ldots, w_{n-1}$; and

                 capacity M;

Find: the maximum value of

$$P = \sum_{i=0}^{n-1} p_i x_i$$

subject to

$$\sum_{i=0}^{n-1} w_i x_i \leq M$$

and $[x_0, \ldots, x_{n-1}] \in \{0, 1\}^n$.

Problem 1.3 is called an *optimal value problem*. In this version of the problem, there is no target profit specified in the problem instance. Instead, it is desired to find the largest target profit, P, such that the decision problem will have the answer "yes".

Problem 1.4:   Knapsack (optimization)

Instance: *profits* $p_0, p_1, p_2, \ldots, p_{n-1}$;

                 weights $w_0, w_1, w_2, \ldots, w_{n-1}$; and

                 capacity M

Find: an n-tuple $[x_0, \ldots, x_{n-1}] \in \{0, 1\}^n$ such that

$$P = \sum_{i=0}^{n-1} p_i x_i$$

is maximized, subject to

$$\sum_{i=0}^{n-1} w_i x_i \leq M.$$

Problem 1.4 is an *optimization problem*. It is similar to the optimal value problem, except that it is required to find an n-tuple $[x_0, \ldots, x_{n-1}] \in \{0, 1\}^n$ which yields the optimal profit.

In an optimization problem, we typically have several *constraints* that need to be satisfied. An n-tuple that satisfies the constraints is called a *feasible solution*. In Problem 1.4, the condition $\sum w_i x_i \leq M$ is the constraint that needs to be satisfied: an n-tuple $[x_0, \ldots, x_{n-1}] \in \{0,1\}^n$ is feasible if $\sum w_i x_i \leq M$. Associated with each feasible solution is an *objective function*, which is an integer or real number that is typically thought of as a *cost* or *profit*. In Problem 1.4, we define the profit of a feasible n-tuple by the formula $P = \sum p_i x_i$. The object of the problem in this case is to find a feasible solution that attains the maximum possible profit. For some optimization problems the objective function is defined as a cost measure, and it is desired to find a feasible solution that incurs a minimum cost.

1.4 O-**Notation**

In the next section, we will discuss the mathematical analysis of algorithms. This will involve using a notation $O(\cdot)$ which is called O-*notation*. We give a formal definition now.

Definition 1.5: Suppose $f : \mathbb{Z}^+ \to \mathbb{R}$ and $g : \mathbb{Z}^+ \to \mathbb{R}$. We say that $f(n)$ is $O(g(n))$ provided that there exist constants $c > 0$ and $n_0 \geq 0$ such that $0 \leq f(n) \leq c \times g(n)$ for all $n \geq n_0$.

In other words, $f(n)$ is $O(g(n))$ provided that $f(n)$ is bounded above by a constant factor times $g(n)$ for large enough n.

As a simple illustrative example, we show that the function $2n^2 + 5n + 6$ is $O(n^2)$. For all $n \geq 1$, it is the case that

$$2n^2 + 5n + 6 \leq 2n^2 + 5n^2 + 6n^2 = 13n^2.$$

Hence, we can take $c = 13$ and $n_0 = 1$, and the definition is satisfied.

Two related notations are Ω-*notation* and Θ-*notation*.

Definition 1.6: Suppose $f : \mathbb{Z}^+ \to \mathbb{R}$ and $g : \mathbb{Z}^+ \to \mathbb{R}$. We say that $f(n)$ is $\Omega(g(n))$ provided that there exist constants $c > 0$ and $n_0 \geq 0$ such that $0 \leq c \times g(n) \leq f(n)$ for all $n \geq n_0$.

Definition 1.7: Suppose $f : \mathbb{Z}^+ \to \mathbb{R}$ and $g : \mathbb{Z}^+ \to \mathbb{R}$. We say that $f(n)$ is $\Theta(g(n))$ provided that there exist constants $c, c' > 0$ and $n_0 \geq 0$ such that $0 \leq c \times g(n) \leq f(n) \leq c' \times g(n)$ for all $n \geq n_0$.

If $f(n)$ is $\Theta(g(n))$, then we say that f and g have the same *growth rate*.

Among the useful rules for working with these notations are the following sum and product rules. We state these rules for O-notation; similar rules hold for Ω- and Θ-notation.

THEOREM 1.1 *Suppose that the two functions $f_1(n)$ and $f_2(n)$ are both $O(g(n))$. Then the function $f_1(n) + f_2(n)$ is $O(g(n))$.*

THEOREM 1.2 *Suppose that $f_1(n)$ is $O(g_1(n))$ and $f_2(n)$ is $O(g_2(n))$. Then the function $f_1(n) f_2(n)$ is $O(g_1(n) g_1(n))$.*

As examples of the use of these notations, we have that n^2 is $O(n^3)$, n^3 is $\Omega(n^2)$, and $2n^2 + 3n - \sin n + 1/n$ is $\Theta(n^2)$.

We now collect a few results on growth rates of functions that arise frequently in algorithm analysis. The first of these results says that a polynomial of degree d, in which the high-order coefficient is positive, has growth rate n^d.

THEOREM 1.3 *Suppose that $a_d > 0$. Then the function $a_0 + a_1 n + \ldots + a_d n^d$ is $\Theta(n^d)$.*

The next result says that *logarithmic growth* does not depend on the base to which logarithms are computed. It can be proved easily using the formula $\log_a n = \log_a b \times \log_b n$.

THEOREM 1.4 *The function $\log_a n$ is $\Theta(\log_b n)$ for any $a, b > 1$.*

The next result can be proved using Stirling's formula. It gives the growth rate of the factorial function in terms of exponential functions.

THEOREM 1.5 *The function $n!$ is $\Theta(n^{n+1/2} e^{-n})$.*

1.5   Analysis of algorithms

Most combinatorial problems are big — bigger than can be handled on most computers — and hence the development of fast algorithms is very important. When we design an algorithm to solve a particular problem, we want to know how much resources (i.e., time and space) an implementation will consume. Mathematical methods can often be used to predict the time and space required by an algorithm, without actually implementing it in the form of a computer program. This is important for several reasons. The most important is that we can save work by not having to implement algorithms in order to test their suitability. The analysis done ahead of time can compare several proposed algorithms, or compare new algorithms to old algorithms. Then we can implement the best alternative, knowing that the others will be inferior.

The analysis of an algorithm will describe how the running time of the algorithm behaves as a function of the size of the input. We will illustrate the basic ideas by looking at a simple sorting algorithm called INSERTIONSORT , which we present as Algorithm 1.1. This algorithm sorts an array $A = [A[0], \ldots, A[n-1]]$ of n items in increasing order. To begin we see that an array with a single entry is already sorted. Now suppose that the first i values of the array A are in the correct order. The **while** loop finds the correct position j for $A[i]$ and simultaneously moves the entries $A[j+1], A[j+2], \ldots, A[i-1]$ to make room for it. This puts the first $i + 1$ values of the array in the correct order.

Algorithm 1.1: INSERTIONSORT (A, n)

for $i \leftarrow 1$ **to** $n - 1$

do $\begin{cases} x \leftarrow A[i] \\ j \leftarrow i - 1 \\ \textbf{while } j \geq 0 \textbf{ and } A[j] > x \\ \quad \textbf{do } \begin{cases} A[j+1] \leftarrow A[j] \\ j \leftarrow j - 1 \end{cases} \\ A[j+1] \leftarrow x \end{cases}$

It is not difficult to perform a mathematical analysis of Algorithm 1.1. Within any iteration of the **while** loop, a constant amount of time is spent, say c_1. The number of iterations of the **while** loop is at most $i - 1$ (where i is the index of the **for** loop). The amount of time spent in iteration i of the **for** loop is bounded above by $c_2 + c_1 (i - 1)$, where c_2 is the amount of time used within the **for** loop, excluding the time spent in the **while** loop. Thus, the running time of the entire algorithm is at most

$$T(n) = \sum_{i=2}^{n} (c_2 + c_1 (i - 1)) = c_2(n - 1) + \frac{c_1 n(n - 1)}{2}.$$

Note that the running time of the algorithm can, in fact, be this bad. If the array A happens to be initially sorted in decreasing order, then the **while** loop will require $i - 1$ iterations during iteration i of the **for** loop, for all i, $1 \leq i \leq n - 1$. In this case, the running time will be $T(n)$, while for other permutations of the n items in the array A, the running time will be less than $T(n)$.

The function $T(n)$ defined above is a quadratic in n, and hence, by Theorem 1.3, $T(n)$ is $\Theta(n^2)$. The growth rate of an algorithm's running time is called the *complexity* of the algorithm. For example, the above argument shows that INSERTIONSORT has quadratic complexity.

The actual coefficients of the quadratic function $T(n)$ are determined by c_1 and c_2, and these depend on the implementation of the algorithm (i.e., the programming language that is used and the machine that the program is run on). In other

words, analysis will reveal the complexity of an algorithm, but it will not tell us
the exact running time.

Let us now consider how we can use a complexity analysis to compare two
algorithms. We have already said that the algorithm INSERTIONSORT has complexity $\Theta(n^2)$. Suppose that we have another sorting algorithm with complexity $\Theta(n \log n)$ (such algorithms do in fact exist; one example is the well-known
HEAPSORT algorithm). We can recognize that $\Theta(n \log n)$ is a lower growth rate
than $\Theta(n^2)$. What this means is that when n is large enough, HEAPSORT will be
faster than INSERTIONSORT .

How big is "large enough"? To answer this, we would require some more
specific knowledge about the constants. (Usually this would be obtained from
a particular implementation of the algorithm.) As an example, suppose that we
know that the running time of INSERTIONSORT is at least cn^2 for $n \geq n_0$, and the
running time of HEAPSORT is at most $c'n \log n$ for $n \geq n_1$. It is certainly the case
that $c'n \log n < cn^2$ if $n \geq \max\{n_0, n_1, c/c'\}$; so, we can identify a particular
value for n beyond which HEAPSORT will run faster than INSERTIONSORT .

1.5.1   Average-case complexity

In the discussion of INSERTIONSORT , we determined the *worst-case complexity*,
by looking at the maximum running time of the algorithm, which we denoted
by the function $T(n)$. As mentioned above, this turns out to correspond to the
situation where the array A is sorted in decreasing order. Another alternative is
to consider *average-case complexity*. This would be determined by looking at the
amount of time the algorithm requires for each of the $n!$ possible permutations of
n given items, and then computing the average of these $n!$ quantities.

Suppose without loss of generality that the items to be sorted are the integers
$0, 1, \ldots, n-1$. Let $A = [A[0], \ldots, A[n-1]]$ be a permutation of $\{0, \ldots, n-1\}$,
and consider the number of iterations of the **while** loop that are required in iteration i of the **for** loop of Algorithm 1.1 (recall that $i = 0, 1, \ldots, n-1$). It
is not hard to see that this quantity is in fact the number of elements among
$A[0], \ldots, A[i-1]$ which are greater than $A[i]$. If we denote this quantity by
$N(A, i)$, then we have that

$$N(A, i) = |\{j : 0 \leq j \leq i-1, A[j] > A[i]\}|$$

for all permutations A and for all i, $0 \leq i \leq n-1$. Now, given a permutation A,
the running time of Algorithm 1.1 is seen to be

$$\sum_{i=1}^{n-1} (c_2 + c_1 N(A, i)).$$

For fixed positions $j < i$, let $m(i, j)$ be the number of permutations A that
have $A[j] > A[i]$. It is obvious that exactly half of the $n!$ permutations will have

$A[j] > A[i]$ (and half will have $A[j] < A[i]$). Hence, $m(i, j) = n!/2$. Then the average running time of Algorithm 1.1 over all $n!$ permutations A is

$$\frac{1}{n!} \sum_A \sum_{i=1}^{n-1} (c_2 + c_1 N(A, i)) = c_2(n - 1) + \frac{1}{n!} \sum_A \sum_{i=1}^{n-1} c_1 N(A, i)$$

$$= c_2(n - 1) + \frac{c_1}{n!} \sum_{i=1}^{n-1} \sum_A N(A, i)$$

$$= c_2(n - 1) + \frac{c_1}{n!} \sum_{i=1}^{n-1} \sum_{j=0}^{i-1} m(i, j)$$

$$= c_2(n - 1) + \frac{c_1}{n!} \sum_{i=1}^{n-1} \frac{n! \, i}{2}$$

$$= c_2(n - 1) + \frac{c_1}{2} \sum_{i=1}^{n-1} i$$

$$= c_2(n - 1) + \frac{c_1 n(n - 1)}{4}.$$

It follows that the average running time of Algorithm 1.1 is roughly half as long as the maximum running time. The average-case complexity is quadratic, as was the worst-case complexity.

1.6 Complexity classes

In general, we hope to find algorithms having *polynomial complexity*, i.e., $\Theta(n^d)$ for some positive integer d. Algorithms with *exponential complexity* (i.e., $\Theta(c^n)$ for some $c > 1$) will have a significantly higher growth rate and often become impractical for values of n that are not too large. There is a considerable amount of theory that has been developed to try to determine which problems can be solved by algorithms having polynomial complexity. This has given rise to definitions of various *complexity classes*. We will briefly discuss in an informal way some of the basic concepts and terminology.

Much of the terminology refers to decision problems. A *decision problem* can be described as a problem that requires a "yes" or "no" answer. The class P refers to the set of all decision problems for which polynomial-time algorithms exist. (P stands for "polynomial".)

There is a larger class of decision problems called NP. These problems have the property that, for any problem instance for which the answer is "yes", there exists a proof that the answer is "yes" that can be verified by a polynomial-time algorithm. (NP stands for "non-deterministic polynomial".) Note that we do not stipulate that there is any efficient method to find the given proof; we require only that a given hypothetical proof can be checked for validity by a polynomial-time algorithm.

It is not too difficult to show that $P \subseteq NP$. It seems likely that NP is a much larger class than P, but this is unproven at present.

Here is an example to illustrate the concepts defined above.

Problem 1.5:   Vertex Coloring

Instance: A graph $\mathcal{G} = (\mathcal{V}, \mathcal{E})$ and a positive integer k.

Question: Does there exist a function color : $\mathcal{V} \to \{0, \ldots, k-1\}$ such that color(u) $\neq$ color(v) for all $\{u, v\} \in \mathcal{E}$? (Such a function is called a *vertex k-coloring* or *k-coloring* of $\mathcal{G}$.)

Note that Problem 1.5 is phrased as a decision problem. An algorithm that "solves" it is required only to give the correct answer "yes" or "no." Let us first observe that this problem is in the class NP. If the graph $\mathcal{G}$ does have a k-coloring, color, then color itself can serve as the desired proof. To verify that a given color is in fact a k-coloring, it suffices to consider every edge of $\mathcal{G}$ and check to see if the two endpoints have received different colors.

Although verifying that a given function color is a k-coloring is easy, it does not seem so easy to find color in the first place. Even if we fix $k = 3$, then there is no known polynomial-time algorithm for this problem. (On the other hand, if $k = 2$, then the resulting problem is in the class P. A graph $\mathcal{G}$ has a 2-coloring if and only if it is bipartite.)

In fact, Vertex Coloring is generally believed to be in the class NP\P. There is further evidence to support this conjecture, beyond the fact that no one has managed to find a polynomial-time algorithm to solve it. Problem 1.5 turns out to be one of the so-called NP-complete problems.

The concept of NP-completeness is based on the idea of a polynomial transformation, which we define now. Suppose D_1 and D_2 are both decision problems. A *polynomial transformation* from D_1 to D_2 is a polynomial-time algorithm, TRANSFORM , which, when given any instance I of the problem D_1, will construct an instance TRANSFORM(I) of the problem D_2, in such a way that I is a yes-instance of D_1 if and only if TRANSFORM(I) is a yes-instance of D_2. We use the notation $D_1 \propto D_2$ to indicate that there is a polynomial transformation from D_1 to D_2.

We give a very simple example of a polynomial transformation. An *independent set* in a graph $\mathcal{G} = (\mathcal{V}, \mathcal{E})$ is a subset $S \subseteq \mathcal{V}$ such that $\{x, y\} \notin \mathcal{E}$ for all $x, y \in S$. The Maximum Independent Set and Maximum Clique decision problems are defined in the obvious way, as follows.

Problem 1.6:	Maximum Independent Set (decision)
Instance:	a graph $\mathcal{G}$; and
	an integer K
Question:	does $\mathcal{G}$ have an independent set of size at least K?

Problem 1.7:	Maximum Clique (decision)
Instance:	a graph $\mathcal{G}$; and
	an integer K
Question:	does $\mathcal{G}$ have a clique of size at least K?

It is easy to show that Maximum Independent Set (decision) $\propto$ Maximum Clique (decision) . Let $I = (\mathcal{G}, K)$ be an instance of Maximum Independent Set (decision) , where $\mathcal{G} = (\mathcal{V}, \mathcal{E})$ is a graph and K is an integer. The algorithm TRANSFORM constructs the instance $I' = (\mathcal{G}^c, K)$ of Maximum Clique (decision) , where $\mathcal{G}^c = (\mathcal{V}, \mathcal{F})$ is the graph in which the edge set $\mathcal{F}$ is defined by the rule

$$\{x, y\} \in \mathcal{F} \Leftrightarrow \{x, y\} \notin \mathcal{E},$$

for all $x, y \in \mathcal{V}$, $x \neq y$. Clearly $\mathcal{G}^c$ can be constructed in time $O(n^2)$, where $n = |\mathcal{V}|$. It is also easy to see that $\mathcal{G}^c$ has a clique of size K if and only if $\mathcal{G}$ has an independent set of size K. Thus the algorithm TRANSFORM is a polynomial transformation.

Suppose we have a polynomial transformation TRANSFORM from D_1 to D_2. Further, suppose that we have a polynomial-time algorithm A to solve D_2. Then we can construct a polynomial-time algorithm B to solve D_1, as follows. Given any instance I of D_1, construct the instance $J = \text{TRANSFORM}(I)$ of D_2. Then run the algorithm B on the instance J. Take the resulting answer to be the output of the algorithm A on input I.

A decision problem D is said to be NP-*complete* provided that $D \in$ NP, and for any problem $D' \in$ NP, $D' \propto D$. It follows that if $D \in$ P (i.e., if there is a polynomial-time algorithm for D), then there is a polynomial-time algorithm for any problem in NP, and hence P = NP.

Over the years, many decision problems have been shown to be NP-complete. Maximum Independent Set, Maximum Clique and Knapsack are all examples of NP-complete problems. We showed above that if any NP-complete problem can be solved in polynomial time, then they all can. However, it is generally believed that no NP-complete problem can be solved in polynomial-time. For such problems, this means that we will be forced to look at slower, exponential-time algorithms, which we will do in later chapters.

1.6.1   Reductions between problems

The concept of an NP-complete problem is a very useful and powerful idea, but NP-complete problems are, by definition, decision problems. The combinatorial problems of greatest practical interest tend to be search or optimization problems. To classify these problems, we need to first introduce the idea of a *Turing reduction*, or more simply, *reduction*. Informally, a Turing reduction is a method of using an algorithm A for one problem, say D_1, as a "subroutine" to solve another problem, say D_2. Note that the two problems D_1 and D_2 need not be decision problems. The algorithm A can be invoked one or many times, but the resulting algorithm, say B, should have the property that A is polynomial-time if and only if B is polynomial-time. Informally, a Turing reduction establishes that D_1 is no more difficult to solve than D_2. The existence of a Turing reduction is written notationally as $D_1 \propto_T D_2$. Note that a polynomial transformation provides a particularly simple type of Turing reduction, i.e., $D_1 \propto D_2$ implies $D_1 \propto_T D_2$.

It is an interesting exercise to find Turing reductions between the different flavors of the Knapsack problem. An easy example of a Turing reduction is to show that Knapsack (decision) $\propto_T$ Knapsack (optimization). Suppose that A is an algorithm that solves the Knapsack (optimization) problem, and let $I = (p_0, \ldots, p_{n-1}; w_0, \ldots, w_{n-1}; M; P)$ be an instance of the Knapsack (decision) problem. We construct an algorithm B as follows. Define $I' = (p_0, \ldots, p_{n-1}; w_0, \ldots, w_{n-1}; M)$. Then run A on I', obtaining an optimal n-tuple, $[x_0, \ldots, x_{n-1}]$, and compute

$$Q = \sum_{i=0}^{n-1} p_i x_i.$$

The algorithm B returns the answer "yes" if $Q \geq P$, and it returns the answer "no", otherwise.

The above example establishes the intuitively obvious fact that the optimization version of the Knapsack problem is at least as difficult as the decision version. It is more interesting (and more difficult) to find a converse reduction, i.e., to prove that Knapsack (optimization) $\propto_T$ Knapsack (decision). Here, we will prove that Knapsack (search) $\propto_T$ Knapsack (decision). This reduction is presented as Algorithm 1.2. In this algorithm, the hypothetical algorithm A is assumed to be an algorithm to solve the Knapsack (decision) problem.

Algorithm 1.2: KNAPREDUCTION $(p_0, \ldots, p_{n-1}; w_0, \ldots, w_{n-1}; M; P)$

external $A()$
if $A(p_0, \ldots, p_{n-1}; w_0, \ldots, w_{n-1}; M; P) = $ "no"
  then return ("no")
  else $\begin{cases} \\ \\ \mathbf{do} \begin{cases} \mathbf{for}\ i \leftarrow n-1\ \mathbf{downto}\ 0 \\ \qquad \begin{cases} \mathbf{if}\ A(p_0, \ldots, p_{i-1}; w_0, \ldots, w_{i-1}; M; P) = \text{"no"} \\ \qquad \mathbf{then} \begin{cases} x_i \leftarrow 1 \\ P \leftarrow P - p_i \\ W \leftarrow W - w_i \end{cases} \\ \qquad \mathbf{else}\ x_i \leftarrow 0 \end{cases} \\ \mathbf{return}\ ([x_0, \ldots, x_{n-1}]) \end{cases} \end{cases}$

The analog of NP-complete problems among search and optimization problems are the NP-hard problems. A problem D_2 is NP-*hard* if there exists an NP-complete prob-lem D_1 such that $D_1 \propto_T D_2$. As an example, since Knapsack (decision) $\propto_T$ Knapsack (optimization) and Knapsack (decision) is NP-complete, it follows that Knapsack (optimization) is NP-hard. In general, optimization and search versions of NP-complete decision problems are NP-hard. Note that an NP-hard problem can be solved in polynomial time only if P = NP.

1.7 Data structures

A data structure is an implementation or machine representation of a mathematical structure. In combinatorial algorithms, the choice of data structure can greatly affect the efficiency of an algorithm. In this section, we briefly discuss some useful data structures that we will use in combinatorial algorithms in the remaining chapters. More thorough discussions of data structures are given in the many available textbooks on data structures and algorithms.

1.7.1 Data structures for sets

Many combinatorial problems involve the manipulation of one or more subsets of a finite ground set X. To discuss the algorithms presented in this section, we will assume that $X = \{0, \ldots, n-1\}$ for some integer $n = |X|$. (Only minor modifications would be required for ground sets not of this form.) Among the operations that we want to perform on subsets of X are the following:

1. test membership of an element $x \in X$ in a subset $S \subseteq X$;
2. insert and delete elements from a subset S;

3. compute intersections and unions of subsets;

4. compute cardinality of subsets; and

5. list the elements of a subset.

One obvious way to store a subset $S \subseteq X$ is as a sorted array. That is, we write $S = [S[0], S[1], \ldots]$, where $S[0] < S[1] < \ldots$. We can keep track of the value $|S|$ in a separate auxiliary variable. Since $|S|$ is kept up-to-date every time an element is inserted or deleted from S, it is clear that no separate computation is required to determine $|S|$. Thus $|S|$ is computed in time $O(1)$, i.e., in a constant amount of time. Listing the elements of S can be done in time $O(|S|)$. Testing membership in S can be accomplished using a binary search, which takes time $O(\log |S|)$ (the binary search algorithm is described in Section 1.8.3). For the insertion or deletion of an element y, a binary or linear search, followed by a shift of the elements later than y, will accomplish the task in time $O(|S|)$. Intersection or union of two subsets S_1 and S_2 can be computed in time $O(|S_1| + |S_2|)$ by a simple merging algorithm.

If a subset is instead maintained as an unordered array, then it is easy to see that testing membership in a subset S requires time $O(|S|)$, since a linear search will be required. Computation of intersections and unions will also be less efficient.

There are other implementations of sets, based on balanced trees, in which all the operations can be performed in time $O(\log |S|)$ (except for listing all the elements of the set, which must take time $\Omega(|S|)$). One of the most popular data structures to do this is the so-called red-black tree, which is discussed in various textbooks.

All of the running times of the set operations in the implementations described above depend on the sizes of the subsets involved, and not on the size of the ground set. For many practical combinatorial algorithms, however, the ground set may be relatively small, say $|X| = n < 1000$. For small ground sets, an efficient alternative method is to represent subsets of X using a *bit array*.

Let $S \subseteq X$, and construct a bit array $B = [B[0], \ldots, B[n-1]]$ in which $B[i] = 1$ if $i \in S$, and $B[i] = 0$ if $i \notin S$. This representation requires one bit for each element in the ground set X. Bit arrays can be stored in the computer as unsigned integers, and we can take advantage of the bit operations that are available in programming languages such as C. (Note that we will not be performing any arithmetic operations on the bit arrays.) This approach can often improve the speed of the many algorithms.

Let $m \wedge n$ denote the "*bitwise boolean and*" and let $m \vee n$ denote the "*bitwise boolean or*" of the unsigned integers m and n. Let $m \ll j$ denote the *shift left* of the integer m by j bits, filling the rightmost bits with 0's. Similarly, define $m \gg j$ to be a *shift right* by j bits. Finally, we will use $\neg m$ to denote the *bitwise complement* of m.

Suppose β is the number of bits in an unsigned integer word. In general, this is a machine-dependent quantity. Currently $\beta = 32$ on most machines, and $\beta = 60$ or 64 on a few special machines. Thus, a bit array representation of a subset

$S \subseteq X$ will actually be an array A of

$$\omega = \left\lceil \frac{n}{\beta} \right\rceil$$

unsigned integers. The elements of the array A are defined as follows:

$$u \in S \text{ if and only if the } j\text{th bit of } A[i] \text{ is a 1;}$$

where

$$i = \left\lfloor \frac{u}{\beta} \right\rfloor$$

and

$$j = \beta - 1 - (u \bmod \beta).$$

Recall that in a β-bit integer, the 0th bit is the rightmost (least significant) bit, and the $(\beta - 1)$st bit is the leftmost (most significant) bit. If we think of A as the concatenation of the *bit strings* $A[0]$, $A[1]$, etc., then this representation of S has the effect that the bits (from left to right) of A correspond to the elements $0, 1, \ldots, n - 1$, in that order.

As an example, suppose that $n = 20$, $\beta = 8$ and $S = \{1, 3, 11, 16\}$. The bit string representation of S is

01010000000100001000.

The elements of A (represented as bit strings) are as follows:

$$A[0] = 01010000$$
$$A[1] = 00010000$$
$$A[2] = 10000000.$$

Here, the element $16 \in S$ corresponds to the seventh bit of $A[2]$, since

$$i = \left\lfloor \frac{16}{8} \right\rfloor = 2$$

and

$$j = 8 - 1 - (16 \bmod 8) = 7.$$

The seventh bit of $A[2]$ is the high-order (leftmost) bit.

Of course, there are other reasonable representations of a set as an array of integers. We have chosen this one since it seems fairly natural.

Now we look at how the various set operations would be implemented using bit operations. First, we consider the SETINSERT operation, where we want to replace S with $S \cup \{u\}$ (note that u may or may not be in the set S before this operation is performed). The following algorithm can be used to perform a SETINSERT operation.

Algorithm 1.3: SETINSERT (u, A)

$j \leftarrow \beta - 1 - (u \bmod \beta)$
$i \leftarrow \lfloor \frac{u}{\beta} \rfloor$
$A[i] \leftarrow A[i] \vee (1 \ll j)$

The SETDELETE operation, $S \leftarrow S \setminus \{u\}$, is accomplished with the following algorithm.

Algorithm 1.4: SETDELETE (u, A)

$j \leftarrow \beta - 1 - (u \bmod \beta)$
$i \leftarrow \lfloor \frac{u}{\beta} \rfloor$
$A[i] \leftarrow A[i] \wedge \neg(1 \ll j)$

Testing membership in a set is also easy.

Algorithm 1.5: MEMBEROFSET (u, A)

$j \leftarrow \beta - 1 - (u \bmod \beta)$
$i \leftarrow \lfloor \frac{u}{\beta} \rfloor$
if $A[i] \wedge (1 \ll j)$
   then return (**true**)
   else return (**false**)

Algorithms 1.3, 1.4, and 1.5 each use $O(1)$ operations. The union of two sets can be accomplished with the "bitwise boolean or" operation, $\vee$.

Algorithm 1.6: UNION (A, B)

global ω
for $i \leftarrow 0$ **to** $\omega - 1$
   do $C[i] \leftarrow A[i] \vee B[i]$
return (C)

For intersection, we use the "bitwise boolean and" operation, $\wedge$. Algorithm 1.7 computes the intersection of two sets.

Algorithm 1.7: INTERSECTION (A, B)

global ω
for $i \leftarrow 0$ **to** $\omega - 1$
  do $C[i] \leftarrow A[i] \wedge B[i]$
return (C)

Algorithms 1.6 and 1.7 each use $O(\omega)$ operations. To compute the cardinality of a set S, we could run over all the elements of X and count which ones are in S using Algorithm 1.5. This will require $\Omega(n)$ operations. A more efficient approach is to precompute an array *look*, whose ith entry is the number of 1 bits in the unsigned integer with value i, for $i = 0, 1, 2, \ldots, 2^{\alpha} - 1$. For example, if $\alpha = 4$, then

$$look = [0, 1, 1, 2, 1, 2, 2, 3, 1, 2, 2, 3, 2, 3, 3, 4].$$

Let *mask* be the unsigned integer whose rightmost α bits are all 1's, and whose remaining $\beta - \alpha$ bits are 0s. For example, when $\alpha = 4$ and $\beta = 32$, we have

$$mask = \underbrace{000 \cdots 0}_{28} 1111 = 15.$$

Then the number of 1 bits in an unsigned integer can be computed by using the "bitwise boolean and" and "right shift" operators to obtain small chunks of α bits. The numerical value of each chunk of α bits can be used to index the array *look* and thus obtain the number of 1 bits in the given chunk. The resulting algorithm is as follows.

Algorithm 1.8: SETORDER (A)

global $\alpha, \omega, look, mask$
$ans \leftarrow 0$
for $i \leftarrow 0$ **to** $\omega - 1$
  do $\begin{cases} x \leftarrow A[i] \\ \textbf{while } x \neq 0 \\ \quad \textbf{do } \begin{cases} ans \leftarrow ans + look[x \wedge mask] \\ x \leftarrow (x \gg \alpha) \end{cases} \end{cases}$
return (ans)

There is of course a space-time tradeoff with this approach. The algorithm runs faster as α becomes larger, but an increase in the size of α gives an exponential increase in the size of the array *look*. We usually use $\alpha = 8$ as a convenient compromise.

1.7.2 Data structures for lists

Recall that a list consists of a sequence of items given in a specific order. In this book, we will always use an array to store the elements of a list. Note that if we have a list of distinct items, and the order of the items in the list is irrelevant for our intended application, then we can think of the list as being a set, and use any of the data structures in the previous section to implement it.

An alternative data structure that can be used to store a set is a linked list. These are described in most textbooks on data structures.

1.7.3 Data structures for graphs and set systems

There are several convenient data structures for storing graphs. The most popular are the following:

1. a list (or set) of edges;

2. an incidence matrix;

3. an adjacency matrix; or

4. an adjacency list.

We illustrate the different methods for the graph that was presented in Figure 1.1. The original description of this graph presented the set of edges explicitly. This is the first of the four representations.

The next representation is as an incidence matrix. The *incidence matrix* of a graph $G = (V, \mathcal{E})$ is the $|V|$ by $|\mathcal{E}|$ matrix whose $[x, e]$-entry is a 1 if $x \in e$, and 0, otherwise. $e \in \mathcal{E}$. Every column of this matrix contains exactly two 1's, and the sum of the entries in row x is equal to the degree of vertex x.

The *adjacency matrix* of a graph $G = (V, \mathcal{E})$ is the $|V|$ by $|V|$ matrix whose $[x, y]$-entry is a 1 if $\{x, y\} \in \mathcal{E}$, and 0, otherwise.

An *adjacency list* of a graph $G = (V, \mathcal{E})$ is a list A of $|V|$ items, corresponding the vertices $x \in V$. Each item $A[x]$ is itself a list, consisting of the vertices incident with vertex x. Usually, the order of the items within each list $A[x]$ is irrelevant. Hence each $A[x]$ can be represented as a set, if desired. If this is done, then A is a list of sets.

In Figure 1.2, these representations are illustrated for the graph presented in Figure 1.1.

An incidence matrix is also a suitable data structure for an arbitrary set system, $(\mathcal{X}, \mathcal{A})$. The columns of the incidence matrix will be labeled by the blocks of the set system, and a column A will contain $|A|$ 1's, for each block $A \in \mathcal{A}$. Of course, a set system can also be stored as a list of blocks. However, adjacency matrices and adjacency lists do not have obvious analogs for set systems.

	01	02	04	13	15	23	26	37	45	46	57	67
0	1	1	1	0	0	0	0	0	0	0	0	0
1	1	0	0	1	1	0	0	0	0	0	0	0
2	0	1	0	0	0	1	1	0	0	0	0	0
3	0	0	0	1	0	1	0	1	0	0	0	0
4	0	0	1	0	0	0	0	0	1	1	0	0
5	0	0	0	0	1	0	0	0	1	0	1	0
6	0	0	0	0	0	0	1	0	0	1	0	1
7	0	0	0	0	0	0	0	1	0	0	1	1

	0	1	2	3	4	5	6	7
0	0	1	1	0	1	0	0	0
1	1	0	0	1	0	1	0	0
2	1	0	0	1	0	0	1	0
3	0	1	1	0	0	0	0	1
4	1	0	0	0	0	1	1	0
5	0	1	0	0	1	0	0	1
6	0	0	1	0	1	0	0	1
7	0	0	0	1	0	1	1	0

$$[\{1,2,4\}, \{0,3,5\}, \{0,3,6\}, \{1,2,7\}, \{1,5,6\}, \{1,4,7\}, \{2,4,7\}, \{3,5,6\}]$$

FIGURE 1.2
The incidence matrix, adjacency matrix and adjacency list for the cube.

1.8 Algorithm design techniques

It is often useful to classify algorithms by the design techniques. In this section, we describe three popular and useful design techniques for combinatorial algorithms.

1.8.1 Greedy algorithms

Many optimization problems can be solved by a *greedy strategy*. Greedy algorithms do not, in general, always find optimal solutions, but they are easy to apply and may succeed in finding solutions that are reasonably close to being optimal solutions. For some very special problems, it can be proved that greedy algorithms will always find an optimal solution.

The idea of a greedy algorithm is to build up a particular feasible solution by improving the objective function, or some other measure, as much as possible during each stage of the algorithm. This can be done, for example, when a feasible solution is defined as a list of items chosen from an underlying set. A feasible solution is constructed step by step, making sure at each stage that none of the constraints are violated. At the end of the algorithm, we should have constructed a feasible solution, which may or may not be optimal.

As a simple illustration, consider the Maximum Clique (optimization) problem, in which we are required to find a clique of maximum size in a graph. Suppose we are given a graph $\mathcal{G} = (\mathcal{V}, \mathcal{E})$, where $\mathcal{V} = \{0, \ldots, n - 1\}$. A greedy algorithm could construct a clique S by initializing S to be the empty set. Then each vertex x is considered in turn. We insert the vertex x into S if and only if $S \cup \{x\}$ is also a clique. At the end of the algorithm, a clique has been constructed.

The clique constructed by the greedy algorithm might turn out to be a maximum clique, or it could be much smaller than optimal, depending on the order in which the vertices of the graph are considered in the algorithm. This illustrates a common feature of greedy algorithms: since only one feasible solution is constructed in the course of the algorithm, the initial ordering of the items under consideration may have a drastic effect on the outcome of the algorithm. In the Maximum Clique (optimization) problem, for example, it might be a good strategy to first reorder the vertices in decreasing order of their degrees. This is reasonable because we might expect that vertices of large degree are more likely to occur in large cliques (note that the size of any clique containing a vertex x cannot exceed the degree of x).

It is usually fairly easy to devise greedy algorithms for combinatorial optimization problems. Even though the outcome of a greedy algorithm may not be very close to an optimal solution, greedy algorithms are still useful since they can provide nontrivial bounds on optimal solutions. We will see examples of this in Chapter 4.

1.8.2 Dynamic programming

Another method of solving optimization problems is *dynamic programming*. It requires being able to express or compute the optimal solution to a given problem instance I in terms of optimal solutions to smaller instances of the same problem. This is called a *problem decomposition*. The optimal solutions to all the relevant smaller problem instances are then computed and stored in tabular form. The smallest instances are solved first, and at the end of the algorithm, the optimal solution to the original instance I is obtained. Dynamic programming can thus be thought of as a "bottom-up" design strategy.

We give a simple illustration of a dynamic programming algorithm using the Knapsack (optimal value) problem, which was defined in Problem 1.3. Suppose we are given a problem instance $I = (p_0, \ldots, p_{n-1}; w_0, \ldots, w_{n-1}; M)$. For $0 \leq m \leq M$ and $0 \leq j \leq n - 1$, define $P[j, m]$ to be the optimal solution to the Knapsack (optimal value) problem for the instance $(p_0, \ldots, p_j; w_0,$

$\ldots, w_j; m)$.

The basis of the dynamic programming algorithm is the following recurrence relation for the values $P[j, m]$:

$$
P[j, m] = \begin{cases}
P[j-1, m] & \text{if } j \geq 1 \text{ and } w_j > m \\
\max\{P[j-1, m], P[j-1, m-w_j] + p_j\} & \text{if } j \geq 1 \text{ and } w_j \leq m \\
0 & \text{if } j = 0 \text{ and } w_0 > m \\
p_0 & \text{if } j = 0 \text{ and } w_0 \leq m.
\end{cases}
$$

Note that the recurrence relation is based on considering two possible cases for $P[j, m]$ that can arise, depending on whether $x_j = 0$ or $x_j = 1$.

The dynamic programming algorithm proceeds to compute the following table of values:

$$
\begin{bmatrix}
P[0, 0] & P[0, 1] & \cdots & P[0, M] \\
P[1, 0] & P[1, 1] & \cdots & P[1, M] \\
\vdots & \vdots & & \vdots \\
P[n-1, 0] & P[n-1, 1] & \cdots & P[n-1, M]
\end{bmatrix}
$$

The elements in the 0th row are computed first, then the elements in the first row are computed, etc. The value $P[n-1, M]$ is the solution to the problem instance I. Note that each entry in the table is computed in time $O(1)$ using the recurrence relation, and hence the running time of the algorithm is $O(nM)$. A particularly interesting aspect of this algorithm is that the running time grows linearly as a function of n, even though the optimal solution is computed as the maximum profit attained by one of 2^n possible n-tuples. However, the running time also is a linear function of M, and thus the algorithm will not be practical if M is too large.

1.8.3 Divide-and-conquer

The *divide-and-conquer* design strategy also utilizes a problem decomposition. In general, a solution to a problem instance I should be obtained by "combining" in some way solutions to one or more smaller instances of the same problem. Divide-and-conquer algorithms are often implemented as recursive algorithms.

Many familiar algorithms employ the divide-and-conquer method. One example is a binary search algorithm which can be used to determine if a desired item occurs in a sorted list. Suppose $X = [X[0], X[1], \ldots, X[n-1]]$ is a list of integers, where $X[0] < X[1] < \cdots < X[n-1]$, and we are given an integer y. If y is an item in the list X, then we want to find the index i such that $X[i] = y$, and if y is not an element in the list, then we should report that fact.

In the BINARYSEARCH algorithm, we compare the integer y to the item in the midpoint of the list X. If y is less than this item, then we can restrict the search to the first half of the list, while if y is greater than this item, then we can restrict the search to the second half of the list. (If this item has the value y, then the search is successful and we're done.) This is easily implemented as a recursive algorithm, Algorithm 1.9. To get started, we invoke Algorithm 1.9 with $lo = 0$ and $hi = n - 1$.

Algorithm 1.9: BINARYSEARCH (X, y, lo, hi)

if $lo > hi$
    then return $(y$ does not occur in the list $X)$
    else $\begin{cases} mid \leftarrow \lfloor \frac{lo+hi}{2} \rfloor \\ \textbf{if } X[mid] = y \\ \quad \textbf{then return } (mid) \\ \quad \textbf{else } \begin{cases} \textbf{if } X[mid] < y \\ \quad \textbf{then } \text{BINARYSEARCH}(X, y, mid + 1, hi) \\ \quad \textbf{else } \text{BINARYSEARCH}(X, y, lo, mid - 1) \end{cases} \end{cases}$

Performing a binary search of a sorted list of length n involves testing the midpoint of the list, and then recursively performing a binary search of a list of size $< n/2$. The complexity of the algorithm can be shown to be $O(\log n)$. Note that other divide-and-conquer algorithms may require solving more than one smaller problem in order to solve the original problem.

1.9   Notes

Section 1.1

There are quite a number of books on combinatorial algorithms, but many of them were written in the 1970s and are somewhat out of date. The following books are good general references for material and techniques on combinatorial algorithms that we discuss in this book: Even [28], Hu [44], Kučera [61], Reingold, Nievergelt and Deo [90], Stanton and White [101], Wells [111], and Wilf [80] and Wilf [114].

Section 1.2

There are many general textbooks on combinatorics, as well as textbooks on certain types of combinatorial structures. Good general textbooks include Brualdi [14], Cameron [16], van Lint and Wilson [67], Roberts [92], Straight [105] and Tucker [107]. Some good textbooks on graph theory are Bondy and Murty [7], and West [112]. Good references for material on combinatorial designs are Colbourn and Dinitz [20], Lindner and Rodger [66], and Wallis [109].

Section 1.3

Some recommended books on combinatorial optimization problems include Nemhauser and Wolsey [79] and Papadimitriou and Steiglitz [83].

Section 1.5

A good book discussing the analysis of algorithms is Purdom and Brown [84].

Section 1.6

Garey and Johnson [31] provides a very readable treatment of NP-completeness and related topics. See also the book by Papadimitriou [82].

Section 1.7

Data structures and algorithms are discussed in numerous textbooks. Cormen, Leiserson and Rivest [22] is a good general reference. Other recommended books include Baase [3], Kozen [56], Mehlhorn [72], Sedgewick [97] and Wilf [113].

Section 1.8

Textbooks that emphasize algorithm design techniques include Brassard and Bratley [9] and Stinson [103].

Exercises

1.1 Enumerate all the Hamiltonian circuits of the graph in Example 1.1.

1.2 Describe the transversal design $(\mathcal{X}, \mathcal{B})$ given below as a Latin square.

$$\mathcal{X} = \{1, 2, 3, 4, 5, 6, 7, 8, 9\}$$

$$\mathcal{X}_1 = \{1, 8, 9\}$$

$$\mathcal{X}_2 = \{2, 3, 4\}$$

$$\mathcal{X}_3 = \{5, 6, 7\}$$

$$\mathcal{B} = \left\{ \begin{array}{l} \{1, 2, 7\}, \{1, 3, 6\}, \{1, 4, 5\}, \{2, 5, 9\}, \{2, 6, 8\}, \\ \{3, 5, 8\}, \{3, 7, 9\}, \{4, 6, 9\}, \{4, 7, 8\} \end{array} \right\}$$

1.3 For all positive integers n, give a construction for a $TD(n)$.

1.4 Prove Theorem 1.1.

1.5 Prove Theorem 1.2.

1.6 Prove Theorem 1.3.

1.7 Prove Theorem 1.4.

1.8 Assuming Stirling's formula, which states that

$$\sqrt{2\pi n}\; e^{\frac{1}{12n+1}} \left(\frac{n}{e}\right)^n \leq n! \leq \sqrt{2\pi n}\; e^{\frac{1}{12n}} \left(\frac{n}{e}\right)^n,$$

prove Theorem 1.5.

1.9 For all permutations A of $\{0, 1, 2, 3\}$ and for $1 \leq i \leq 3$, compute $N(A, i)$, as it is defined in Section 1.5.1.

1.10 Prove that the problem **Maximum Independent Set (decision)** is in the class NP.

1.11 Prove that **Knapsack (optimal value)** $\propto_T$ **Knapsack (decision)** using a binary search technique. Then prove that **Knapsack (optimization)** $\propto_T$ **Knapsack (decision)**.

1.12 Let $A \subseteq \{0, \ldots, 31\}$ denote the subset of all the prime numbers in this interval. Use the algorithm SETORDER with $\alpha = 4$ and $\omega = 8$ to compute $|A|$.

1.13 Let $\mathcal{G}$ be the graph on vertex set $\{0, 1, \ldots, 9\}$ consisting of the following 15 edges:

$$\{01\}\{12\}\{23\}\{34\}\{04\}\{05\}\{16\}\{27\}\{38\}\{49\}\{56\}\{67\}\{78\}\{89\}\{59\}.$$

Give the incidence matrix and adjacency matrix for this graph, which is called the *Petersen graph*.

1.14 Describe an optimization version of the **Vertex Coloring** problem. Construct a greedy algorithm for this problem, and determine the result when the algorithm is run on the Petersen graph from the previous exercise.

1.15 Use a dynamic programming algorithm to solve the following instance of the **Knapsack (optimal value)** problem:

profits $1, 2, 3, 5, 7, 10$;
weights $2, 3, 5, 8, 13, 16$;
capacity 30.

Then, using the table of values $P[j, m]$, solve the **Knapsack (optimization)** problem for the same problem instance.

1.16 The algorithm MERGESORT , given below, is a divide-and-conquer algorithm that will sort the array

$$X = [X[0], X[1], \ldots, X[n-1]]$$

in increasing order. Give a worst-case analysis of the running time $T(n)$ for this algorithm.

Hint: suppose $n = 2^k$ and let $T(n)$ be the running time of MERGESORT . Show that $T(n) \leq c f(n)$ where c is a constant and f satisfies the following recurrence relation.

$$f(n) = \begin{cases} 2f(n/2) + n & \text{if } n > 2 \\ 2 & \text{if } n \leq 2. \end{cases}$$

Then show that $f(n)$ is $O(n \log n)$.

MERGESORT (n, X)

if $n = 1$
  then return
if $n = 2$
  then $\left\{\begin{array}{l} \textbf{if } X[0] > X[1] \\ \quad \textbf{then } \left\{\begin{array}{l} T \leftarrow X[0] \\ X[0] \leftarrow X[1] \\ X[1] \leftarrow T \end{array}\right. \end{array}\right.$
  else $\left\{\begin{array}{l} m \leftarrow \lfloor n/2 \rfloor \\ \textbf{for } i \leftarrow 0 \textbf{ to } m - 1 \\ \quad \textbf{do } A[i] \leftarrow X[i] \\ \textbf{for } i \leftarrow m \textbf{ to } n - 1 \\ \quad \textbf{do } B[i] \leftarrow X[i] \\ \text{MERGESORT}(m, A) \\ \text{MERGESORT}(n - m, B) \\ i \leftarrow 0 \\ j \leftarrow 0 \\ \textbf{for } k \leftarrow 0 \textbf{ to } n - 1 \\ \quad \textbf{do } \left\{\begin{array}{l} \textbf{if } A[i] \leq B[j] \\ \quad \textbf{then } \left\{\begin{array}{l} X[k] \leftarrow A[i] \\ i \leftarrow i + 1 \end{array}\right. \\ \quad \textbf{else } \left\{\begin{array}{l} X[k] \leftarrow B[j] \\ j \leftarrow j + 1 \end{array}\right. \end{array}\right. \end{array}\right.$

2

Generating Elementary Combinatorial Objects

2.1   Combinatorial generation

Often it is necessary to find nice algorithms to solve problems such as generating all the subsets of a given set S of size n, say. Related problems include generating all the permutations of S, or all the k-subsets of S.

Among the generation algorithms we will study are those that generate the desired objects in a *lexicographic order*, and the so-called *minimal change algorithms*, in which each object is generated from the previous one by performing a very small change. Both of these types of *sequential generation* are often accomplished by means of a *successor* algorithm, which is used to find the next object following a given one, with respect to a given ordering.

As well as sequential generation, we will be interested in *ranking* and *unranking* algorithms. A ranking algorithm determines the position (or rank) of a combinatorial object among all the objects (with respect to a given order); an unranking algorithm finds the object having a specified rank. Thus, ranking and unranking can be considered as inverse operations.

Here are slightly more formal mathematical descriptions of these concepts. Suppose that $\mathcal{S}$ is a finite set and $N = |\mathcal{S}|$. A ranking function will be a bijection

$$\mathsf{rank} : \mathcal{S} \to \{0, \dots, N-1\}.$$

A rank function defines a total ordering on the elements of $\mathcal{S}$, by the obvious rule

$$s < t \Leftrightarrow \mathsf{rank}(s) < \mathsf{rank}(t).$$

Conversely, there is a unique rank function associated with any total ordering defined on $\mathcal{S}$.

If rank is a ranking function defined on $\mathcal{S}$, then there is a unique unranking function associated with the function rank. This function unrank is also a bijection,

$$\mathsf{unrank} : \{0, \dots, N-1\} \to \mathcal{S}.$$

unrank is the inverse function of the function rank, i.e., we have

$$\text{rank}(s) = i \Leftrightarrow \text{unrank}(i) = s,$$

for all $s \in S$ and all $i \in \{0, \ldots, N-1\}$.

Efficient ranking and unranking algorithms have several potential uses. We mention a couple now. One application is the generation of random objects from a specified set S. This can be done easily by generating a random integer $i \in \{0, \ldots, N-1\}$, where $N = |S|$, and then unranking i. This algorithm ensures that every element of S is chosen with equal probability $1/N$, assuming that the random number generator being used is unbiased. For an example of the use of such an algorithm, see Algorithm 4.20.

Another use of ranking and unranking algorithms is in storing combinatorial objects in the computer. Instead of storing a combinatorial structure, which could be quite complicated, an alternative would be to simply store its rank, which of course is just an integer. If the structure is needed at any time, then it can be recovered by using the unranking algorithm.

Given a ranking function, rank, defined on S, the successor function, which we name successor, satisfies the following rule:

$$\text{successor}(s) = t \Leftrightarrow \text{rank}(t) = \text{rank}(s) + 1.$$

Thus, successor(s) is the next element following s in the total ordering. We will use the convention that successor(s) is undefined if rank$(s) = N - 1$ (i.e., if s is the last (largest) element in S).

The function successor can easily be constructed from the functions rank and unrank, according to the following formula:

$$\text{successor}(s) = \begin{cases} \text{unrank}(\text{rank}(s) + 1) & \text{if rank}(s) < N - 1 \\ \text{undefined} & \text{if rank}(s) = N - 1. \end{cases}$$

However, for a given set S under consideration, it may be that there is a more efficient way to construct a successor function.

Once we have constructed a successor function, it is a simple matter to generate all the elements in S. We would do this by beginning with the first element of S, and applying the function successor $N - 1$ times.

2.2 Subsets

2.2.1 Lexicographic ordering

Suppose that n is a positive integer, and $S = \{1, \ldots, n\}$. Define S to consist of the 2^n subsets of $S = \{1, \ldots, n\}$. We begin by describing how to generate the subsets in S in lexicographic order.

Given a subset $T \subseteq S$, let us define the *characteristic vector* of T to be the n-tuple

$$\chi(T) = [x_{n-1}, \ldots, x_0],$$

where

$$x_i = \begin{cases} 1 & \text{if } n - i \in T \\ 0 & \text{if } n - i \notin T. \end{cases}$$

This method of labeling the coordinates $x_{n-1}, \ldots, x_0$ is convenient for the purposes of the algorithms we are going to describe. It is essentially the same method that we used in Section 1.7.1, except that here we are taking the base set to be $\{1, \ldots, n\}$.

Next, define the *lexicographic ordering* on the set of subsets of S to be that induced by the lexicographic ordering of the characteristic vectors. If we think of these characteristic vectors as being the binary representations of the integers from 0 to $2^n - 1$, then this ordering corresponds to the usual ordering of the integers. With respect to this ordering, the rank of a subset T, denoted $\text{rank}(T)$, is just the integer whose binary representation is $\chi(T)$. That is,

$$\text{rank}(T) = \sum_{i=0}^{n-1} x_i 2^i.$$

We illustrate by taking $n = 3$, and tabulating the eight subsets of $S = \{1, 2, 3\}$:

T	$\chi(T) = [x_2, x_1, x_0]$	$\text{rank}(T)$
$\emptyset$	$[0, 0, 0]$	0
$\{3\}$	$[0, 0, 1]$	1
$\{2\}$	$[0, 1, 0]$	2
$\{2, 3\}$	$[0, 1, 1]$	3
$\{1\}$	$[1, 0, 0]$	4
$\{1, 3\}$	$[1, 0, 1]$	5
$\{1, 2\}$	$[1, 1, 0]$	6
$\{1, 2, 3\}$	$[1, 1, 1]$	7

We now present ranking and unranking algorithms for lexicographic generation of subsets. These are very simple. As mentioned above, ranking a subset $T \subseteq \{1, \ldots, n\}$ consists of computing the integer whose binary representation is $\chi(T)$. Unranking an integer r, where $0 \leq r \leq 2^n - 1$, requires the computation of the subset T having rank r. These algorithms are described below without reference to the characteristic vectors $\chi(T)$.

Algorithm 2.1: SUBSETLEXRANK (n, T)

$r \leftarrow 0$
for $i \leftarrow 1$ **to** n
$\quad$ **do** $\begin{cases} \textbf{if } i \in T \\ \quad \textbf{then } r \leftarrow r + 2^{n-i} \end{cases}$
return (r)

Algorithm 2.2: SUBSETLEXUNRANK (n, r)

$T \leftarrow \emptyset$
for $i \leftarrow n$ **downto** 1
$\quad$ **do** $\begin{cases} \textbf{if } r \bmod 2 = 1 \\ \quad \textbf{then } T \leftarrow T \cup \{i\} \\ r \leftarrow \lfloor \frac{r}{2} \rfloor \end{cases}$
return (T)

As an example, suppose that $n = 8$ and $T = \{1, 3, 4, 6\}$. Then Algorithm 2.1 computes

$$\text{rank}(T) = 2^7 + 2^5 + 2^4 + 2^2$$

$$= 128 + 32 + 16 + 4$$

$$= 180.$$

Conversely, if we run Algorithm 2.2 with $n = 8$ and $r = 180$, then we obtain the following:

i	r	$r \bmod 2$	T
8	180	0	$\emptyset$
7	90	0	$\emptyset$
6	45	1	$\{6\}$
5	22	0	$\{6\}$
4	11	1	$\{4, 6\}$
3	5	1	$\{3, 4, 6\}$
2	2	0	$\{3, 4, 6\}$
1	1	1	$\{1, 3, 4, 6\}$.

This example illustrates that ranking and unranking are inverse operations, since the subset $\{1, 3, 4, 6\}$ has rank 180 and unranking 180 produces the set $\{1, 3, 4, 6\}$.

We have assumed in this section that our base set is $S = \{1, \ldots, n\}$. What would we do if we wanted to rank and unrank the subsets of some other n-element set, say S'? We could of course design algorithms for ranking and unranking subsets of S', but a different approach is usually more convenient. It suffices to construct a bijection $\phi : S' \to S$. Now we can rank any subset $X \subseteq S'$, using a rank function for subsets of S, from the following formula:

$$\text{rank}(X) = \text{rank}(\phi(X)).$$

Similarly, we can unrank r to a subset of S', using the following formula:

$$\text{rank}(r) = \phi^{-1}(\text{unrank}(r)).$$

In the formula above, ϕ^{-1} is the inverse function of ϕ, i.e., $\phi(X) = Y$ if and only if $\phi^{-1}(Y) = X$, where $X \subseteq S'$ and $Y \subseteq S$.

For example if we want to rank and unrank subsets of $S' = \{0, \ldots, n-1\}$, then we can use the bijections ϕ and ϕ^{-1} defined by the following formulas:

$$\phi(X) = \{i + 1 : i \in X\}$$

and

$$\phi^{-1}(Y) = \{i - 1 : i \in Y\}.$$

2.2.2 Gray codes

The lexicographic ordering defined above makes ranking and unranking very simple, but the ordering is not well suited to the sequential generation of all 2^n subsets of an n-set. This is because subsets that are consecutive with respect to the ordering can be very "different." For example, in the case $n = 3$ considered above, $\text{rank}(\{2,3\}) = 3$ and $\text{rank}(\{1\}) = 4$. Hence, we have two consecutive subsets that are in fact complements of each other (so they are as different as they could possibly be).

Given two subsets $T_1, T_2 \subseteq S$, we define the *symmetric difference* of T_1 and T_2, denoted $T_1 \Delta T_2$, to be

$$T_1 \Delta T_2 = (T_1 \backslash T_2) \cup (T_2 \backslash T_1).$$

The *distance* between T_1 and T_2 is defined to be

$$\text{dist}(T_1, T_2) = |T_1 \Delta T_2|.$$

Alternatively, $\text{dist}(T_1, T_2)$ is equal to the number of coordinates in which $\chi(T_1)$ and $\chi(T_2)$ have different entries, which is called the *Hamming distance* between the vectors $\chi(T_1)$ and $\chi(T_2)$. The relevance of the distance between two subsets is that it represents the number of elements that need to be added to and/or deleted from one subset in order to obtain the other.

If we are going to generate all 2^n subsets sequentially, it might be desirable to do so in such a way that any two consecutive subsets have distance one (the smallest possible). This means that any subset can be obtained from the previous one by either deleting a single element or adding a single element. Such an ordering of the 2^n subsets of an n-set will be called a *minimal change ordering*.

As an example in the case $n = 3$, the ordering

$$\emptyset, \{3\}, \{2, 3\}, \{2\}, \{1, 2\}, \{1, 2, 3\}, \{1, 3\}, \{1\}$$

is a minimal change ordering.

The characteristic vectors of the subsets in a minimal change ordering form a structure that is known as a *Gray code*. Thus, a Gray code is an ordering of the 2^n binary vectors of length n in such a way that any two consecutive vectors have Hamming distance equal to one.

From the minimal change ordering presented above, the following Gray code is obtained:

$$000, 001, 011, 010, 110, 111, 101, 100.$$

There is another way to formulate the concept of minimal change coverings or Gray codes. Consider the n-dimensional unit cube, whose 2^n vertices are labeled by the 2^n binary vectors. The edges of this cube join vertices having Hamming distance equal to one. Thus, a Gray code is nothing more than a *Hamiltonian path* in the n-dimensional unit cube, i.e., a method traversing the edges of the cube so that each vertex is visited exactly once. Examples are given in Figure 2.1.

There has been a considerable amount of study done on different constructions for Gray codes. We will look at a particularly nice class of Gray codes called the *binary reflected Gray codes*. G^n will denote the binary reflected Gray code for the 2^n binary n-tuples, and it will be written as a list of 2^n vectors, as follows:

$$G^n = [G_0^n, G_1^n, \ldots, G_{2^n-1}^n].$$

The codes G^n are defined recursively. The first one, G^1, is defined to be

$$G^1 = [0, 1].$$

Given G^{n-1}, the Gray code G^n is defined to be

$$G^n = \left[0G_0^{n-1}, \ldots, 0G_{2^{n-1}-1}^{n-1}, 1G_{2^{n-1}-1}^{n-1}, \ldots, 1G_0^{n-1}\right].$$

Equivalently, we have that

$$G_i^n = \begin{cases} 0G_i^{n-1} & \text{if } 0 \le i \le 2^{n-1} - 1 \\ 1G_{2^n-1-i}^{n-1} & \text{if } 2^{n-1} \le i \le 2^n - 1. \end{cases}$$

The code G^n is constructed from G^{n-1} in two steps. First, we take a copy of G^{n-1} with a "0" prepended to each vector. Then we take a copy of G^{n-1} in reverse order, with a "1" prepended to each vector. The fact that the second copy of G^{n-1} is in reverse order is the reason for the name "reflected."

The next two Gray codes produced by this recipe are

$$G^2 = [00, 01, 11, 10]$$

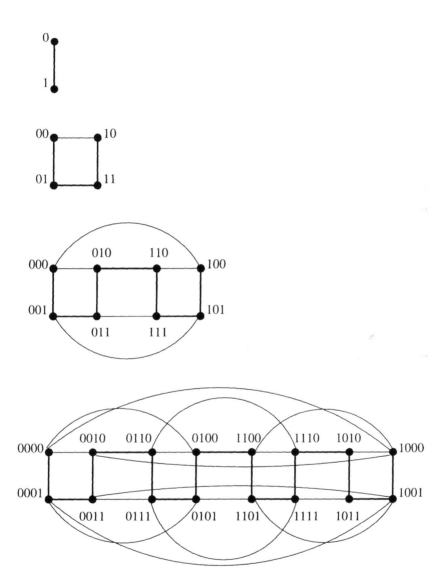

FIGURE 2.1
The evolution of the binary reflected Gray code.

and
$$G^3 = [000, 001, 011, 010, 110, 111, 101, 100].$$

Figure 2.1 depicts the binary reflected Gray codes $G^1, \ldots, G^4$.

Our first result is to prove that any G^n is a Gray code. Since the codes G^n are defined recursively, it is most natural to prove the result by induction on n.

THEOREM 2.1   *For any integer $n \geq 1$, G^n is a Gray code.*

PROOF   The proof is by induction on n. For $n = 1$, the result is easily seen to be true. As an induction hypothesis, suppose that G^{n-1} is a Gray code, for some integer $n \geq 2$. We will prove that G^n is a Gray code.

First, it is clear that G^n contains all 2^n binary n-tuples. By induction, the 2^{n-1} n-tuples that begin with "0" are contained in the first half of G^n, and the 2^{n-1} n-tuples that begin with "1" are contained in the second half of G^n.

It remains to verify the minimal change property. Consider two consecutive n-tuples in G^n, say G_i^n and G_{i+1}^n. There are in fact three cases to consider, depending on the value of i.

First, if $0 \leq i \leq 2^{n-1} - 2$, then G_i^n and G_{i+1}^n are formed from two consecutive $(n-1)$-tuples in G^{n-1} by prepending a "0" to each of them. Therefore, by induction, G_i^n and G_{i+1}^n have Hamming distance equal to 1.

The second case is $2^{n-1} \leq i \leq 2^n - 2$. This is similar to the first case. This time, G_i^n and G_{i+1}^n are formed from two consecutive $(n-1)$-tuples in G^{n-1} by prepending a "1" to each of them. Therefore, by induction, G_i^n and G_{i+1}^n have Hamming distance equal to 1.

The final case is $i = 2^{n-1} - 1$. $G_{2^{n-1}-1}^n$ and $G_{2^{n-1}}^n$ are both formed from $G_{2^{n-1}-1}^{n-1}$, by prepending a "0" and a "1" respectively. Therefore, $G_{2^{n-1}-1}^n$ and $G_{2^{n-1}}^n$ have Hamming distance equal to 1. (Note that this last case works out precisely because of the fact that the second copy of G^{n-1} in G^n is in reverse order.)

By induction on n, the result is true for all integers $n \geq 1$.   ∎

We now present Algorithm 2.3, which computes the successor function for the binary reflected gray Code G^n. Suppose that the binary vector $A = [a_{n-1}, \ldots, a_0]$ represents the set $T \subseteq \{1, \ldots, n\}$. Then $a_i = 1$ if $n - i \in T$ and $a_i = 0$ otherwise. Let $w(A)$ denote the *Hamming weight* of A (i.e., the number of "1"s in the vector A); note that $w(A) = |T|$.

Algorithm 2.3 works as follows. If $w(A)$ is even, then the last bit of A (namely, a_0) is flipped; if $w(A)$ is odd, then we find the first "1" from the right, and flip the next bit (to the left). The last vector in G^n, which has no successor, is $[1, 0, \ldots, 0]$. This corresponds to the set $\{1\}$.

Algorithm 2.3 is described in terms of the set T. It could alternatively be presented in terms of the binary vector A, if desired. The operation "Δ", which denotes the symmetric difference of two sets, was defined earlier.

Algorithm 2.3: GRAYCODESUCCESSOR (n, T)

if $|T|$ is even
   then $U \leftarrow T\Delta\{n\}$
   else $\begin{cases} j \leftarrow n \\ \textbf{while } j \notin T \textbf{ and } j > 0 \\ \quad \textbf{do } j \leftarrow j - 1 \\ \textbf{if } j = 1 \\ \quad \textbf{then return } (\text{"undefined"}) \\ U \leftarrow T\Delta\{j - 1\} \\ \textbf{return } (U) \end{cases}$

The following theorem can be proved by induction on n.

THEOREM 2.2 *Algorithm 2.3 computes the function* successor *for the Gray code* G^n.

We now proceed to develop ranking and unranking algorithms for the binary reflected Gray code. These algorithms depend on certain relationships between the binary representations of the integers $r = 0, \ldots, 2^n - 1$ and the corresponding vectors G_r^n. Let us begin by tabulating these in the case $n = 3$:

r	binary representation of r	G_r^3
0	000	000
1	001	001
2	010	011
3	011	010
4	100	110
5	101	111
6	110	101
7	111	100

For an integer r such that $0 \le r \le 2^n - 1$, suppose that its binary representation is written as

$$b_n b_{n-1} \ldots b_1 b_0.$$

In other words,

$$r = \sum_{i=0}^{n} b_i 2^i,$$

and $b_n = 0$ since $r \le 2^n - 1$. Also, suppose we write the vector G_r^n in the form

$$\mathsf{G}_r^n = a_{n-1} \ldots a_1 a_0,$$

as in Algorithm 2.3.

The relations in the next lemma will form the basis of the ranking and unranking algorithms.

LEMMA 2.3   *Suppose that $n \geq 1$ is an integer, $0 \leq r \leq 2^n - 1$, and suppose that $b_n, \ldots, b_0$ and $a_{n-1}, \ldots, a_0$ are as defined above. Then*

$$a_j \equiv (b_j + b_{j+1}) \bmod 2 \tag{2.1}$$

and

$$b_j \equiv \sum_{i=j}^{n-1} a_i \bmod 2, \tag{2.2}$$

for $j = 0, 1, \ldots, n - 1$.

PROOF   We begin by proving that Equation (2.1) is true for all $n \geq 1$ and $j = 0, 1, \ldots, n - 1$. The proof is by induction on n. The induction can be started with $n = 1$, where Equation (2.1) can be verified easily.

For some integer $i \geq 2$ assume that Equation (2.1) is true for $n = i - 1$ and $0 \leq j \leq i - 2$. We now consider $n = i$ and $0 \leq j \leq i - 1$. Let r be an integer such that $0 \leq r \leq 2^i - 1$. We divide the proof into two cases, depending on the value of r.

The first case is when $0 \leq r \leq 2^{i-1} - 1$. In this case, we have $b_{i-1} = 0$ and $a_{i-1} = 0$. For $0 \leq j \leq i - 2$, Equation (2.1) is true by induction. For $j = i - 1$, we have

$$b_{i-1} + b_i \equiv 0 \bmod 2$$

and $a_{i-1} = 0$, so Equation (2.1) is true here as well.

Now, we proceed to the second case, $2^{i-1} \leq r \leq 2^i - 1$. In this case, we have that $a_{i-1} = 1$, $b_{i-1} = 1$,

$$G^{i-1}_{2^i-1-r} = a_{n-2} \ldots a_1 a_0,$$

and the binary representation of $2^i - 1 - r$ is

$$0(1 - b_{n-2}) \ldots (1 - b_0).$$

Since Equation (2.1) is true for $n = i - 1$ by induction, we have

$$a_j \equiv (1 - b_j) + (1 - b_{j+1}) \bmod 2$$

for $j = 0, 1 \ldots, i - 2$. Since

$$(1 - b_j) + (1 - b_{j+1}) \equiv (b_j + b_{j+1}) \bmod 2,$$

Equation (2.1) is true for $n = i$ and $j = 0, 1 \ldots, i - 2$. For $j = i - 1$, we have

$$b_{i-1} + b_i \equiv 1 \bmod 2$$

and $a_{i-1} = 1$, so Equation (2.1) is true here as well.

By induction, Equation (2.1) is true for $j = 0, 1, \ldots, n - 1$, for all integers $n \geq 1$.

To complete the proof, we show that, for any $n \geq 1$, the truth of Equation (2.1) for $j = 0, 1, \ldots, n - 1$ implies the truth of Equation (2.2) for $j = 0, 1, \ldots, n - 1$. This is an easy computation:

$$\sum_{i=j}^{n-1} a_i \equiv \sum_{i=j}^{n-1} (b_i + b_{i+1}) \bmod 2$$

$$\equiv (b_j + b_n) \bmod 2$$

$$\equiv b_j \bmod 2,$$

since $b_n = 0$. ∎

The relations in Lemma 2.3 give rise to the ranking and unranking algorithms for the binary reflected Gray code which are presented as Algorithms 2.4 and 2.5. We provide brief explanations.

First, consider unranking. In iteration i of the **for** loop of Algorithm 2.5, b corresponds to b_{i+1} and b' corresponds to b_i. The algorithm successively computes $b_{n-1}, \ldots, b_0$, which are the bits in the binary representation of r. Recalling that $a_i \equiv b_i + b_{i+1} \bmod 2$, we see that

$$n - i \in T \Leftrightarrow a_i = 1 \Leftrightarrow b \neq b'.$$

Now we look at ranking. In iteration i of the **for** loop of Algorithm 2.4, b corresponds to b_i. Initially, $b = 0$ (corresponding to $b_n = 0$). Since

$$b_i = b_{i+i} + a_i \bmod 2,$$

we can update b during each iteration of the **for** loop by checking if $n - i \in T$ (since $a_i = 1$ if $n - i \in T$ and $a_i = 0$ if $n - i \notin T$). Whenever $b = 1$, we add 2^i to r, since $b = b_i$ is just bit i in the binary representation of r.

Algorithm 2.4: GRAYCODERANK (n, T)

$r \leftarrow 0$
$b \leftarrow 0$
for $i \leftarrow n - 1$ **downto** 0
  do $\begin{cases} \textbf{if } n - i \in T \\ \quad \textbf{then } b \leftarrow 1 - b \\ \textbf{if } b = 1 \\ \quad \textbf{then } r \leftarrow r + 2^i \end{cases}$
return (r)

Algorithm 2.5: GRAYCODEUNRANK (n, r)

$T \leftarrow \emptyset$
$b' \leftarrow 0$
for $i \leftarrow n - 1$ **downto** 0
$\quad$ **do** $\begin{cases} b \leftarrow \lfloor \frac{r}{2^i} \rfloor \\ \textbf{if } b \neq b' \\ \quad \textbf{then } T \leftarrow T \cup \{n - i\} \\ b' \leftarrow b \\ r \leftarrow r - b\,2^i \end{cases}$
return (T)

Let's work out a couple of examples to illustrate Algorithms 2.4 and 2.5. Suppose that $n = 8$ and $T = \{1, 2, 3, 4, 5, 7, 8\}$. We first use Algorithm 2.4 to compute rank(T).

i	2^i	$n - i \in T$?	b	r
7	128	yes	1	128
6	64	yes	0	128
5	32	yes	1	160
4	16	yes	0	160
3	8	yes	1	168
2	4	no	1	172
1	2	yes	0	172
0	1	yes	1	173

Thus, rank$(T) = 173$. The inverse algorithm, Algorithm 2.5, can be used to compute unrank(173). It executes as follows:

b'	r	i	2^i	b	T
0	173	7	128	1	$\{1\}$
1	45	6	64	0	$\{1, 2\}$
0	45	5	32	1	$\{1, 2, 3\}$
1	13	4	16	0	$\{1, 2, 3, 4\}$
0	13	3	8	1	$\{1, 2, 3, 4, 5\}$
1	5	2	4	1	$\{1, 2, 3, 4, 5\}$
1	1	1	2	0	$\{1, 2, 3, 4, 5, 7\}$
0	1	0	1	1	$\{1, 2, 3, 4, 5, 7, 8\}$

Hence, unrank$(173) = \{1, 2, 3, 4, 5, 7, 8\}$.

2.3 *k*-**Element subsets**

2.3.1 Lexicographic ordering

Suppose that n is a positive integer, and $S = \{1, \ldots, n\}$. Define $\mathcal{S}$ to consist of the $\binom{n}{k}$ k-element subsets of S. We begin by describing how to generate the subsets in $\mathcal{S}$ in lexicographic order.

A k-element subset $T \subseteq S$ can be written in a natural way as a list

$$\overrightarrow{T} = [t_1, t_2, \ldots, t_k],$$

where

$$t_1 < t_2 < \cdots < t_k.$$

The lexicographic ordering on $\mathcal{S}$ is induced by the lexicographic ordering on the sequences $\overrightarrow{T}$ ($T \in \mathcal{S}$).

We illustrate with a small example. Let $n = 5$ and $k = 3$. The lexicographic ordering of the ten 3-element subsets $T \subseteq \{1, \ldots, 5\}$ is as follows:

T	$\overrightarrow{T}$	rank(T)
$\{1, 2, 3\}$	$[1, 2, 3]$	0
$\{1, 2, 4\}$	$[1, 2, 4]$	1
$\{1, 2, 5\}$	$[1, 2, 5]$	2
$\{1, 3, 4\}$	$[1, 3, 4]$	3
$\{1, 3, 5\}$	$[1, 3, 5]$	4
$\{1, 4, 5\}$	$[1, 4, 5]$	5
$\{2, 3, 4\}$	$[2, 3, 4]$	6
$\{2, 3, 5\}$	$[2, 3, 5]$	7
$\{2, 4, 5\}$	$[2, 4, 5]$	8
$\{3, 4, 5\}$	$[3, 4, 5]$	9

It is fairly straightforward to describe a successor algorithm for $\mathcal{S}$. This algorithm is presented in Algorithm 2.6.

Algorithm 2.6: KSUBSETLEXSUCCESSOR $(\overrightarrow{T}, k, n)$

$\overrightarrow{U} \leftarrow \overrightarrow{T}$
$i \leftarrow k$
while $(i \geq 1)$ **and** $(t_i = n - k + i)$
  do $i \leftarrow i - 1$
if $i = 0$
  then return ("undefined")
  else $\begin{cases} \textbf{for } j \leftarrow i \textbf{ to } k \\ \quad \textbf{do } u_j \leftarrow t_i + 1 + j - i \\ \textbf{return } (\overrightarrow{U}) \end{cases}$

To construct a ranking algorithm, we need to count the number of k-element subsets preceding a given set T in this ordering. Suppose that t_1 is an integer such that $1 \leq t_1 \leq n$. It is easy to see that there are exactly $\binom{n-t_1}{k-1}$ subsets $X \in \mathcal{S}$ such that $x_1 = t_1$, where $\overrightarrow{X} = [x_1, \ldots, x_k]$. More generally, for any $i \leq k$ integers $t_1, \ldots, t_i$ such that $1 \leq t_1 < \cdots < t_i \leq n$, there are exactly $\binom{n-t_i}{k-i}$ subsets $X \in \mathcal{S}$ such that $x_1 = t_1, \ldots$, and $x_i = t_i$.

Now, suppose that $T \in \mathcal{S}$, and $\overrightarrow{T} = [t_1, t_2, \ldots, t_k]$ is defined as above. The k-element subsets X preceding T in lexicographic order are the following:

- The subsets X with $1 \leq x_1 \leq t_1 - 1$.
- The subsets X with $x_1 = t_1$ and $t_1 + 1 \leq x_2 \leq t_2 - 1$.
- The subsets X with $x_1 = t_1$, $x_2 = t_2$, and $t_2 + 1 \leq x_3 \leq t_3 - 1$.
- etc.
- The subsets X with $x_1 = t_1, x_2 = t_2, \ldots, x_{k-1} = t_{k-1}$ and $t_{k-1} + 1 \leq x_k \leq t_k - 1$.

From these facts, we can write down a formula for $\mathsf{rank}(T)$, where $\overrightarrow{T} = [t_1, t_2, \ldots, t_k]$. We get the following formula, where we define $t_0 = 0$ for convenience:

$$\mathsf{rank}(T) = \sum_{i=1}^{k} \sum_{j=t_{i-1}+1}^{t_i-1} \binom{n-j}{k-i}.$$

This formula immediately yields a ranking algorithm, which we present as Algorithm 2.7.

Algorithm 2.7: KSUBSETLEXRANK $(\overrightarrow{T}, k, n)$

$r \leftarrow 0$
$t_0 \leftarrow 0$
for $i \leftarrow 1$ **to** k

$\qquad$ **do** $\begin{cases} \textbf{if } t_{i-1} + 1 \leq t_i - 1 \\ \qquad \textbf{then } \begin{cases} \textbf{for } j \leftarrow t_{i-1} + 1 \textbf{ to } t_i - 1 \\ \qquad \textbf{do } r \leftarrow r + \binom{n-j}{k-i} \end{cases} \end{cases}$

return (r)

Now we unravel Algorithm 2.7 to obtain an unranking algorithm. Suppose that $0 \leq r \leq \binom{n}{k} - 1$, and suppose that $T = \mathsf{unrank}(r)$ with $\overrightarrow{T} = [t_1, \ldots, t_k]$. The smallest element in T, t_1, can be determined by the observation that

$$t_1 = x \Leftrightarrow \sum_{j=1}^{x-1} \binom{n-j}{k-1} \leq r < \sum_{j=1}^{x} \binom{n-j}{k-1}.$$

Having determined t_1, we can compute t_2 in a similar way:

$$t_2 = x \Leftrightarrow \sum_{j=t_1+1}^{x-1} \binom{n-j}{k-2} \le r - \sum_{j=1}^{t_1-1} \binom{n-j}{k-1} < \sum_{j=t_1+1}^{x} \binom{n-j}{k-2}.$$

The pattern continues, and the entire algorithm is presented as Algorithm 2.8.

Algorithm 2.8: KSUBSETLEXUNRANK (r, k, n)

$x \leftarrow 1$
for $i \leftarrow 1$ **to** k

$\mathbf{do} \begin{cases} \textbf{while } \binom{n-x}{k-i} \le r \\ \quad \mathbf{do} \begin{cases} r \leftarrow r - \binom{n-x}{k-i} \\ x \leftarrow x + 1 \end{cases} \\ t_i \leftarrow x \\ x \leftarrow x + 1 \end{cases}$

return $(\overrightarrow{T})$

2.3.2   Co-lex ordering

There is a useful alternative to the lexicographic ordering for k-element subsets of an n-set. The ordering is called the *co-lex ordering*, and it is defined as follows. A k-element subset $T \subseteq S$ is written as a list

$$\overleftarrow{T} = [t_1, t_2, \dots, t_k],$$

where

$$t_1 > t_2 > \cdots > t_k.$$

The co-lex ordering is induced by the lexicographic ordering on the sequences $\overleftarrow{T}$ $(T \in \mathcal{S})$.

We illustrate the co-lex ordering when $n = 5$ and $k = 3$. The co-lex ordering of the ten 3-element subsets of $\{1, \dots, 5\}$ is as follows:

T	$\overleftarrow{T}$	rank(T)
$\{1, 2, 3\}$	$[3, 2, 1]$	0
$\{1, 2, 4\}$	$[4, 2, 1]$	1
$\{1, 3, 4\}$	$[4, 3, 1]$	2
$\{2, 3, 4\}$	$[4, 3, 2]$	3
$\{1, 2, 5\}$	$[5, 2, 1]$	4
$\{1, 3, 5\}$	$[5, 3, 1]$	5
$\{2, 3, 5\}$	$[5, 3, 2]$	6
$\{1, 4, 5\}$	$[5, 4, 1]$	7
$\{2, 4, 5\}$	$[5, 4, 2]$	8
$\{3, 4, 5\}$	$[5, 4, 3]$	9

It is straightforward to find a successor algorithm for the co-lex ordering. This is left as an exercise.

We proceed to develop ranking and unranking algorithms, which are much simpler in the co-lex ordering than in the lexicographic ordering. Suppose that t_1 is an integer such that $1 \le t_1 \le n$. It is easy to see that there are exactly $\binom{t_1 - 1}{k}$ subsets $X \in S$ such that $x_1 < t_1$, where $\overleftarrow{X} = [x_1, \ldots, x_k]$. More generally, for any $i \le k$ integers $t_1, \ldots, t_i$ such that $1 \le t_1 < \ldots < t_i \le n$, there are exactly $\binom{t_i - 1}{k+1-i}$ subsets $X \in S$ such that $x_1 = t_1, \ldots, x_{i-1} = t_{i-1}$ and $x_i < t_i$.

Now, suppose that $T \in S$, and $\overleftarrow{T} = [t_1, t_2, \ldots, t_k]$ is defined as above. The k-element subsets X preceding T in lexicographic order are the following:

- The subsets X with $x_1 < t_1$.
- The subsets X with $x_1 = t_1$ and $x_2 < t_2$.
- The subsets X with $x_1 = t_1$, $x_2 = t_2$, and $x_3 < t_3$.
- etc.
- The subsets X with $x_1 = t_1$, $x_2 = t_2, \ldots, x_{k-1} = t_{k-1}$ and $x_k < t_k$.

These facts permit us to state a formula for the rank of a subset T, where $\overleftarrow{T} = [t_1, t_2, \ldots, t_k]$:

$$\operatorname{rank}(T) = \sum_{i=1}^{k} \binom{t_i - 1}{k + 1 - i}.$$

It is interesting to observe that this formula does not depend on the value of n.

The ranking algorithm presented in Algorithm 2.9 is an immediate consequence of the formula presented above.

Algorithm 2.9: KSUBSETCOLEXRANK $(\overleftarrow{T}, k)$

$r \leftarrow 0$
for $i \leftarrow 1$ **to** k
$\quad$ **do** $r \leftarrow r + \binom{t_i - 1}{k+1-i}$
return (r)

The unranking algorithm for co-lex ordering is presented as Algorithm 2.10.

Algorithm 2.10: KSUBSETCOLEXUNRANK (r, k, n)

$x \leftarrow n$
for $i \leftarrow 1$ **to** k

$\quad$ **do** $\begin{cases} \textbf{while } \binom{x}{k+1-i} > r \\ \quad \textbf{do } x \leftarrow x - 1 \\ t_i \leftarrow x + 1 \\ r \leftarrow r - \binom{x}{k+1-i} \end{cases}$

return $(\overleftarrow{T})$

Our final observation in this section is a relationship between the co-lex ordering and the lexicographic ordering. Given a k-element subset $T \subseteq \{1, \ldots, n\}$, define

$$T' = \{n + 1 - i : i \in T\}.$$

Suppose we take $k = 3$ and $n = 5$, and recall the lexicographic ordering:

$$\begin{array}{cccc} \{1,2,3\} & \{1,2,4\} & \{1,2,5\} & \{1,3,4\} \\ \{1,3,5\} & \{1,4,5\} & \{2,3,4\} & \{2,3,5\} \\ \{2,4,5\} & \{3,4,5\} \end{array}$$

Now, replace every set T by T':

$$\begin{array}{cccc} \{3,4,5\} & \{2,4,5\} & \{1,4,5\} & \{2,3,5\} \\ \{1,3,5\} & \{1,2,5\} & \{2,3,4\} & \{1,3,4\} \\ \{1,2,4\} & \{1,2,3\} \end{array}$$

The result is just the reverse of the co-lex ordering! The property which we have observed in this example can be proved to hold in general, as stated in the following theorem. The proof is left as an exercise.

THEOREM 2.4 *Let* S *consist of all* k-element subsets of the n-set $S = \{1, \ldots, n\}$. *Suppose that* rank_L *denotes rank in the lexicographic ordering and* rank_C *denotes rank in the co-lex ordering. Then, for any* k-set $T \subseteq S$, *we have*

$$\mathrm{rank}_L(T) + \mathrm{rank}_C(T') = \binom{n}{k} - 1,$$

where $T' = \{n + 1 - i : i \in T\}$.

Theorem 2.4 provides an alternative method of computation of lexicographic rank. Given T, we would first compute T', then find the co-lex rank of T', and finally subtract the result from $\binom{n}{k} - 1$. This would, in general, be more efficient than Algorithm 2.7. A similar strategy could be employed to do unranking, too.

2.3.3   Minimal change ordering

It is easy to see that if T_1 and T_2 are two k-element subsets of $S = \{1, \ldots, n\}$, and $T_1 \neq T_2$, then $\text{dist}(T_1, T_2) \geq 2$. Hence, if $\mathcal{S}_k^n$ denotes the set of all $\binom{n}{k}$ k-element subsets of S, then a minimal change ordering on $\mathcal{S}_k^n$ will be one in which any two consecutive subsets have distance two. In this section, we study a minimal change ordering called the *revolving door ordering*. The revolving door ordering for $\mathcal{S}_k^n$ will be written as a list $\mathsf{A}^{n,k}$ of $\binom{n}{k}$ k-element sets, as follows:

$$\mathsf{A}^{n,k} = \left[\mathsf{A}_0^{n,k}, \mathsf{A}_1^{n,k}, \ldots, \mathsf{A}_{\binom{n}{k}-1}^{n,k} \right].$$

The revolving door algorithm is motivated by *Pascal's identity* for binomial coefficients:

$$\binom{n}{k} = \binom{n-1}{k-1} + \binom{n-1}{k}.$$

This identity can be proved by observing that the set of all $\binom{n}{k}$ k-element subsets of S can be partitioned into two disjoint subcollections: the $\binom{n-1}{k-1}$ k-element subsets that contain the element n, and the $\binom{n-1}{k}$ k-element subsets that do not contain the element n.

The definition of $\mathsf{A}^{n,k}$ follows a similar recursive pattern. Given $\mathsf{A}^{n-1,k-1}$ and $\mathsf{A}^{n-1,k}$, the list $\mathsf{A}^{n,k}$ is defined as follows:

$$\mathsf{A}^{n,k} = \left[\mathsf{A}_0^{n-1,k}, \ldots, \mathsf{A}_{\binom{n-1}{k}-1}^{n-1,k}, \mathsf{A}_{\binom{n-1}{k-1}-1}^{n-1,k-1} \cup \{n\}, \ldots, \mathsf{A}_0^{n-1,k-1} \cup \{n\} \right].$$

This recursive definition can be applied whenever $1 \leq k \leq n-1$. The lists $\mathsf{A}^{n,0}$ and $\mathsf{A}^{n,n}$ are given as initial conditions to start the recursion. They are as follows:

$$\mathsf{A}^{n,0} = [\emptyset]$$

and

$$\mathsf{A}^{n,n} = [\{1, \ldots, n\}].$$

This construction is very reminiscent of the method used to construct the binary reflected Gray codes in Section 2.2.2. Similar features include the construction of a list by gluing two smaller lists together, and the reversed order of the second list.

Let's construct some of the small lists $\mathsf{A}^{n,k}$ before proceeding further.

$\mathsf{A}^{2,1} = [\{1\}, \{2\}]$

$\mathsf{A}^{3,1} = [\{1\}, \{2\}, \{3\}]$

$\mathsf{A}^{3,2} = [\{1,2\}, \{2,3\}, \{1,3\}]$

$\mathsf{A}^{4,1} = [\{1\}, \{2\}, \{3\}, \{4\}]$

$$A^{4,2} = [\{1,2\}, \{2,3\}, \{1,3\}, \{3,4\}, \{2,4\}, \{1,4\}]$$

$$A^{4,3} = [\{1,2,3\}, \{1,3,4\}, \{2,3,4\}, \{1,2,4\}]$$

$$A^{5,1} = [\{1\}, \{2\}, \{3\}, \{4\}, \{5\}]$$

$$A^{5,2} = [\{1,2\}, \{2,3\}, \{1,3\}, \{3,4\}, \{2,4\}, \{1,4\}, \{4,5\}, \{3,5\}, \{2,5\}, \{1,5\}]$$

$$A^{5,3} = [\{1,2,3\}, \{1,3,4\}, \{2,3,4\}, \{1,2,4\}, \{1,4,5\},$$
$$\{2,4,5\}, \{3,4,5\}, \{1,3,5\}, \{2,3,5\}, \{1,2,5\}]$$

$$A^{5,4} = [\{1,2,3,4\}, \{1,2,4,5\}, \{2,3,4,5\}, \{1,3,4,5\}, \{1,2,3,5\}]$$

Our first task is to prove that $A^{n,k}$ is a minimal change ordering of $\mathcal{S}_k^n$. As one would expect, the proof will be by induction on n, similar to the proof of Theorem 2.1. Induction will automatically ensure that the minimal change property holds within the two sublists that are pasted together to form $A^{n,k}$. The tricky part is to prove that the minimal change property holds at the transition between the two sublists. In order to handle this part of the proof, we need to know what the last subset is in any $A^{n,k}$.

From the lists computed above, it appears that

$$A^{n,k}_{\binom{n}{k}-1} = \{1, \dots, k-1, n\}, \tag{2.3}$$

for any integer k such that $1 \leq k \leq n$. If we can prove that Equation (2.3) holds for all relevant k and n, then we will be in good shape. It also seems natural to try to prove that Equation (2.3) is valid by induction on n. However, we quickly encounter another snag if we do this, since the last set in the list $A^{n,k}$ is described in terms of the first set in the list $A^{n-1,k-1}$.

This forces us to also obtain a description of the first set in any list $A^{n,k}$. Again, we can guess a formula by inspection:

$$A^{n,k}_0 = \{1, \dots, k\}, \tag{2.4}$$

for any integer k such that $1 \leq k \leq n$.

We begin by proving that Equations (2.3) and (2.4) are valid.

LEMMA 2.5 *Suppose that $1 \leq k \leq n$. Then Equations (2.3) and (2.4) hold.*

PROOF The proof is by induction on n. For $n = 1$, we must have $k = 1$ and

$$A^{1,1}_0 = \{1\}.$$

Thus both equations hold for $n = 1$.

As an induction hypothesis, suppose that Equations (2.3) and (2.4) hold when $n = j - 1$, for all k such that $1 \leq k \leq j - 1$, where $j \geq 2$ is an integer.

Now, we consider $n = j$, and let $1 \leq k \leq j$. If $k = j$, then we recall that $A^{j,j}_0 = \{1, \dots, j\}$ by definition. Hence, Equations (2.3) and (2.4) hold for $k = j$.

Thus, we may assume that $1 \leq k \leq j - 1$. First, we consider $A_0^{j,k}$. By definition,

$$A_0^{j,k} = A_0^{j-1,k}.$$

By induction, therefore, we have

$$A_0^{j,k} = \{1, \ldots, k\}.$$

We turn to $A_{\binom{n}{k}-1}^{j,k}$. By definition, we have

$$A_{\binom{j}{k}-1}^{j,k} = A_0^{j-1,k-1} \cup \{j\}.$$

Hence, by induction, we have

$$A_{\binom{j}{k}-1}^{j,k} = \{1, \ldots, k-1, j\},$$

as desired.

By induction on n, the proof is complete.   ∎

We are now in a position to prove our main result.

THEOREM 2.6   *For any integers k and n such that $1 \leq k \leq n$, $A^{n,k}$ is a minimal change ordering of S_k^n.*

PROOF   The proof will be induction on n. For $n = 1$, we have $k = 1$ and the result is true. As an induction hypothesis, suppose that $A^{n-1,k}$ is a minimal change ordering for all k such that $1 \leq k \leq n - 1$, where $n \geq 2$ is an integer. We will prove that $A^{n,k}$ is a minimal change ordering for all k such that $1 \leq k \leq n$.

It is trivial that $A^{n,k}$ is a minimal change ordering for $k \in \{1, n\}$ (since there is only one set in the list in these two cases). Hence, we may assume that $2 \leq k \leq n - 1$.

Consider two consecutive sets in $A^{n,k}$, say $A_i^{n,k}$ and A_{i+1}^n. There are three cases to consider, depending on the value of i.

First, if $0 \leq i \leq \binom{n-1}{k} - 2$, then $A_i^{n,k}$ and $A_{i+1}^{n,k}$ are two consecutive k-element sets in $A^{n-1,k}$. Therefore, by induction, $A_i^{n,k}$ and $A_{i+1}^{n,k}$ have distance equal to two.

The second case is $\binom{n-1}{k} \leq i \leq \binom{n}{k} - 2$. This is similar to the first case. This time, $A_i^{n,k}$ and $A_{i+1}^{n,k}$ are formed from two consecutive $(k-1)$-element sets in $A^{n-1,k-1}$ by inserting the new element n into each of them. Therefore, by induction, $A_i^{n,k}$ and $A_{i+1}^{n,k}$ have distance equal to two.

The final case is $i = \binom{n-1}{k} - 1$. We have

$$A_{\binom{n-1}{k}-1}^{n,k} = A_{\binom{n-1}{k}-1}^{n-1,k}$$

$$= \{1, \ldots, k-1, n-1\},$$

by Lemma 2.5. Also,

$$A^{n,k}_{\binom{n-1}{k}} = A^{n-1,k-1}_{\binom{n-1}{k-1}-1} \cup \{n\}$$
$$= \{1,\ldots,k-2,n-1,n\},$$

by Lemma 2.5. Therefore $A^{n,k}_{\binom{n-1}{k}-1}$ and $A^{n,k}_{\binom{n-1}{k}}$ have distance equal to two.

By induction, the proof is complete. ∎

Ranking and unranking turn out to be quite simple in the revolving door ordering. As before, we will write a subset T in increasing order as $\overrightarrow{T} = [t_1, t_2, \ldots, t_k]$, where $t_1 < t_2 < \cdots < t_k$. The following formula holds:

$$\text{rank}(T) = \sum_{i=1}^{k}(-1)^{k-i}\left(\binom{t_i}{i} - 1\right) = \begin{cases} \displaystyle\sum_{i=1}^{k}(-1)^{k-i}\binom{t_i}{i} & \text{if } k \text{ is even} \\[2ex] \displaystyle\sum_{i=1}^{k}(-1)^{k-i}\binom{t_i}{i} - 1 & \text{if } k \text{ is odd.} \end{cases}$$

The ranking algorithm presented in Algorithm 2.11 is an immediate consequence of the formula given above.

Algorithm 2.11: KSUBSETREVDOORRANK $(\overrightarrow{T}, k)$

if $k \equiv 0 \bmod 2$
   then $r \leftarrow 0$
   else $r \leftarrow -1$
$s \leftarrow 1$
for $i \leftarrow k$ **downto** 1
   do $\begin{cases} r \leftarrow r + s\binom{t_i}{i} \\ s \leftarrow -s \end{cases}$
return (r)

The corresponding unranking algorithm is displayed in Algorithm 2.12.

Algorithm 2.12: KSUBSETREVDOORUNRANK (r, k, n)

$x \leftarrow n$
for $i \leftarrow k$ **downto** 1
   do $\begin{cases} \textbf{while } \binom{x}{i} > r \\ \quad \textbf{do } x \leftarrow x - 1 \\ t_i \leftarrow x + 1 \\ r \leftarrow \binom{x+1}{i} - r - 1 \end{cases}$
return $(\overrightarrow{T})$

Finally, we present a successor algorithm for the revolving door ordering, in Algorithm 2.13. In this algorithm, the successor of the last k-subset is the first one. In other words, we think of the list $A^{n,k}$ as being ordered cyclicly, and therefore we define

$$\text{successor}(\{1,\ldots,k-1,n\}) = \{1,\ldots,k\}.$$

Note that this is also a minimal change.

Algorithm 2.13 begins by defining t_{k+1} to be $n+1$. This means that we do not have to handle the situation $j = k$ as a special case.

Algorithm 2.13: KSUBSETREVDOORSUCCESSOR $(\overrightarrow{T}, k, n)$

$t_{k+1} \leftarrow n + 1$
$j \leftarrow 1$
while $(j \le k)$ **and** $(t_j = j)$
  do $j \leftarrow j + 1$
if $k \not\equiv j \bmod 2$

$\text{then} \begin{cases} \textbf{if } j = 1 \\ \quad \textbf{then } t_1 \leftarrow t_1 - 1 \\ \quad \textbf{else } \begin{cases} t_{j-1} \leftarrow j \\ t_{j-2} \leftarrow j - 1 \end{cases} \end{cases}$

$\text{else} \begin{cases} \textbf{if } t_{j+1} \ne t_j + 1 \\ \quad \textbf{then } \begin{cases} t_{j-1} \leftarrow t_j \\ t_j \leftarrow t_j + 1 \end{cases} \\ \quad \textbf{else } \begin{cases} t_{j+1} \leftarrow t_j \\ t_j \leftarrow j \end{cases} \end{cases}$

return $(\overrightarrow{T})$

2.4   Permutations

2.4.1   Lexicographic ordering

We now look at the generation of all $n!$ permutations of the set $\{1,\ldots,n\}$. A *permutation* is a bijection from a set to itself. One way to represent a permutation $\pi : \{1,\ldots,n\} \to \{1,\ldots,n\}$ is by listing its values, as follows:

$$[\pi[1],\ldots,\pi[n]].$$

We call this the *list representation* of the permutation π. Saying that π is a permutation is equivalent to saying that each element in $\{1,\ldots,n\}$ occurs exactly once in this list.

First, we will look at the lexicographic ordering of permutations. The lexicographic ordering is defined in terms of the list representation. As an example, when $n = 3$, the lexicographic ordering of the six permutations of $\{1, 2, 3\}$ is as follows:

$$[1, 2, 3], [1, 3, 2], [2, 1, 3], [2, 3, 1], [3, 1, 2], [3, 2, 1].$$

We begin by describing an algorithm for generating permutations in lexicographic order. This generation algorithm depends on a successor algorithm that finds the permutation that immediately follows a given permutation (in lexicographic order). In Algorithm 2.14, π is a permutation of $\{1, \ldots, n\}$ given in list representation.

Algorithm 2.14 has four steps. In the first **while** loop, we find i such that

$$\pi[i] < \pi[i + 1] > \pi[i + 2] > \cdots > \pi[n].$$

Note that by setting $\pi[0]$ to 0, we ensure that the **while** loop terminates with $0 \leq i \leq n - 1$. If $i = 0$, then

$$\pi = [n, n - 1, \ldots, 1]$$

is the last permutation lexicographically and has no successor. Otherwise, we proceed to the second **while** loop, where we find the integer j such that $\pi[j] > \pi[i]$ and $\pi[k] < \pi[i]$ for $j < k \leq n$ (i.e., j is the position of the last element among $\pi[i+1], \ldots, \pi[n]$ that is greater than $\pi[i]$). The third step is to interchange $\pi[i]$ and $\pi[j]$, and the fourth step is to reverse the sublist

$$[\pi[i + 1], \ldots, \pi[n]].$$

Algorithm 2.14: PERMLEXSUCCESSOR (n, π)

$\pi[0] \leftarrow 0$
$i \leftarrow n - 1$
while $\pi[i + 1] < \pi[i]$
   do $i \leftarrow i - 1$
if $i = 0$
   then return ("undefined")
$j \leftarrow n$
while $\pi[j] < \pi[i]$
   do $j \leftarrow j - 1$
$t \leftarrow \pi[j]$
$\pi[j] \leftarrow \pi[i]$
$\pi[i] \leftarrow t$
for $h \leftarrow i + 1$ **to** n
   do $\rho[h] \leftarrow \pi[h]$
for $h \leftarrow i + 1$ **to** n
   do $\pi[h] \leftarrow \rho[n + i + 1 - h]$
return (π)

As an example, suppose that $n = 7$ and

$$\pi = [3, 6, 2, 7, 5, 4, 1].$$

Then, after the first **while** loop, we have $i = 3$, since

$$2 < 7 > 5 > 4 > 1.$$

After the second **while** loop, we have $j = 6$ since $4 > 2$ and $1 < 2$. In the third step, we interchange π_3 and π_6, producing

$$[3, 6, 4, 7, 5, 2, 1].$$

Finally, we reverse the sublist
$$[7, 5, 2, 1],$$

producing the permutation
$$[3, 6, 4, 1, 2, 5, 7],$$

which is the successor of π.

It is now easy to generate all $n!$ permutations of $\{1, \dots, n\}$. We can begin with the permutation $[1, 2, \dots, n]$ (which is the first permutation lexicographically) and invoke Algorithm 2.14 a total of $n! - 1$ times.

We next turn to ranking and unranking permutations in lexicographic order. In the lexicographic ordering of permutations of $\{1, \dots, n\}$, we first have the $(n-1)!$ permutations that begin with a "1", followed by the $(n - 1)!$ permutations that begin with a "2", etc. Hence, if π is a permutation of $\{1, \dots, n\}$, it is clear that

$$(\pi[1] - 1)(n - 1)! \leq \mathsf{rank}(\pi) \leq \pi[1](n - 1)! - 1.$$

Let r' denote the rank of π within the group of $(n - 1)!$ permutations that begin with $\pi[i]$. Then r' is the rank of $[\pi[2], \dots, \pi[n]]$ when it is considered as a permutation of $\{1, \dots, n\} \backslash \{\pi[1]\}$. If we decrease every element of $[\pi[2], \dots, \pi[n]]$ that is greater than $\pi[1]$ by one, then we obtain a permutation π' of $\{1, \dots, n-1\}$ that also has rank r'.

This observation leads to a recursive formula for lexicographic rank of permutations of $\{1, \dots, n\}$. For $n > 1$, we have

$$\mathsf{rank}(\pi, n) = (\pi[1] - 1)(n - 1)! + \mathsf{rank}(\pi', n - 1),$$

where
$$\pi'[i] = \begin{cases} \pi[i+1] - 1 & \text{if } \pi[i+1] > \pi[1] \\ \pi[i+1] & \text{if } \pi[i+1] < \pi[1]. \end{cases}$$

Initial conditions for this recurrence relation are given by

$$\mathsf{rank}([1], 1) = 0.$$

We work out a small example to illustrate:

$$\text{rank}([2,4,1,3],4) = 6 + \text{rank}([3,1,2],3)$$
$$= 6 + 4 + \text{rank}([1,2],2)$$
$$= 6 + 4 + 0 + \text{rank}([1],1)$$
$$= 6 + 4 + 0 + 0$$
$$= 10.$$

It is easy to convert this recursive formula into a non-recursive algorithm, which we present as Algorithm 2.15.

Algorithm 2.15: PERMLEXRANK (n, π)

$r \leftarrow 0$
$\rho \leftarrow \pi$
for $j \leftarrow 1$ **to** n
$\quad$**do** $\begin{cases} r \leftarrow r + (\rho[j] - 1)\,(n - j)! \\ \textbf{for } i \leftarrow j + 1 \textbf{ to } n \\ \quad \textbf{do} \begin{cases} \textbf{if } \rho[i] > \rho[j] \\ \quad \textbf{then } \rho[i] \leftarrow \rho[i] - 1 \end{cases} \end{cases}$
return (r)

Now suppose we want to unrank the integer r, where $0 \le r \le n! - 1$. Unranking can be done fairly easily if we first determine the *factorial representation* of r, by expressing r in the form

$$r = \sum_{i=1}^{n-1} (d_i \cdot i!),$$

where $0 \le d_i \le i$ for $i = 1, \ldots, n - 1$. (We leave it as an exercise to prove that any non-negative integer r such that $0 \le r \le n! - 1$ has a unique factorial representation of this form.)

Suppose that $\pi = \text{unrank}(r)$ in the lexicographic ordering. It is easy to see that

$$\pi[1] = d_{n-1} + 1.$$

Thus the first element of π is determined immediately from the factorial representation of r. Now, denote

$$r' = r - d_{n-1} \cdot (n - 1)!,$$

and suppose that $\pi' = \text{unrank}(r')$, where π' is a permutation of $\{1, \ldots, n - 1\}$. (This could be done recursively, for example.) Suppose we increment by one all elements of π' that are greater than d_{n-1}. Finally, define

$$\pi[i] = \pi'[i + 1]$$

for $2 \leq i \leq n$. Then it will be the case that $\pi = \text{unrank}(r)$.

As an example, suppose that $n = 4$ and $r = 10$. The factorial representation of r is

$$1 \cdot 3! + 2 \cdot 2! + 0 \cdot 1!.$$

Hence, $\pi[1] = d_3 + 1 = 2$. Now, compute $r' = r - 6 = 4$. It can be verified that $\pi' = \text{unrank}(4) = [3, 1, 2]$. Then we increment the first and third elements by one, so $\pi' = [4, 1, 3]$. Hence, we obtain

$$\text{unrank}(10) = [2, 4, 1, 3].$$

Algorithm 2.16 is a non-recursive implementation of this unranking algorithm. In this algorithm, we use a function mod which performs modular reduction according to the following rule:

$$\text{mod}(x, m) = r \Leftrightarrow x \equiv r \bmod m \text{ and } 0 \leq r \leq m - 1.$$

Algorithm 2.16: PERMLEXUNRANK (n, r)

$\pi[n] \leftarrow 1$
for $j \leftarrow 1$ **to** $n - 1$
$\quad$**do** $\begin{cases} d \leftarrow \dfrac{\text{mod}(r, (j+1)!)}{j!} \\ r \leftarrow r - d \cdot j! \\ \pi[n - j] \leftarrow d + 1 \\ \textbf{for } i \leftarrow n - j + 1 \textbf{ to } n \\ \quad \textbf{do} \begin{cases} \textbf{if } \pi[i] > d \\ \quad \textbf{then } \pi[i] \leftarrow \pi[i] + 1 \end{cases} \end{cases}$
return (π)

We illustrate Algorithm 2.16 by recomputing unrank(10). Initially, we set

$$\pi[4] = 1.$$

When $j = 1$, we compute

$$d = \frac{\text{mod}(10, 2)}{1} = 0,$$

$$\pi[3] = 1 \text{ and } \pi[4] = 2.$$

When $j = 2$, we have

$$d = \frac{\text{mod}(10, 6)}{2} = 2,$$

$$r = 10 - 2 \cdot 2 = 6,$$

and

$$\pi[2] = 3.$$

Finally, when $j = 3$, we have

$$d = \frac{\mod(6, 24)}{6} = 1,$$

$$r = 6 - 1 \cdot 6 = 0,$$

$$\pi[1] = 2, \pi[2] = 4 \text{ and } \pi[4] = 3.$$

Hence, we obtain

$$\text{unrank}(10) = [2, 4, 1, 3],$$

as before.

2.4.2 Minimal change ordering

First we need to give some thought as to what a minimal change would be in the context of permutations. It is certainly the case that any two distinct permutations π and π' of $\{1, \ldots, n\}$ must differ in at least two positions. Further, if π and π' differ in exactly two positions, then one can be obtained from the other by a single *transposition* (i.e., by exchanging the elements in the two given positions). It may even happen that the two positions are adjacent; so, we in fact transpose two adjacent elements in order to transform π into π'. This is equivalent to saying that there exists an integer i, $1 \leq i \leq n - 1$, such that

$$\pi'[j] = \begin{cases} \pi[j+1] & \text{if } j = i \\ \pi[j-1] & \text{if } j = i + 1 \\ \pi[j] & \text{if } j \neq i, i+1. \end{cases}$$

This is in fact the definition we will take for a minimal change for permutations.

The *Trotter-Johnson algorithm* is a nice example of a minimal change algorithm for generating the $n!$ permutations. It can be most easily described recursively. Suppose we have a listing of the $(n-1)!$ permutations of $\{1, \ldots, n-1\}$ in minimal change order, say

$$\mathsf{T}^{n-1} = [\pi_0, \pi_1, \ldots, \pi_{(n-1)!-1}].$$

Form a new list by repeating each permutation in the list T^{n-1} n times. Now insert the element n into each of the n copies of each permutation π_i, as follows. If i is even, then we first insert element n after the element in position $n - 1$, then after the element in position $n - 2$, etc., and finally preceding the element in position 1. If i is odd, then we proceed in the opposite order, inserting element n into the n copies of π_i from the beginning to the end of π.

We illustrate the procedure for $n = 1, 2, 3$ and 4. We begin with $n = 1$, where we have

$$\mathsf{T}^1 = [1].$$

Next, we obtain

$$\mathsf{T}^2 = [[1, 2], [2, 1]].$$

The next list, T^3, would be produced by taking three copies of each permutation in T^2, and inserting the element 3 as follows:

```
            1       2  3
            1   3   2
        3   1       2
        3   2       1
            2   3   1
            2       1  3
```

This gives the following:

$$T^3 = [[1, 2, 3], [1, 3, 2], [3, 1, 2], [3, 2, 1], [2, 3, 1], [2, 1, 3]].$$

To do the next case ($n = 4$), we would repeat each permutation in T^3 four times, and insert the element 4 as follows:

```
            1       2       3  4
            1       2   4   3
            1   4   2       3
        4   1       2       3
        4   1       3       2
            1   4   3       2
            1       3   4   2
            1       3       2  4
            3       1       2  4
            3       1   4   2
            3   4   1       2
        4   3       1       2
        4   3       2       1
            3   4   2       1
            3       2   4   1
            3       2       1  4
            2       3       1  4
            2       3   4   1
            2   4   3       1
        4   2       3       1
        4   2       1       3
            2   4   1       3
            2       1   4   3
            2       1       3  4
```

This yields T^4:

$$T^4 = [[1,2,3,4], [1,2,4,3], [1,4,2,3], [4,1,2,3],$$
$$[4,1,3,2], [1,4,3,2], [1,3,4,2], [1,3,2,4],$$
$$[3,1,2,4], [3,1,4,2], [3,4,1,2], [4,3,1,2],$$
$$[4,3,2,1], [3,4,2,1], [3,2,4,1], [3,2,1,4],$$
$$[2,3,1,4], [2,3,4,1], [2,4,3,1], [4,2,3,1],$$
$$[4,2,1,3], [2,4,1,3], [2,1,4,3], [2,1,3,4]].$$

We now develop ranking and unranking algorithms for the Trotter-Johnson ordering. Let's look first at ranking. Suppose

$$\pi = [\pi[1], \ldots, \pi[n]],$$

where $\pi[k] = n$. Define a permutation of π' of $\{1, \ldots, n-1\}$ as follows:

$$\pi' = [\pi[1], \ldots, \pi[k-1], \pi[k+1], \ldots, \pi[n]].$$

Observe that π was constructed from π' by inserting the element n. Since each permutation in T^{n-1} was replicated n times in the construction of T^n, we see that

$$n \, \mathsf{rank}(\pi') \leq \mathsf{rank}(\pi) \leq n \, \mathsf{rank}(\pi') + n - 1.$$

The exact rank of π is determined from the position of π' into which the element n was inserted (there are in fact two cases, depending on whether $\mathsf{rank}(\pi')$ is even or odd).

In this way, we obtain the following recursive formula for rank:

$$\mathsf{rank}(\pi) = n \, \mathsf{rank}(\pi') + \epsilon,$$

where

$$\epsilon = \begin{cases} n - k & \text{if } \mathsf{rank}(\pi', n-1) \text{ is even} \\ k - 1 & \text{if } \mathsf{rank}(\pi', n-1) \text{ is odd.} \end{cases}$$

As an example, we compute $\mathsf{rank}([3,4,2,1])$. We have $n = 4$, $k = 2$ and $\pi' = [3,2,1]$. Now, $\mathsf{rank}([3,2,1]) = 3$ could be computed recursively, using the same method. Since 3 is odd, we see that

$$\mathsf{rank}([3,4,2,1]) = 4 \times 3 + 2 - 1 = 13.$$

This recursive formula can be modified to produce a non-recursive algorithm for computing rank, which we present as Algorithm 2.17.

Algorithm 2.17: TROTTERJOHNSONRANK (π)

$r \leftarrow 0$
for $j \leftarrow 2$ **to** n

$\quad$ **do** $\begin{cases} k \leftarrow 1 \\ i \leftarrow 1 \\ \textbf{while } \pi[i] \neq j \\ \quad \textbf{do} \begin{cases} \textbf{if } \pi[i] < j \\ \quad \textbf{then } k \leftarrow k+1 \\ i \leftarrow i+1 \end{cases} \\ \textbf{if } r \equiv 0 \bmod 2 \\ \quad \textbf{then } r \leftarrow jr+j-k \\ \quad \textbf{else } r \leftarrow jr+k-1 \end{cases}$

return (r)

Now we study unranking. We again approach the problem recursively, using essentially the same approach that we did for ranking. Suppose we are given n, and $0 \leq r \leq n! - 1$. Define

$$r' = \left\lfloor \frac{r}{n} \right\rfloor.$$

Suppose that the permutation π' of $\{1, \ldots, n-1\}$ has rank r'. Then compute $k = r - nr'$. Finally, insert n into π' in position $k+1$ if r' is odd; and into position $n - k$ if r' is even. (When we insert the new element n into a given position of π', the element in that position, and all elements to the right, are shifted one position to the right.)

As an example, suppose we want to unrank $r = 13$ when $n = 4$. We first compute $r' = 3$. The permutation of $\{1, 2, 3\}$ that has rank 3 is $\pi' = [3, 2, 1]$. Now, $k = 13 - 4 \times 3 = 1$. Since $r' = 3$ is odd, we insert the element 4 into π' in position 2. Thus we obtain $\pi = [3, 4, 2, 1]$.

Algorithm 2.18 is a non-recursive unranking algorithm that is based on this method.

Now we develop a successor algorithm. This is somewhat more complicated. Suppose we are given π, a permutation of $\{1, \ldots, n\}$, and we want to compute the successor of π in the list T^n. Suppose that $\pi[k] = n$, and let π' be constructed as we did in our discussion of unranking, by deleting the element n from π. There are essentially four cases that arise:

Algorithm 2.18: TROTTERJOHNSONUNRANK (n, r)

$\pi[1] \leftarrow 1$
$r_2 \leftarrow 0$
for $j \leftarrow 2$ **to** n

$\text{do} \begin{cases} r_1 \leftarrow \left\lfloor \frac{r\,j!}{n!} \right\rfloor \\ k \leftarrow r_1 - j\,r_2 \\ \textbf{if } r_2 \textbf{ is even} \\ \quad \textbf{then} \begin{cases} \textbf{for } i \leftarrow j - 1 \textbf{ downto } j - k \\ \quad \textbf{do } \pi[i+1] \leftarrow \pi[i] \\ \pi[j - k] \leftarrow j \end{cases} \\ \quad \textbf{else} \begin{cases} \textbf{for } i \leftarrow j - 1 \textbf{ downto } k + 1 \\ \quad \textbf{do } \pi[i+1] \leftarrow \pi[i] \\ \pi[k + 1] \leftarrow j; \end{cases} \\ r_2 \leftarrow r_1 \end{cases}$

return (π)

---

1. Suppose that $\text{rank}(\pi')$ is even and $k \neq 1$. Then the successor of π is constructed by interchanging $\pi[k]$ and $\pi[k-1]$.

2. Suppose that $\text{rank}(\pi')$ is odd and $k \neq n$. Then the successor of π is constructed by interchanging $\pi[k]$ and $\pi[k+1]$.

3. Suppose that $\text{rank}(\pi')$ is even and $k = 1$. Suppose that the successor of π' (in the list T^{n-1}) is constructed by interchanging $\pi'[j]$ and $\pi'[j+1]$. Then the successor of π is constructed by interchanging $\pi[j+1]$ and $\pi[j+2]$. (Note that $\pi'[j] = \pi[j+1]$ and $\pi'[j+1] = \pi[j+2]$.)

4. Suppose that $\text{rank}(\pi')$ is odd and $k = n$. Suppose that the successor of π' in T^{n-1} is constructed by interchanging $\pi'[j]$ and $\pi'[j+1]$. Then the successor of π is constructed by interchanging $\pi[j]$ and $\pi[j+1]$. (Note that $\pi'[j] = \pi[j]$ and $\pi'[j+1] = \pi[j+1]$.)

Observe that cases 1 and 2 occur most of the time. (In fact, if π is chosen at random from the set of all permutations of $\{1, \ldots, n\}$, then the probability that one of case 1 or case 2 occurs is $(n-1)/n$.) The pair of elements of π to be interchanged is computed immediately in these two cases. Cases 3 and 4 require determining the successor of π' recursively, within the list T^{n-1}. The successor of π' is formed by interchanging two adjacent elements, say x and y. Then the successor of π is also constructed by interchanging x and y.

The above four cases assume that $\text{rank}(\pi')$ is known. We would prefer to find an algorithm to compute successors that does not require a rank computation. However, we never need to know the precise value of $\text{rank}(\pi')$ — we just need to know whether it is even or odd. It turns out that we can compute the parity of $\text{rank}(\pi')$ fairly easily.

A permutation π of $\{1, \ldots, n\}$ is called an *even permutation* if π can be transformed into the permutation $[1, 2, \ldots, n]$ by performing an even number of interchanges of two elements (recall that this operation is called a transposition). A permutation is an *odd permutation*, otherwise. For example, the permutation $\pi = [5, 1, 3, 4, 2]$ is an even permutation since it can be transformed into $[1, 2, 3, 4, 5]$ by first interchanging the elements 5 and 1, and then interchanging the elements 5 and 2. It can be shown that exactly half of the $n!$ permutations of an n-set are even permutations.

Each permutation $\pi_i \in \mathbf{T}^n$ is obtained from the previous one, π_{i-1}, by a single transposition. Also, the permutation having rank 0 is the identity permutation, which is an even permutation. Thus, the parity of a permutation π (i.e., even or odd) is the same as the parity of $\mathsf{rank}(\pi)$.

There are some relatively easy ways to compute the parity of a permutation π of $\{1, \ldots, n\}$. One way is to count the number of N of ordered pairs (i, j) such that $\pi[i] > \pi[j]$ (where $1 \le i < j \le n$). Then the parity of π is the same as the parity of N. A straightforward computation of N requires time $\Theta(n^2)$.

Another, more efficient, approach is to represent π by a directed graph which we name D_π. A *directed graph* consists of a set of vertices $\mathcal{V}$ and a set $\mathcal{E}$ of *directed edges* or *arcs*. Each arc $e \in \mathcal{E}$ is an ordered pair (u, v), where $u, v \in \mathcal{V}$. If $u = v$, then the arc (u, v) is a *loop*.

The directed graph D_π is defined to have vertex set $\mathcal{V} = \{1, \ldots, n\}$ and arc set $\mathcal{E} = \{(i, \pi[i]) : 1 \le i \le n\}$. The graph D_π consists of a union of disjoint directed circuits. (Note that some of these circuits may be of size 1. These are loops in D_π, which occur whenever $\pi[i] = i$.) If D_π contains exactly c directed circuits, then it can be shown that the parity of π is the same as the parity of $n - c$. This method of computing parity has complexity $\Theta(n)$; so, it is more efficient computationally, at least for large n.

Algorithm 2.19 computes the parity of a permutation using this method.

We are now in a position to present a successor algorithm. Algorithm 2.20 is based on the four cases enumerated above. It is implemented in a non-recursive manner, however.

We illustrate Algorithm 2.20 by computing successor($[4, 3, 1, 2]$). Here, we have $n = 4$. Initially, $st = 0$, $\rho = [4, 3, 1, 2]$, *done* = **true** and $m = 4$. We next compute $d = 1$, $\rho = [3, 1, 2]$ and PERMPARITY($3, [3, 1, 2]$) = 0. Therefore we set $m = 2$ and $st = 1$, and execute the outer **while** loop again. In the second iteration of the **while** loop, we get $d = 1$, $\rho = [1, 2]$, and PERMPARITY($2, [1, 2]$) = 0. Hence, we set $m = 1$ and $st = 2$, and execute the outer **while** loop again. In the third iteration of the **while** loop, we get $d = 2$, $\rho = [1]$, and PERMPARITY($1, [1]$) = 0. Hence, we interchange $\pi[4]$ and $\pi[3]$, and successor($[4, 3, 1, 2]$) = $[4, 3, 2, 1]$.

Algorithm 2.19: PERMPARITY (n, π)

$$\textbf{for } i \leftarrow 1 \textbf{ to } n \textbf{ do } a[i] \leftarrow 0$$
$$c \leftarrow 0$$
$$\textbf{for } j \leftarrow 1 \textbf{ to } n$$
$$\textbf{do} \begin{cases} \textbf{if } a[j] = 0 \\ \quad \textbf{then} \begin{cases} c \leftarrow c + 1 \\ a[j] \leftarrow 1 \\ i \leftarrow j \\ \textbf{while } \pi[i] \neq j \textbf{ do } \begin{cases} i \leftarrow \pi[i] \\ a[i] \leftarrow 1 \end{cases} \end{cases} \end{cases}$$
$$\textbf{return } ((n - c) \bmod 2)$$

Algorithm 2.20: TROTTERJOHNSONSUCCESSOR (n, π)

external PERMPARITY()
$st \leftarrow 0$
for $i \leftarrow 1$ **to** n **do** $\rho[i] \leftarrow \pi[i]$
$done \leftarrow$ **false**
$m \leftarrow n$
while $m > 1$ **and not** $done$

$$\textbf{do} \begin{cases} d \leftarrow 1 \\ \textbf{while } \rho[d] \neq m \textbf{ do } d \leftarrow d + 1 \\ \textbf{for } i \leftarrow d \textbf{ to } m - 1 \textbf{ do } \rho[i] \leftarrow \rho[i+1] \\ par \leftarrow \text{PERMPARITY}(m - 1, \rho) \\ \textbf{if } par = 1 \\ \quad \textbf{then} \begin{cases} \textbf{if } d = m \\ \quad \textbf{then } m \leftarrow m - 1 \\ \quad \textbf{else} \begin{cases} temp \leftarrow \pi[st + d] \\ \pi[st + d] \leftarrow \pi[st + d + 1] \\ \pi[st + d + 1] \leftarrow temp \\ done \leftarrow \textbf{true} \end{cases} \end{cases} \\ \quad \textbf{else} \begin{cases} \textbf{if } d = 1 \\ \quad \textbf{then} \begin{cases} m \leftarrow m - 1 \\ st \leftarrow st + 1 \end{cases} \\ \quad \textbf{else} \begin{cases} temp \leftarrow \pi[st + d] \\ \pi[st + d] \leftarrow \pi[st + d - 1] \\ \pi[st + d - 1] \leftarrow temp \\ done \leftarrow \textbf{true} \end{cases} \end{cases} \end{cases}$$

if $m = 1$
  then return ("undefined")

2.5 Notes

Section 2.1

Several books which contain information on combinatorial generation include
Bogart [6], Cameron [16], Nijenhuis and Wilf [80], Reingold, Nievergelt and
Deo [90], Stanton and White [101], Tucker [107], Wells [111], Wilf [114] and
Williamson [115].

Section 2.2

The topic of Gray codes has been an active research area, particularly in the last
10 years. Good starting points for learning more about this area are Savage [94]
and Wilf [114, Chapters 1 and 2].

Section 2.3

The revolving door algorithm is due to Nijenhuis and Wilf; see [80, 114].

Section 2.4

The Trotter-Johnson algorithm was described independently in [106] and [50].

Exercises

2.1 Let $S = \{2, 3, 5, 7, 11, 13\}$. Determine the rank of the subset $\{3, 7, 13\}$ among the
subsets of S in lexicographic order, and verify that

$$\text{unrank}(\text{rank}(\{3, 7, 13\})) = \{3, 7, 13\}.$$

2.2 Find all possible Gray codes for $n = 4$.

2.3 Prove Theorem 2.2.

2.4 Find the successor and the rank of the binary vector 01010110 in the Gray code G^8.

2.5 What three-element set has rank equal to 1000 in co-lex order?

2.6 Find a successor algorithm for the co-lex ordering of k-subsets of an n-element set.

2.7 What is the rank of $\{3, 6, 7, 9\}$ considered as a 4-subset of $\{0, \ldots, 12\}$, in lexicographic, co-lex and revolving door order? What is its successor in each of these
orders?

2.8 Prove Theorem 2.4.

2.9 Suppose $1 \le k \le n$, and we delete all vectors in the binary reflected Gray code G^n
that do not correspond to subsets of cardinality k. Prove that the vectors that remain
comprise a minimal change ordering for the k-element subsets of an n-set (in fact,
it is precisely the revolving door ordering).

2.10 Another way to order the subsets of an n-element set is to order them first in increasing size, and then in lexicographic order for each fixed size. For example,

when $n = 3$, this ordering for the subsets of $S = \{1, 2, 3\}$ is:

$$\emptyset, \{1\}, \{2\}, \{3\}, \{1, 2\}, \{1, 3\}, \{2, 3\}, \{1, 2, 3\}.$$

Develop unranking, ranking and successor algorithms for the subsets with respect to this ordering.

2.11 Find the rank and successor of the permutation $[2, 4, 6, 7, 5, 3, 1]$ in lexicographic and Trotter-Johnson order.

2.12 A *derangement* is a permutation $[\pi[1], \pi[2], \ldots, \pi[n]]$ of the set $\{1, 2, 3, \ldots, n\}$ such that $\pi[i] \neq i$, for all $i = 1, 2, \ldots, n$. Let D_n denote the number of derangements of an n-element set. Prove the recurrence relation $D_n = (n - 1)(D_{n-1} + D_{n-2})$. Then, use this recurrence relation to develop an algorithm to generate all the derangements.

2.13 A *multiset* is a set with (possibly) repeated elements. A k-multiset is one that contains k elements (counting repetitions). Thus, for example, $\{1, 2, 3, 1, 1, 3\}$ is a 6-multiset. The k-multisets of an n-set can be ordered lexicographically, by sorting the elements in each multiset in non-decreasing order and storing the result as a list of length k. Develop unranking, ranking and successor algorithms for the k-multisets of an n-set.

2.14 k-permutations were defined in Section 1.2.1. Assuming that $k < n$, develop a minimal change algorithm to generate the k-permutations of an n-set with a minimal change algorithm. At each step, this algorithm should change exactly one element.

3

More Topics in Combinatorial Generation

3.1 Integer partitions

Let m be a positive integer. A *partition* of m is a representation of m as a sum of positive integers, say $m = a_1 + \ldots + a_n$. The summands $a_1, \ldots, a_n$ are called the *parts* of the partition, and their order is ignored. The notation $P(m)$ is used to denote the number of partitions of m; $P(m)$ is called a *partition number*.

The first few partition numbers are $P(1) = 1$, $P(2) = 2$, $P(3) = 3$, $P(4) = 5$, $P(5) = 7$ and $P(6) = 11$. As an example, we list the 11 different partitions of the integer 6:

$$6$$
$$5 + 1$$
$$4 + 2$$
$$4 + 1 + 1$$
$$3 + 3$$
$$3 + 2 + 1$$
$$3 + 1 + 1 + 1$$
$$2 + 2 + 2$$
$$2 + 2 + 1 + 1$$
$$2 + 1 + 1 + 1 + 1$$
$$1 + 1 + 1 + 1 + 1 + 1$$

Although partitions have been studied by mathematicians for hundreds of years and many interesting results are known, there is no known formula for the values $P(m)$. The growth rate of $P(m)$ is known however; it can be shown that $P(m)$ is $\Theta\left(e^{\pi\sqrt{2m/3}}/m\right)$.

A partition $m = a_1 + \ldots + a_n$ is said to be in *standard form* if $a_1 \geq a_2 \geq \ldots \geq a_n$. (Note that the 11 partitions of 6 given above are all in standard form.) We will sometimes write a partition in standard form as a list, i.e., $[a_1, a_2, \ldots, a_n]$, particularly in algorithms.

Our first algorithm, Algorithm 3.1, is a simple recursive algorithm that can be used to generate all the partitions of m in standard form.

Algorithm 3.1: GENPARTITIONS (m)

procedure RECPARTITION(m, B, N)
  if $m = 0$
    then output $([a_1, \ldots, a_N])$
    else $\begin{cases} \textbf{for } i \leftarrow 1 \textbf{ to } \min(B, m) \\ \quad \textbf{do } \begin{cases} a_{N+1} \leftarrow i \\ \text{RECPARTITION}(m - i, i, N + 1) \end{cases} \end{cases}$

main
RECPARTITION$(m, m, 0)$

In the procedure RECPARTITION , the values for $a_1, \ldots, a_N$ have been chosen already. The parameter B is an upper bound on the size of the next part to be chosen, and m will be the sum of the values $a_{N+1}, a_{N+2}, \ldots$ (which have not been chosen yet). Thus a_{N+1} can take on any value between 1 and B. If we define a_{N+1} to have the value i, then i becomes an upper bound on the values of the remaining parts (since we are constructing the partition in standard form). Also, the value of m will be decreased by i, and N is increased by one, when the procedure RECPARTITION is called recursively. Algorithm 3.1 simply calls the procedure RECPARTITION with $B = m$ and $N = 0$ in order to get things started.

The *Ferrers-Young diagram* of a partition is formed by first writing the partition in standard form, say $m = a_1 + \ldots + a_n$, and then constructing an array of dots, say, where the ith row contains a_i dots ($1 \leq i \leq n$) and the rows of dots are all left-justified.

For example, the partition $7 = 4 + 2 + 1$ has the following Ferrers-Young diagram:

$$\mathbf{D} = \begin{matrix} \bullet & \bullet & \bullet & \bullet \\ \bullet & \bullet & & \\ \bullet & & & \end{matrix}$$

Suppose we have a Ferrers-Young diagram $\mathbf{D}$, and we construct the diagram $\mathbf{D}^*$ in which the rows of $\mathbf{D}$ become the columns of $\mathbf{D}^*$. Then $\mathbf{D}^*$ is called the *conjugate diagram* of $\mathbf{D}$, and the corresponding partitions are called *conjugate partitions*.

For example, the conjugate of the diagram $\mathbf{D}$ displayed above is the following:

$$\mathbf{D}^* = \begin{matrix} \bullet & \bullet & \bullet \\ \bullet & \bullet & \\ \bullet & & \\ \bullet & & \end{matrix}$$

Hence, the two partitions $7 = 4 + 2 + 1$ and $7 = 3 + 2 + 1 + 1$ are conjugates.

Suppose we consider the set $\mathcal{P}(m)$ of all partitions of m. It is easy to see that the operation of conjugation is a bijection of $\mathcal{P}(m)$ to itself. Further, if $n \leq m$ is a positive integer, then the operation of conjugation is a bijection of the set of partitions of m having n parts with the set of partitions of m in which the largest part has size n. Hence, we have the following fundamental result.

THEOREM 3.1 *The number of partitions of m having n parts is the same as the number of partitions of m in which the largest part has size n.*

Since conjugation is such an important operation in the study of integer partitions, it seems useful to describe an algorithm to compute the conjugate of a partition given in standard form. This is done in Algorithm 3.2.

Algorithm 3.2: CONJPARTITION $([a_1, \ldots, a_n])$

> **for** $i \leftarrow 1$ **to** a_1
>   **do** $b_i \leftarrow 1$
> $n' \leftarrow a_1$
> **for** $j \leftarrow 2$ **to** n
>   **do** $\begin{cases} \textbf{for } i \leftarrow 1 \textbf{ to } a_j \\ \quad \textbf{do } b_i \leftarrow b_i + 1 \end{cases}$
> **return** $([b_1, \ldots, b_{n'}])$

We will use the notation $P(m, n)$ to denote the number of partitions of m having n parts. Clearly $P(m, 1) = P(m, m) = 1$ for all integers $m \geq 1$. It will be convenient to define $P(m, 0) = 0$ for all $m \geq 1$, and $P(0, 0) = 1$.

It is obvious that the following equation holds:

$$P(m) = \sum_{n=1}^{m} P(m, n). \tag{3.1}$$

No general formula for $P(m, n)$ is known. However, for any fixed value of n, it is possible to compute a formula for $P(m, n)$. For example, it is easy to prove that

$$P(m, 2) = \left\lfloor \frac{m}{2} \right\rfloor.$$

With a bit more work, it can also be shown that

$$P(m, 3) = \left\lfloor \frac{m^2 + 4}{12} \right\rfloor.$$

In general, it is known that $P(m, n)$ is $\Theta(m^{n-1})$ for any fixed integer m.

It is straightforward to generate all the partitions of m in which the largest part has size n. Algorithm 3.3 in fact does this.

Algorithm 3.3: GENPARTITIONS2 (m, n)

procedure RECPARTITION(m, B, N)
  if $m = 0$
    then output $([a_1, \ldots, a_N])$
    else $\begin{cases} \textbf{for } i \leftarrow 1 \textbf{ to } \min(B, m) \\ \quad \textbf{do } \begin{cases} a_{N+1} \leftarrow i \\ \text{RECPARTITION}(m - i, i, N + 1) \end{cases} \end{cases}$

main
  $a_1 \leftarrow n$
  RECPARTITION$(m - n, n, 1)$

The recursive procedure RECPARTITION is the same as in Algorithm 3.1. The only difference is in how we invoke this procedure. Now we set the first (largest) part to have size n, and then generate all partitions of $m - n$ in which all parts have size at most n.

Now, suppose that we want to generate all the partitions of m into n parts. Algorithm 3.4 does this by using Algorithm 3.3 to generate the partitions of m in which the largest part has size n, and then conjugating each partition that is produced, using Algorithm 3.2. (Alternatively, it is not difficult to construct a recursive algorithm to generate the desired partitions directly. This is left as an exercise for the reader.)

Algorithm 3.4: GENPARTITIONS3 (m, n)

external CONJPARTITION()
procedure RECPARTITION2(m, B, N)
  if $m = 0$
    then $\begin{cases} [b_1, \ldots, b_{n'}] \leftarrow \text{CONJPARTITION}([a_1, \ldots, a_n]) \\ \textbf{output } ([b_1, \ldots, b_N]) \end{cases}$
    else $\begin{cases} \textbf{for } i \leftarrow 1 \textbf{ to } \min\{B, m\} \\ \quad \textbf{do } \begin{cases} a_{N+1} \leftarrow i \\ \text{RECPARTITION2}(m - i, i, N + 1) \end{cases} \end{cases}$

main
  $a_1 \leftarrow n$
  RECPARTITION2$(m - n, n, 1)$

We next look at some interesting relations involving the numbers $P(m, n)$. In the proofs of the next results, we will use the notation $\mathcal{P}(m, n)$ to denote the set of all partitions of m having n parts (so $|\mathcal{P}(m, n)| = P(m, n)$). Here is one basic result involving the numbers $P(m, n)$.

THEOREM 3.2   *For positive integers m, n with $m \geq n$,*

$$P(m, n) = P(m - 1, n - 1) + P(m - n, n).$$

PROOF   The idea of the proof is to show that $P(m - 1, n - 1)$ is equal to the number of partitions in $\mathcal{P}(m, n)$ having at least one part of size 1; and $P(m-n, n)$ is equal to the number of partitions in $\mathcal{P}(m, n)$ having no part of size 1. This is done by defining two mappings

$$\phi_1 : \mathcal{P}(m, n) \to \mathcal{P}(m - 1, n - 1)$$

and

$$\phi_2 : \mathcal{P}(m, n) \to \mathcal{P}(m - n, n),$$

as follows:

$$\phi_1([a_1, \ldots, a_n]) = [a_1, \ldots, a_{n-1}], \quad \text{and}$$
$$\phi_2([a_1, \ldots, a_n]) = [a_1 - 1, \ldots, a_n - 1].$$

In the above, partitions are in standard form.

Clearly ϕ_1 is a bijection between the set of partitions in $\mathcal{P}(m, n)$ having at least one part of size 1 and $\mathcal{P}(m - 1, n - 1)$. It is also easy to see that ϕ_2 is a bijection between the set of partitions in $\mathcal{P}(m, n)$ having no parts of size 1 and $\mathcal{P}(m - n, n)$. This completes the proof. ∎

Note that Theorem 3.2 holds when $m = n$ since $P(m, m) = 1$ and

$$P(m - 1, m - 1) + P(0, m) = 1 + 0 = 1.$$

Similarly, when $n = 1$ and $m > 1$, we have $P(m, 1) = 1$ and

$$P(m - 1, 0) + P(m - 1, 1) = 0 + 1 = 1.$$

Another relation that can be proved easily is the following.

THEOREM 3.3

$$P(m, n) = \sum_{i=1}^{n} P(m - n, i).$$

PROOF   For $1 \leq i \leq n$, we show that $P(m-n, i)$ is equal to the number of partitions in $\mathcal{P}(m, n)$ in which there are $n - i$ parts equal to 1. This is done as follows. First, define $\mathcal{P}(m, n)_i$ to consist of all the partitions in $\mathcal{P}(m, n)$ having exactly $n - i$ parts equal to 1. Observe that a partition in standard form, $[a_1, \ldots, a_n]$, is in the set $\mathcal{P}(m, n)_i$ if and only if

$$a_i > 1 \text{ and } a_{i+1} = \ldots = a_n = 1.$$

For $1 \leq i \leq n$, we define a mapping

$$\phi_i : \mathcal{P}(m,n)_i \to \mathcal{P}(m-n,i),$$

as follows:

$$\phi_i([a_1,\ldots,a_n]) = [a_1 - 1,\ldots,a_i - 1].$$

It is easy to see that each ϕ_i is a bijection, and the result follows. $\blacksquare$

The recurrence given in Theorem 3.2 provides a convenient method of computing the value of $P(m,n)$ without generating all the relevant partitions. The idea is to compute all the values $P(i,j)$ such that $1 \leq i \leq m$ and $1 \leq j \leq \min\{i,n\}$, in order, using this recurrence relation. This is done in Algorithm 3.5.

Algorithm 3.5: ENUMPARTITIONS (m,n)

$P(0,0) \leftarrow 1$
for $i \leftarrow 1$ **to** m
  do $P(i,0) \leftarrow 0$
for $i \leftarrow 1$ **to** m
  do $\begin{cases} \textbf{for } j \leftarrow 1 \textbf{ to } \min\{i,n\} \\ \qquad \textbf{do} \begin{cases} \textbf{if } i < 2j \\ \qquad \textbf{then } P(i,j) \leftarrow P(i-1,j-1) \\ \qquad \textbf{else } P(i,j) \leftarrow P(i-1,j-1) + P(i-j,j) \end{cases} \end{cases}$
return (P)

Algorithm 3.5 allows $P(m,n)$ to be determined by computing $\Theta(mn)$ additions of integers. Notice that this bottom-up method is much more efficient than the top-down approach of computing the numbers $P(m,n)$ recursively. The algorithmic design technique being applied is dynamic programming, which was discussed in Section 1.8.2. Table 3.1 records the values of $P(m,n)$ and $P(m)$, for $1 \leq n \leq m \leq 15$.

Algorithm 3.5 can easily be adapted to compute $P(m)$. This method of computing $P(m)$ will require $\Theta(m^2)$ additions of integers. We now describe a method in which the number of additions is reduced to $\Theta(m^{3/2})$. The method makes use of the following recursive formula for $P(m)$ that was first discovered by Euler.

THEOREM 3.4

$$P(m) = \sum_{\{j \geq 1 : 3j^2 - j < 2m\}} (-1)^{j+1} P\left(m - \frac{3j^2 - j}{2}\right)$$

$$+ \sum_{\{j \geq 1 : 3j^2 + j < 2m\}} (-1)^{j+1} P\left(m - \frac{3j^2 + j}{2}\right).$$

TABLE 3.1
A table of partition numbers

						$P(m,n)$					
m	$P(m)$	$n=1$	2	3	4	5	6	7	8	9	10
1	1	1									
2	2	1	1								
3	3	1	1	1							
4	5	1	2	1	1						
5	7	1	2	2	1	1					
6	11	1	3	3	2	1	1				
7	15	1	3	4	3	2	1	1			
8	22	1	4	5	5	3	2	1	1		
9	30	1	4	7	6	5	3	2	1	1	
10	42	1	5	8	9	7	5	3	2	1	1
11	55	1	5	10	11	10	7	5	3	2	1
12	75	1	6	12	15	13	11	7	5	3	2
13	97	1	6	14	18	18	14	11	7	5	3
14	128	1	7	16	23	23	20	15	11	7	5
15	164	1	7	19	27	30	26	21	15	11	7
16	212	1	8	21	34	37	35	28	22	15	11
17	267	1	8	24	39	47	44	38	29	22	15
18	340	1	9	27	47	57	58	49	40	30	22
19	423	1	9	30	54	70	71	65	52	41	30
20	530	1	10	33	64	84	90	82	70	54	42
21	653	1	10	37	72	101	110	105	89	73	55
22	807	1	11	40	84	119	136	131	116	94	75
23	984	1	11	44	94	141	163	164	146	123	97
24	1204	1	12	48	108	164	199	201	186	157	128
25	1455	1	12	52	120	192	235	248	230	201	164
26	1761	1	13	56	136	221	282	300	288	252	212
27	2112	1	13	61	150	255	331	364	352	318	267
28	2534	1	14	65	169	291	391	436	434	393	340
29	3015	1	14	70	185	333	454	522	525	488	423
30	3590	1	15	75	206	377	532	618	638	598	530

Theorem 3.4 allows $P(m)$ to be computed as a sum of $\Theta(m^{1/2})$ smaller values $P(j)$. If we compute all the values $P(1), \ldots, P(m)$ by this method, the total number of additions performed is $\Theta(m^{3/2})$. Algorithm 3.6 does this in such a way that no multiplications need to be performed (except for multiplication by the constant 3).

Algorithm 3.6: ENUMPARTITIONS2 (m)

$P(1) \leftarrow 1$
for $i \leftarrow 2$ **to** m

$\qquad$ **do** $\begin{cases} sign \leftarrow 1 \\ sum \leftarrow 0 \\ \omega \leftarrow 1 \\ j \leftarrow 1 \\ \omega' \leftarrow \omega + j \\ \textbf{while } \omega < m \\ \qquad \textbf{do} \begin{cases} \textbf{if } sign = 1 \\ \quad \textbf{then } sum \leftarrow sum + P(i - \omega) \\ \quad \textbf{else } sum \leftarrow sum - P(i - \omega) \\ \textbf{if } \omega' < i \\ \quad \textbf{then} \begin{cases} \textbf{if } sign = 1 \\ \quad \textbf{then } sum \leftarrow sum + P(i - \omega') \\ \quad \textbf{else } sum \leftarrow sum - P(i - \omega') \end{cases} \\ \omega \leftarrow \omega + 3j + 1 \\ j \leftarrow j + 1 \\ \omega' \leftarrow \omega + j \\ sign \leftarrow -sign \end{cases} \\ P(i) \leftarrow sum \end{cases}$

return (P)

3.1.1 Lexicographic ordering

In this section, we develop ranking, unranking and successor algorithms for $\mathcal{P}(m, n)$, based on a modified lexicographic ordering. Given a partition $m = a_1 + \ldots + a_n$ in $\mathcal{P}(m, n)$ is said to be in *reverse standard form* if $a_1 \leq \ldots \leq a_n$. We will use the lexicographic ordering of $\mathcal{P}(m, n)$, where the partitions are written in reverse standard form as lists of length n. We will call this the *rsf-lex ordering*, for short.

As an example, we tabulate the nine partitions in $\mathcal{P}(10, 4)$ in rsf-lex order:

standard form	reverse standard form
$[7, 1, 1, 1]$	$[1, 1, 1, 7]$
$[6, 2, 1, 1]$	$[1, 1, 2, 6]$
$[5, 3, 1, 1]$	$[1, 1, 3, 5]$
$[4, 4, 2, 1]$	$[1, 1, 4, 4]$
$[5, 2, 2, 1]$	$[1, 2, 2, 5]$
$[4, 3, 2, 1]$	$[1, 2, 3, 4]$
$[3, 3, 3, 1]$	$[1, 3, 3, 3]$
$[4, 2, 2, 2]$	$[2, 2, 2, 4]$
$[3, 3, 2, 2]$	$[2, 2, 3, 3]$

It can be shown that Algorithm 3.4 generates the partitions in $\mathcal{P}(m, n)$ in rsf-lex order.

We will first develop a successor algorithm for $\mathcal{P}(m, n)$ in rsf-lex order. Observe that the last partition in rsf-lex order is characterized by the property that all of its parts are equal to $\lfloor \frac{m}{n} \rfloor$ or $\lceil \frac{m}{n} \rceil$. This is equivalent to saying that $a_1 \leq a_n + 1$ when the partition is written in standard form. (In the example above, the partition $[3, 3, 2, 2]$ is the last partition, since $a_1 = 3$, $a_4 = 2$ and $3 \leq 2 + 1$.)

If we are given a partition $m = a_1 + \ldots + a_n$ in standard form and it is not the last partition in rsf-lex order, then we define i to be the smallest integer such that $a_1 > a_i + 1$. Note that $2 \leq i \leq n$. Then the partition

$$m' = a_1 + \ldots + a_{i-1}$$

is the last partition in $\mathcal{P}(m', i - 1)$ in rsf-lex order.

The successor of the partition $m = a_1 + \ldots + a_n$ is then computed as follows. We find the first partition in $\mathcal{P}(m' - 1, i - 1)$ (in rsf-lex order) having $a_{i-1} \geq a_i + 1$. This is easily seen to be

$$m' = a'_1 + \ldots + a'_{i-1},$$

where

$$a'_2 = \ldots = a'_{i-1} = a_i + 1$$

and

$$a'_1 = m' - (i - 2)(a_i + 1) - 1.$$

Now we replace the parts $a_1, \ldots, a_{i-1}$ by $a'_1, \ldots, a'_{i-1}$. Finally, we increase the value of a_i by one.

Algorithm 3.7 is based on the ideas discussed above. As an example, we compute the successor of the partition $[5, 5, 4, 2, 1]$ in $\mathcal{P}(17, 5)$. The value i defined above is computed to be $i = 4$. Since $a_4 = 2$ and $m' = 5 + 5 + 4 = 14$, we have that $a'_1 = 7$ and $a'_2 = a'_3 = 3$. The successor of the partition $[5, 5, 4, 2, 1]$ is therefore $[7, 3, 3, 3, 1]$.

Algorithm 3.7: PARTITIONLEXSUCCESSOR $(m, n, [a_1, \ldots, a_n])$

$i \leftarrow 2$
while $i \leq n$ **and** $a_1 \leq a_i + 1$
  do $i \leftarrow i + 1$
if $i = (n + 1)$
  then return ("undefined")
  else $\begin{cases} a_i \leftarrow a_i + 1 \\ d \leftarrow -1 \\ \textbf{for } j \leftarrow i - 1 \textbf{ downto } 2 \\ \quad \textbf{do } \begin{cases} d \leftarrow d + a_j - a_i \\ a_j \leftarrow a_i \end{cases} \\ a_1 \leftarrow a_1 + d \\ \textbf{return } ([a_1, \ldots, a_n]) \end{cases}$

Our ranking and unranking algorithms are based on the proof of Theorem 3.2. In this proof, $\mathcal{P}(m, n)$ is partitioned into two sets. The first set consists of all partitions in $\mathcal{P}(m, n)$ that contain at least one part equal to 1, and the second set consists of all partitions in $\mathcal{P}(m, n)$ that do not contain a part equal to 1. Notice that all the partitions in the first set will precede all the partitions in the second set in rsf-lex order. A partition in the first set will have rank between 0 and $P(m - 1, n - 1) - 1$, and a partition in the first set will have rank between $P(m - 1, n - 1)$ and $P(m, n) - 1$.

In general, the function rank can be described by means of a simple recursive formula that is suggested by the two bijections ϕ_1 and ϕ_2 used in the proof of Theorem 3.2. The formula is as follows:

$$\text{rank}([a_1, \ldots, a_n]) = \begin{cases} \text{rank}([a_1, \ldots, a_{n-1}]) & \text{if } a_n = 1 \\ \text{rank}([a_1', \ldots, a_n']) + P(m - 1, n - 1) & \text{if } a_n > 1, \end{cases}$$

where $a_i' = a_i - 1$, $1 \leq i \leq n$. This formula is easily iterated, and can be converted into a non-recursive algorithm. The result is presented as Algorithm 3.8. Note that the first step in this algorithm is to first precompute and store all the partition numbers $P(i, j)$ with $i \leq m$ and $j \leq n$, using Algorithm 3.5. Then these numbers will be available for use throughout the algorithm.

Algorithm 3.8: PARTITIONLEXRANK $(m, n, [a_1, \ldots, a_n])$

external ENUMPARTITIONS()
ENUMPARTITIONS(n, m)
for $i \leftarrow 1$ **to** n
  do $b_i \leftarrow a_i$
$r \leftarrow 0$
while $m > 0$

$\mathbf{do} \begin{cases} \text{if } b_n = 1 \\ \qquad \mathbf{then} \begin{cases} m \leftarrow m - 1 \\ n \leftarrow n - 1 \end{cases} \\ \qquad \mathbf{else} \begin{cases} \mathbf{for}\ i \leftarrow 1\ \mathbf{to}\ n \\ \quad \mathbf{do}\ b_i \leftarrow b_i - 1 \\ r \leftarrow r + P(m-1, n-1) \\ m \leftarrow m - n \end{cases} \end{cases}$

return (r)

To illustrate, we evaluate rank$[5, 5, 4, 2, 1]$ using Algorithm 3.8. Initially, we have $r = 0$, $m = 17$, $n = 5$ and $[b_1, b_2, b_3, b_4, b_5] = [5, 5, 4, 2, 1]$. In the various iterations of the **while** loop, the following computations are done:

- $b_5 = 1$, so we set $m = 16$ and $n = 4$.
- $b_4 > 1$, so we set $[b_1, b_2, b_3, b_4] = [4, 4, 3, 1]$, $r = P(15, 3) = 19$, and $m = 12$.
- $b_4 = 1$, so we set $m = 11$ and $n = 3$.
- $b_3 > 1$, so we set $[b_1, b_2, b_3] = [3, 3, 2]$, $r = 19 + P(10, 2) = 24$, and $m = 8$.
- $b_3 > 1$, so we set $[b_1, b_2, b_3] = [2, 2, 1]$, $r = 24 + P(7, 2) = 27$, and $m = 5$.
- $b_3 = 1$, so we set $m = 4$ and $n = 2$.
- $b_2 > 1$, so we set $[b_1, b_2] = [1, 1]$, $r = 27 + P(3, 1) = 28$, and $m = 2$.
- $b_2 = 1$, so we set $m = 1$ and $n = 1$.
- $b_1 = 1$, so we set $m = 0$ and $n = 0$.

Thus, the result is that rank$([5, 5, 4, 2, 1]) = 28$.

The behavior of the unranking function can be analyzed in a similar manner, and an algorithm developed to implement it. We present such an algorithm as Algorithm 3.9.

Algorithm 3.9: PARTITIONLEXUNRANK (m, n, r)

external ENUMPARTITIONS()

ENUMPARTITIONS(n, m)

for $i \leftarrow 1$ **to** n

 do $a_1 \leftarrow 0$

while $m > 0$

$\mathbf{do} \begin{cases} \mathbf{if}\ r < P(m-1, n-1) \\ \quad \mathbf{then} \begin{cases} a_n \leftarrow a_n + 1 \\ m \leftarrow m - 1 \\ n \leftarrow n - 1 \end{cases} \\ \quad \mathbf{else} \begin{cases} \mathbf{for}\ i \leftarrow 1\ \mathbf{to}\ n \\ \quad \mathbf{do}\ a_i \leftarrow a_i + 1 \\ r \leftarrow r - P(m-1, n-1) \\ m \leftarrow m - n \end{cases} \end{cases}$

return $([a_1, \ldots, a_n])$

3.2   Set partitions, Bell and Stirling numbers

In the last section, we discussed partitions of integers. Equivalently, we could have defined $P(m)$ to be the number of ways of partitioning a collection of m identical elements into non-empty subsets. In this section, we consider partitions of a set of m distinct elements into non-empty subsets. (Recall that a partition of a set was defined in Section 1.2.3.)

For a positive integer m, let $\mathcal{S}(m)$ denote the set of all partitions of $\{1, \ldots, m\}$ into non-empty subsets. For positive integers m and n with $n \leq m$, let $\mathcal{S}(m, n)$ denote the set of all partitions of $\{1, \ldots, m\}$ into exactly n non-empty subsets. The *Bell number*, $B(m)$, is defined to be $B(m) = |\mathcal{S}(m)|$, and the *Stirling number of the second kind*, $S(m, n)$, is defined to be $S(m, n) = |\mathcal{S}(m, n)|$. It is clear from the definitions that

$$B(m) = \sum_{n=1}^{m} S(m, n).$$

The first few Bell numbers are $B(1) = 1$, $B(2) = 2$, $B(3) = 5$, $B(4) = 15$ and $B(5) = 52$. As an example, we tabulate the 15 ways to partition the set $\{1, 2, 3, 4\}$ into non-empty subsets.

$$
\begin{array}{lll}
\{\{1\}, \{2\}, \{3\}, \{4\}\} & \{\{1,2\}, \{3\}, \{4\}\} & \{\{1,3\}, \{2\}, \{4\}\} \\
\{\{1,4\}, \{2\}, \{3\}\} & \{\{2,3\}, \{1\}, \{4\}\} & \{\{2,4\}, \{1\}, \{3\}\} \\
\{\{3,4\}, \{1\}, \{2\}\} & \{\{1,2\}, \{3,4\}\} & \{\{1,3\}, \{2,4\}\} \\
\{\{1\,4\}, \{2,3\}\} & \{\{1,2,3\}, \{4\}\} & \{\{1,2,4\}, \{3\}\} \\
\{\{1,3,4\}, \{2\}\} & \{\{2,3,4\}, \{1\}\} & \{\{1,2,3,4\}\}
\end{array}
$$

From this we see that $S(4,4) = 1$, $S(4,3) = 6$, $S(4,2) = 7$ and $S(4,1) = 1$.

The first result we prove in this section is an explicit formula for $S(m,n)$. The proof will make use of a result known as the *principle of inclusion-exclusion*. We state the following without proof.

THEOREM 3.5 *(Principle of inclusion-exclusion)* *Suppose that X is a finite set, and $X_1, \ldots, X_n \subseteq X$. For any $I \subseteq \{1, \ldots, n\}$, define*

$$X_I = \bigcup_{i \in I} X_i.$$

Then

$$\left| X \setminus \bigcup_{i=1}^{n} X_i \right| = \sum_{I \subseteq \{1,\ldots,n\}} (-1)^{|I|} |X_I|.$$

Now, suppose X denotes the set of all functions $f : \{1, \ldots, m\} \to \{1, \ldots, n\}$. For $1 \leq i \leq n$, let X_i consist of all functions $f \in X$ such that $f(j) \neq i$ for all j, $1 \leq j \leq m$. It is easy to see that

$$|X_I| = (n - |I|)^m$$

for any $I \subseteq \{1, \ldots, n\}$, where x_I is as defined above. Also, there are exactly $\binom{n}{j}$ choices for $I \subseteq \{1, \ldots, n\}$ with $|I| = j$. Thus, applying Theorem 3.5, we see that

$$\left| X \setminus \bigcup_{i=1}^{n} X_i \right| = \sum_{j=0}^{n} (-1)^j \binom{n}{j} (n-j)^m$$

$$= \sum_{j=1}^{n} (-1)^{n-j} \binom{n}{j} j^m,$$

where, in the last line, we replace j by $n-j$ and make use of the fact that $\binom{n}{n-j} = \binom{n}{j}$.

Since a function $f \in X \setminus \bigcup_{i=1}^{n} X_i$ if and only if it is a surjective function, we have proved the following result.

THEOREM 3.6 *There are exactly*

$$\sum_{j=1}^{n} (-1)^{n-j} \binom{n}{j} j^m$$

surjective functions $f : \{1, \ldots, m\} \to \{1, \ldots, n\}$.

How does this result relate to Stirling numbers? Suppose $f : \{1, \ldots, m\} \to \{1, \ldots, n\}$ is a surjective function. Then it is clear that

$$\{f^{-1}(1), \ldots, f^{-1}(n)\}$$

is a partition in $S(m, n)$. Further, there are exactly $n!$ functions f that give rise in this way to any given partition in $S(m, n)$. Thus we obtain the following formula for $S(m, n)$.

THEOREM 3.7   *For positive integers m and n with $m \geq n$,*

$$S(m, n) = \frac{1}{n!} \sum_{j=1}^{n} (-1)^{n-j} \binom{n}{j} j^m.$$

The numbers $S(m, n)$ also satisfy a simple recurrence relation, which we state and prove now.

THEOREM 3.8   *For positive integers n and m with $n \geq m$,*

$$S(m, n) = n\, S(m - 1, n) + S(m - 1, n - 1).$$

PROOF   We will define a bijection

$$\phi : (S(m - 1, n) \times \{1, \ldots, n\}) \cup S(m\quad 1, n - 1) \to S(m, n).$$

Suppose that $\{A_1, \ldots, A_n\} \in S(m - 1, n)$ and $1 \leq i \leq n$. Then define

$$\phi(\{A_1, \ldots, A_n\}, i) = \{A_1, \ldots, A_{i-1}, A_i \cup \{m\}, A_{i+1}, \ldots, A_n\}.$$

Further, suppose that $\{A_1, \ldots, A_{n-1}\} \in S(m - 1, n - 1)$. Then define

$$\phi(\{A_1, \ldots, A_{n-1}\}) = \{A_1, \ldots, A_{n-1}, \{m\}\}.$$

It is clear that ϕ is the desired bijection, and the result follows.   ∎

Note that, in order for Theorem 3.8 to be valid, we need to define initial conditions as follows:

$$S(m, m + 1) = 0 \quad \text{for all } m \geq 0$$
$$S(m, 0) = 0 \quad \text{for all } m \geq 1, \text{ and}$$
$$S(0, 0) = 1.$$

Likewise, the Bell numbers satisfy a recurrence relation. Note that in the following theorem, we define $B(0) = 1$ as an initial condition.

THEOREM 3.9 *For any integer $m \geq 1$, we have that*

$$B(m) = \sum_{i=0}^{m-1} \binom{m-1}{i} B(i),$$

where $B(0) = 1$.

PROOF Suppose that $1 \leq i \leq m$. Then there are exactly $\binom{m-1}{i-1}$ different sets $Y \subseteq \{1, \ldots, m-1\}$ such that $|Y| = i - 1$. Given Y and i, there are exactly $B(m - i)$ partitions in $\mathcal{S}(m)$ in which $Y \cup \{i\}$ is one of the sets in the partition. Thus we have

$$B(m) = \sum_{i=1}^{m} \binom{m-1}{i-1} B(m-i)$$

$$= \sum_{i=0}^{m-1} \binom{m-1}{i} B(i),$$

where, in the last line, we replace i by $m - i$ and make use of the fact that $\binom{m-1}{m-i-1} = \binom{m-1}{i}$. ∎

Any one of Theorems 3.7, 3.8 or 3.9 can be used to compute the relevant Stirling or Bell numbers. As an example, we present an algorithm to compute a Stirling number that is based on Theorem 3.8. Algorithm 3.10 uses the same dynamic programming strategy that was employed in Algorithm 3.5.

Algorithm 3.10: STIRLINGNUMBERS2 (m, n)

$S(0, 0) \leftarrow 1$
for $i \leftarrow 1$ **to** m
  do $S(i, 0) \leftarrow 0$
for $i \leftarrow 0$ **to** m
  do $S(i, i + 1) \leftarrow 0$
for $i \leftarrow 1$ **to** m
  do $\begin{cases} \textbf{for } j \leftarrow 1 \textbf{ to } \min\{i, n\} \\ \quad \textbf{do } S(i, j) \leftarrow j\, S(i - 1, j) + S(i - j, j - 1) \end{cases}$
return (S)

We close this section by presenting a table of Stirling and Bell numbers; see Table 3.2.

3.2.1 Restricted growth functions

In this subsection, we develop algorithms for generating set partitions in lexicographic order, as well as ranking and unranking algorithms. The first order of

TABLE 3.2
A table of Stirling numbers of the second kind

						$S(m, n)$					
m	$B(m)$	$n = 1$	2	3	4	5	6	7	8	9	10
1	1	1									
2	2	1	1								
3	5	1	3	1							
4	15	1	7	6	1						
5	52	1	15	25	10	1					
6	203	1	31	90	65	15	1				
7	877	1	63	301	350	140	21	1			
8	4140	1	127	966	1701	1050	266	28	1		
9	21147	1	255	3025	7770	6951	2646	462	36	1	
10	115975	1	511	9330	34105	42525	22827	5880	750	45	1

business is to decide how lexicographic ordering should be defined. It is both natural and convenient to consider a different representation of a set partition that is called a *restricted growth function*. Let $m \geq 1$. Define $\mathcal{R}(m)$ to consist of all functions $f : \{1, \ldots, m\} \to \mathbb{Z}^+$ which satisfy the following conditions:

$$f(1) = 1 \tag{3.2}$$

$$f(i) \leq \max\{f(1), \ldots, f(i-1)\} + 1, \quad \text{if } 2 \leq i \leq m. \tag{3.3}$$

A function $f \in \mathcal{R}(m)$ is called a restricted growth function of *length* m. We will represent a function $f \in \mathcal{R}(m)$ as the m-tuple $[f(1), \ldots, f(m)]$. Thus, we will think of $\mathcal{R}(m)$ as being a set of m-tuples of positive integers.

For any integer $m \geq 1$, there is a simple bijection between the two sets $\mathcal{S}(m)$ and $\mathcal{R}(m)$. These bijections are described as Algorithms 3.11 and 3.12. Algorithm 3.11 takes as input $\{A_1, \ldots, A_n\}$, which is a partition of $\{1, \ldots, m\}$, and constructs the corresponding function $[b_1, \ldots, b_m] \in \mathcal{R}(m)$. Algorithm 3.12 reverses the process.

Algorithm 3.11: SETPARTTORGF $(m, n, \{A_1, \ldots, A_n\})$

for $j \leftarrow 1$ **to** m
  do $f_j \leftarrow 0$
$j \leftarrow 1$
$i \leftarrow 1$
for $i \leftarrow 1$ **to** n
  do $\begin{cases} \textbf{while } f_j \neq 0 \\ \quad \textbf{do } j \leftarrow j + 1 \\ h \leftarrow 1 \\ \textbf{while } j \notin A_h \\ \quad \textbf{do } h \leftarrow h + 1 \\ \textbf{for each } g \in A_h \\ \quad \textbf{do } f_g \leftarrow i \end{cases}$
return $([f_1, \ldots, f_m])$

Algorithm 3.12: RGFTOSETPART $(m, n, [f_1, \ldots, f_m])$

$n \leftarrow 1$
for $j \leftarrow 1$ **to** m
  do $\begin{cases} \textbf{if } f_j > n \\ \quad \textbf{then } n \leftarrow f_j \end{cases}$
for $i \leftarrow 1$ **to** n
  do $A_i \leftarrow \emptyset$
for $j \leftarrow 1$ **to** m
  do $A_{f_j} \leftarrow A_{f_j} \cup \{j\}$
return $(\{A_1, \ldots, A_n\})$

As an example, we tabulate the 15 restricted growth functions of length four and the corresponding 15 partitions of $\{1, 2, 3, 4\}$. Note that the 4-tuples in $\mathcal{R}(m)$ are listed in lexicographic order.

RGF	partition	RGF	partition
$[1,1,1,1]$	$\{\{1,2,3,4\}\}$	$[1,1,1,2]$	$\{\{1,2,3\},\{4\}\}$
$[1,1,2,1]$	$\{\{1,2,4\},\{3\}\}$	$[1,1,2,2]$	$\{\{1,2\},\{3,4\}\}$
$[1,1,2,3]$	$\{\{1,2\},\{3\},\{4\}\}$	$[1,2,1,1]$	$\{\{1,3,4\},\{2\}\}$
$[1,2,1,2]$	$\{\{1,3\},\{2,4\}\}$	$[1,2,1,3]$	$\{\{1,3\},\{2\},\{4\}\}$
$[1,2,2,1]$	$\{\{1,4\},\{2,3\}\}$	$[1,2,2,2]$	$\{\{1\},\{2,3,4\}\}$
$[1,2,2,3]$	$\{\{1\},\{2,3\},\{4\}\}$	$[1,2,3,1]$	$\{\{1,4\},\{2\},\{3\}\}$
$[1,2,3,2]$	$\{\{1\},\{2,4\},\{3\}\}$	$[1,2,3,3]$	$\{\{1\},\{2\},\{3,4\}\}$
$[1,2,3,4]$	$\{\{1\},\{2\},\{3\},\{4\}\}$		

Since we have a bijection between the two sets $\mathcal{S}(m)$ and $\mathcal{R}(m)$, they have the same cardinality, and thus the following result holds.

THEOREM 3.10   *There are precisely $B(m)$ restricted growth functions of length m, for any integer $m \geq 1$.*

The next algorithm, Algorithm 3.13, generates the functions in $\mathcal{R}(m)$ in lexicographic order. Here is a bit of explanation as to how the algorithm works. $[f[1], \ldots, f[m]]$ is the "current" RGF. The auxiliary array $[f_{max}[1], \ldots, f_{max}[m]]$ is defined as follows:

$$f_{max}[i] = \begin{cases} 1 + \max\{f(j) : 1 \leq j \leq i - 1\} & \text{if } 2 \leq i \leq m \\ 2 & \text{if } i = 1. \end{cases}$$

For $2 \leq i \leq m$, $f_{max}[i]$ denotes the maximum value that can be assigned to $f[i]$ (given values for $f[1], \ldots, f[i-1]$) without violating the RGF condition.

Initially, we begin with the RGF $[1, \ldots, 1]$. At any point in the algorithm, we find the first position from the right such that $f_{max}[i] \neq f[i]$. The value of $f[i]$ can be increased by 1, and then we set $f[j] = 1$ for all j such that $i + 1 \leq j \leq m$. Finally, the auxiliary array is updated.

This process is iterated until we reach the terminating condition, which is that $f_{max}[i] = f[i]$ for all i such that $2 \leq i \leq m$. The last RGF in lexicographic order is $[1, 2, \ldots, m]$.

Algorithm 3.13: GENERATERGF (m)

for $i \leftarrow 1$ **to** m

do $\begin{cases} f[i] \leftarrow 1 \\ f_{max}[i] \leftarrow 2 \end{cases}$

$done \leftarrow$ **false**

while not $done$

do $\begin{cases} \textbf{output } ([f(1), \ldots, f(m)]) \\ j \leftarrow m + 1 \\ \textbf{repeat} \\ \quad j \leftarrow j - 1 \\ \textbf{until } f[j] \neq f_{max}[j] \\ \textbf{if } j > 1 \\ \quad \textbf{then} \begin{cases} f[j] \leftarrow f[j] + 1 \\ \textbf{for } i \leftarrow j + 1 \textbf{ to } m \\ \quad \textbf{do} \begin{cases} f[i] \leftarrow 1 \\ \textbf{if } f[j] = f_{max}[j] \\ \quad \textbf{then } f_{max}[i] \leftarrow f_{max}[j] + 1 \\ \quad \textbf{else } f_{max}[i] \leftarrow f_{max}[j] \end{cases} \end{cases} \\ \quad \textbf{else } done \leftarrow \textbf{true} \end{cases}$

Now we develop ranking and unranking algorithms for restricted growth functions. In order to accomplish this, we first consider a generalization of this concept, which we define as follows. Let $m \geq 1$ and $j \geq 0$ be integers. Define

$\mathcal{R}(m, j)$ to consist of all functions $f : \{1, \ldots, m\} \to \mathbb{Z}^+$ which satisfy the following conditions:

$$f(1) \leq j + 1$$

$$f(i) \leq \max\{f(1), \ldots, f(i-1), j\} + 1, \quad \text{if } 2 \leq i \leq m.$$

A function $f \in \mathcal{R}(m, j)$ is called a j-*restricted growth function* (or j-RGF) of length m. Note that the case $j = 0$ coincides with our earlier definition, i.e., $\mathcal{R}(m, j) = \mathcal{R}(m)$.

The idea of j-RGF is useful because of the following simple fact, which we state without proof.

LEMMA 3.11 *Suppose that $1 \leq i \leq m$, and $[f[1], \ldots, f[m-i]] \in \mathcal{R}(m-i)$. Denote $j = \max\{f[1], \ldots, f[m-i]\}$. Then $[f[1], \ldots, f[m]] \in \mathcal{R}(m)$ if and only if $[f[m-i+1], \ldots, f[m]] \in \mathcal{R}(m, j)$.*

Now, suppose that we define $d_{i,j} = |\mathcal{R}(i, j)|$ for $i \geq 1$ and $j \geq 0$. Then Lemma 3.11 asserts that the number of ways of completing the "partial" RGF (namely, $[f[1], \ldots, f[m-i]]$) to an RGF (namely, $[f[1], \ldots, f[m]]$) is $d_{i,j}$, where j is the largest value among $f[1], \ldots, f[m-i]$.

We will make use of the following recurrence relation for the values $d_{i,j}$.

THEOREM 3.12 *For all positive integers i and j, it holds that*

$$d_{i,j} = j\, d_{i-1,j} + d_{i-1,j+1}.$$

Note that, in order for Theorem 3.12 to be valid, we should define initial conditions as follows:

$$d_{0,j} = 1$$

for all $j \geq 0$.

Theorem 3.12 leads to a convenient method of computing the values $d_{i,j}$, by applying the same dynamic programming strategy as was used in Algorithm 3.10. Algorithm 3.14 uses this technique to determine the numbers $d_{i,j}$ for all i, j such that $i + j \leq m$. Notice that the last value computed is $d_{m,0} = B(m)$; thus this algorithm provides another method of computing Bell numbers.

Algorithm 3.14: GENERALIZEDRGF (m)

for $j \leftarrow 0$ **to** m
  do $d_{0,j} \leftarrow 1$
for $i \leftarrow 1$ **to** m
  do $\begin{cases} \textbf{for } j \leftarrow 0 \textbf{ to } m - i \\ \quad \textbf{do } d_{i,j} \leftarrow j\, d_{i-1,j} + d_{i-1,j+1} \end{cases}$
return (d)

TABLE 3.3
The number of j-restricted growth functions of length i

	$d_{i,j}$						
i	$j = 0$	1	2	3	4	5	6
0	1	1	1	1	1	1	1
1	1	2	3	4	5	6	
2	2	5	10	17	26		
3	5	15	37	77			
4	15	52	151				
5	52	203					
6	203						

To illustrate, we tabulate the values $d_{i,j}$ for all i, j such that $i + j \leq 6$; see Table 3.3.

Once we have computed the values $d_{i,j}$ for all i, j such that $i + j \leq m$, it is a simple matter to rank or unrank. Both of these tasks are accomplished by traversing an RGF from left to right. Consider first the algorithm for ranking an RGF, say f, which is presented as Algorithm 3.15. The rank, denoted r, is initialized to be 0. During the ith iteration of the **for** loop, we are looking at the partial RGF $[f[1], \ldots, f[i]]$, and $j = \max\{f[1], \ldots, f[i-1]\}$. Thus it must be the case that $1 \leq f[i] \leq j + 1$. For any given value of $f[i]$ such that $1 \leq f[i] \leq j$, there are $d_{m-i,j}$ ways to complete the partial RGF to a complete RGF. Thus we add the quantity $(f[i] - 1) d_{m-i,j}$ to r, and the value of j is updated, in preparation for the next iteration of the loop.

Algorithm 3.15: RANKRGF $(m, [f[1], \ldots, f[m]])$

external GENERALIZEDRGF()
$d \leftarrow$ GENERALIZEDRGF(m)
$r \leftarrow 0$
$j \leftarrow 1$
for $i \leftarrow 2$ **to** m
$\quad$ **do** $\begin{cases} r \leftarrow r + (f[i] - 1) d_{m-i,j} \\ j \leftarrow \max\{j, f[i]\} \end{cases}$
return (r)

As an example, we compute rank$([1, 2, 2, 1])$ using Algorithm 3.15. Initially, $r \leftarrow 0$. When $i = 2$, r is changed to $d_{2,1} = 5$. When $i = 3$, r is changed to $5 + d_{1,2} = 8$. r is not changed when $i = 4$, so rank$([1, 2, 2, 1]) = 8$.

Unranking is only a bit more complicated. The algorithm is presented as Algorithm 3.16. Here the rank r is given, and we are constructing the RGF, f. As in Algorithm 3.15, the value $j = \max\{f[1], \ldots, f[i-1]\}$ during the ith iteration of the **for** loop. At this point, we have to determine the correct value of $f[i]$ (where $1 \le f[i] \le j + 1$). If $r \ge j\, d_{m-i,j}$, then it follows that $f[i] = j + 1$. In this case, j is updated and we subtract the quantity $j\, d_{m-i,j}$ from r. If $r < j\, d_{m-i,j}$, then it follows that

$$f[i] = \left\lfloor \frac{r}{d_{m-i,j}} \right\rfloor + 1,$$

and the value of r is updated appropriately.

Algorithm 3.16: UNRANKRGF (m, r)

external GENERALIZEDRGF()

$d \leftarrow$ GENERALIZEDRGF(m)
$j \leftarrow 1$
$f[1] \leftarrow 1$
for $i \leftarrow 2$ **to** m

$$\textbf{do} \begin{cases} \textbf{if } j\, d_{m-i,j} \le r \\ \quad \textbf{then } \begin{cases} f[i] \leftarrow j + 1 \\ r \leftarrow r - j\, d_{m-i,j} \\ j \leftarrow j + 1 \end{cases} \\ \quad \textbf{else } \begin{cases} f[i] \leftarrow \lfloor \frac{r}{d_{m-i,j}} \rfloor + 1 \\ r \leftarrow r \bmod d_{m-i,j} \end{cases} \end{cases}$$

return $([f[1], \dots, f[m]])$

As an example, suppose we use Algorithm 3.16 to unrank the value $r = 8$ when $m = 4$. First, we set $f[1] \leftarrow 1$. When $i = 2$, we have $1 \times d_{2,1} = 5 \le 8$, so $f[2] \leftarrow 2$, $r \leftarrow 3$ and $j \leftarrow 2$. When $i = 3$, we have $2 \times d_{1,2} = 6 > 3$, so $f[3] \leftarrow \lfloor \frac{3}{3} \rfloor + 1 = 2$ and $r \leftarrow 0$. When $i = 4$, $f[4] \leftarrow 1$. Thus, unrank(8) $= [1, 2, 2, 1]$.

3.2.2 Stirling numbers of the first kind

Recall that a permutation π on a set X is a bijection from X to X. For example,

x	1	2	3	4	5	6	7	8	9	10	11
$\pi(x)$	11	2	4	1	6	5	8	9	7	10	3

is a permutation on $X = \{1, 2, 3, 4, 5, 6, 7, 8, 9, 10, 11\}$. Observe that the bottom row of this table is a permutation in the sense of Section 2.4, and in a computer we usually store the bottom row $[11, 2, 4, 1, 6, 5, 8, 9, 7, 10, 3]$ as an array to represent π. Although this is simple, an even simpler notation is often used, which is called *cycle notation*. The cycle notation for π is

$$\pi = (1, 11, 3, 4)(2)(5, 6)(7, 8, 9)(10).$$

To see how this notation works, we draw the directed graph on vertex set X in which the arcs are $(x, \pi(x))$, for each $x \in X$. (This representation of a permutation was introduced in Section 2.4.2.) For the above example, we obtain the following:

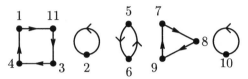

The resulting graph is always a union of directed circuits. A sequence of vertices enclosed between parentheses in the cycle notation for the permutation π is called a *cycle* of π. In the above example, the cycles are

$$(0, 3, 4, 1), \quad (2), \quad (5, 6), \quad (7, 8, 9), \quad \text{and} \quad (10).$$

For positive integers m and n with $n \le m$, let $\Pi(m, n)$ denote the set of all permutations of $\{1, \ldots, m\}$ containing exactly n cycles. Then the *Stirling number of the first kind*, $s(m, n)$, is defined to be

$$s(m, n) = (-1)^{m-n} |\Pi(m, n)|.$$

It is clear that

$$m! = \sum_{n=1}^{m} |s(m, n)|,$$

since $m!$ is the total number of permutations on a set of size m.

As an example, we tabulate the 24 permutations of $\{1, 2, 3, 4\}$, classified according to the number of cycles they contain. There is one permutation having four cycles; so, we see that $\Pi(4, 4) = \{(1)(2)(3)(4)\}$. There are six permutations having three cycles:

$$\Pi(4, 3) = \left\{ \begin{array}{lll} (12)(3)(4), & (13)(2)(4), & (14)(2)(3), \\ (1)(23)(4), & (1)(24)(3), & (1)(2)(34) \end{array} \right\}.$$

There are eleven permutations having two cycles:

$$\Pi(4, 2) = \left\{ \begin{array}{lll} (12)(34), & (13)(24), & (14)(23), \\ (123)(4), & (132)(4), & (124)(3), \\ (142)(3), & (134)(2), & (143)(2), \\ (1)(234), & (1)(243) \end{array} \right\}.$$

Finally, there are six permutations having one cycle:

$$\Pi(4, 1) = \left\{ \begin{array}{lll} (1234), & (1243), & (1324), \\ (1342), & (1423), & (1432) \end{array} \right\}.$$

Hence, $s(4, 1) = -6$, $s(4, 2) = 11$, $s(4, 3) = -6$ and $s(4, 4) = 1$.

We state without proof the following complicated explicit formula for Stirling numbers of the first kind.

THEOREM 3.13 *For positive integers* m *and* n *with* $m \geq n$,

$$s(m,n) = \sum_{k=0}^{m-n} (-1)^k \binom{m-1+k}{m-n+k} \binom{2m-n}{m-n-k} S(m-n+k,k).$$

Note that this formula also involves Stirling numbers of the second kind, which can in turn be computed using Theorem 3.7. Thus, the computation of a Stirling number of the first kind using Theorem 3.13 involves a double sum of products of binomial coefficients. It is therefore much more efficient to use the following recurrence relation, which is similar to Theorem 3.8, to compute these numbers.

THEOREM 3.14 *For positive integers* n *and* m *with* $m \geq n$,

$$s(m,n) = s(m-1,n-1) - (m-1)s(m-1,n).$$

The initial conditions for the above recurrence relation are as follows:

$$s(m, m+1) = 0 \quad \text{for all } m \geq 0$$

$$s(m,0) = 0 \quad \text{for all } m \geq 1, \text{ and}$$

$$s(0,0) = 1.$$

Algorithm 3.17 is a dynamic programming algorithm which computes Stirling numbers of the first kind using Theorem 3.14.

Algorithm 3.17: STIRLINGNUMBERS1 (m,n)

$s(0,0) \leftarrow 1$
for $i \leftarrow 1$ **to** m
    do $s(i,0) \leftarrow 0$
for $i \leftarrow 0$ **to** m
    do $s(i, i+1) \leftarrow 0$
for $i \leftarrow 1$ **to** m
    do $\begin{cases} \textbf{for } j \leftarrow 1 \textbf{ to } \min\{i,n\} \\ \quad \textbf{do } s(i,j) \leftarrow s(i-j,j-1) - (i-1)s(i-1,j) \end{cases}$
return (s)

We present in Table 3.4 a table of Stirling numbers of the first kind.

Stirling numbers of the first and second kind are related by an interesting orthogonality relation, as stated in the following theorem. (A proof will be outlined in the Exercises.)

TABLE 3.4
A table of Stirling numbers of the first kind

			$s(m,n)$							
m	$n=1$	2	3	4	5	6	7	8	9	10
1	1									
2	-1	1								
3	2	-3	1							
4	-6	11	-6	1						
5	24	-50	35	-10	1					
6	-120	274	-225	85	-15	1				
7	720	-1764	1624	-735	175	-21	1			
8	-5040	13068	-13132	6769	-1960	322	-28	1		
9	40320	-109584	118124	-67284	22449	-4536	546	-36	1	
10	-362880	1026576	-1172700	723680	-269325	63273	-9450	870	-45	1

THEOREM 3.15 *For positive integers n and m with $m \geq n$,*

$$\sum_{k=n}^{m} S(m,k)s(k,n) = \sum_{k=n}^{m} s(m,k)S(k,n) = \delta_{mn},$$

where

$$\delta_{mn} = \begin{cases} 1 & \text{if } m = n \\ 0 & \text{otherwise.} \end{cases}$$

Theorem 3.15 can be stated in an equivalent form involving matrices. First, for positive integers i and j, define $S(i,j) = s(i,j) = 0$ if $i < j$. Now, for a fixed positive integer n, define the two n by n matrices S and s, where the (i,j) entry of S is $S(i,j)$ and the (i,j) entry of s is $s(i,j)$. Then we have the following corollary to Theorem 3.15.

COROLLARY 3.16 *For a positive integer n, define the n by n matrices S and s as above. Then $S = s^{-1}$.*

As an example, suppose $n = 4$. Then Corollary 3.16 asserts that the following matrix equation holds:

$$\begin{pmatrix} 1 & 0 & 0 & 0 \\ 1 & 1 & 0 & 0 \\ 1 & 3 & 1 & 0 \\ 1 & 7 & 6 & 1 \end{pmatrix} \begin{pmatrix} 1 & 0 & 0 & 0 \\ -1 & 1 & 0 & 0 \\ 2 & -3 & 1 & 0 \\ -6 & 11 & -6 & 1 \end{pmatrix} = \begin{pmatrix} 1 & 0 & 0 & 0 \\ 0 & 1 & 0 & 0 \\ 0 & 0 & 1 & 0 \\ 0 & 0 & 0 & 1 \end{pmatrix}.$$

3.3 Labeled trees

A *tree* is a connected graph without circuits. A famous theorem of Cayley states that there are exactly n^{n-2} different labeled trees on a given set of n vertices, say $\mathcal{V} = \{1, \ldots, n\}$. (Two trees $(\mathcal{V}, \mathcal{E}_1)$ and $(\mathcal{V}, \mathcal{E}_2)$ are said to be *different* if $\mathcal{E}_1 \neq \mathcal{E}_2$.)

Suppose we denote by $\mathcal{T}(n)$ the set of all trees on vertex set $\mathcal{V}$. There is a nice proof of Cayley's theorem that is accomplished by exhibiting a bijection between the two sets $\mathcal{T}(n)$ and $\mathcal{V}^{n-2}$ (i.e., the set of all $(n-2)$-tuples of vertices). Since $|\mathcal{V}^{n-2}| = n^{n-2}$, this proves the desired result. This bijection is known as the *Prüfer correspondence* .

The Prüfer correspondence is actually described as an algorithm. Given a tree $T = (\mathcal{V}, \mathcal{E}) \in \mathcal{T}(n)$, we will construct a $1 - 1$ function

$$\text{Prüfer} : \mathcal{T}(n) \to \mathcal{V}^{n-2}$$

and the inverse function

$$\text{InvPrüfer} : \mathcal{V}^{n-2} \to \mathcal{T}(n).$$

These are presented in Algorithms 3.18 and 3.19.

Algorithm 3.18: PRUFER $(\mathcal{E}, n)$

for $i \leftarrow 1$ **to** n
  do $d[i] \leftarrow 0$
for each $\{x, y\} \in \mathcal{E}$
  do $\begin{cases} d[x] \leftarrow d[x] + 1 \\ d[y] \leftarrow d[y] + 1 \end{cases}$
for $i \leftarrow 1$ **to** $n - 2$
  do $\begin{cases} x \leftarrow n \\ \textbf{while } d[x] \neq 1 \\ \quad \textbf{do } x \leftarrow x - 1 \\ y \leftarrow n \\ \textbf{while } \{x, y\} \notin \mathcal{E} \\ \quad \textbf{do } y \leftarrow y - 1 \\ L[i] \leftarrow y \\ d[x] \leftarrow d[x] - 1 \\ d[y] \leftarrow d[y] - 1 \\ \mathcal{E} \leftarrow \mathcal{E} \backslash \{\{x, y\}\} \end{cases}$
return $(L = [L[1], \ldots, L[n-2]])$

The list L that is returned by Algorithm 3.18 is defined to be Prüfer(T). Let's see how this list is constructed. Since $(\mathcal{V}, \mathcal{E})$ is a tree on n vertices, it can be shown that $\mathcal{E}$ consists of $n - 1$ edges, and $\mathcal{V}$ has at least two vertices of degree one. During the ith iteration of the **for** loop, we have a tree on $n + 1 - i$ vertices (which therefore contains $n - i$ edges). We find the largest vertex of degree one, which is denoted by x. Then we find the (unique) vertex such that $\{x, y\} \in \mathcal{E}$. We define $L[i]$ to be y, and delete the edge $\{x, y\}$ from the tree. Since x had degree equal to one, we have reduced the number of vertices in the tree by one. At the end of the algorithm (i.e., after iteration $n - 2$) there will remain two vertices and the edge joining them.

An important property of the list L is that the number of occurrences of a vertex x in the list L is equal to $\deg(x) - 1$.

Algorithm 3.19: INVPRUFER (L, n)

$\mathcal{E} \leftarrow \emptyset$
$L[n - 1] \leftarrow 1$
for $i \leftarrow 1$ **to** n
  do $d[i] \leftarrow 1$
for $i \leftarrow 1$ **to** $n - 2$
  do $d[L[i]] \leftarrow d[L[i]] + 1$
for $i \leftarrow 1$ **to** $n - 1$
  do $\begin{cases} x \leftarrow n \\ \textbf{while } d[x] \neq 1 \\ \quad \textbf{do } x \leftarrow x - 1 \\ y \leftarrow L[i] \\ d[x] \leftarrow d[x] - 1 \\ d[y] \leftarrow d[y] - 1 \\ \mathcal{E} \leftarrow \mathcal{E} \cup \{\{x, y\}\} \end{cases}$
return $(\mathcal{E})$

It turns out that Algorithm 3.19 adds edges to $\mathcal{E}$ in the same order that Algorithm 3.18 removes them. Algorithm 3.19 begins by computing the vertex degrees from the list L, as described above. In the ith iteration of the **for** loop, x is determined as the current largest vertex of degree one. Then y is obtained, since $y = L[i]$, and the edge $\{x, y\}$ is included in the tree being constructed. Clearly this is the same edge that was deleted during iteration i of Algorithm 3.18.

We need to consider the "last" edge, i.e., the edge that remains at the termination of Algorithm 3.18. This edge is in fact $\{1, x\}$, for some $2 \leq x \leq n$. It can be shown that x is the maximum vertex of degree one when Algorithm 3.18 terminates. By setting $L[n - 1] \leftarrow 1$ at the beginning of Algorithm 3.19, we ensure that we add this same edge $\{1, x\}$ to the tree in iteration $n - 1$ of the **for** loop.

This discussion establishes that $\mathsf{Prüfer}(T) = L$ if and only if $\mathsf{InvPrüfer}(L) = T$. Thus the two functions are inverses of each other, and hence they are in fact bijections of the two sets.

Let's do an example to illustrate. Consider the following tree T_8 on $n = 8$ vertices:

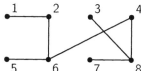

When given the edge set of T_8 as input, Algorithm 3.18 executes as follows:

vertex degrees	i	x	$L[i] = y$	edge deleted
$1, 2, 1, 2, 1, 3, 1, 3$	1	7	8	$\{7, 8\}$
$1, 2, 1, 2, 1, 3, 0, 2$	2	5	6	$\{5, 6\}$
$1, 2, 1, 2, 0, 2, 0, 2$	3	3	8	$\{3, 8\}$
$1, 2, 0, 2, 0, 2, 0, 1$	4	8	4	$\{4, 8\}$
$1, 2, 0, 1, 0, 2, 0, 0$	5	4	6	$\{4, 6\}$
$1, 2, 0, 0, 0, 1, 0, 0$	6	6	2	$\{2, 6\}$
$1, 1, 0, 0, 0, 0, 0, 0$				

Thus $\text{Prüfer}(T_8) = [8, 6, 8, 4, 6, 2]$. At the end of Algorithm 3.18, the edge remaining is $\{1, 2\}$. The reader can check that, when we execute Algorithm 3.19 with input $L = [8, 6, 8, 4, 6, 2]$, the same values of the variables are computed!

We can extend the Prüfer correspondence very easily to obtain ranking and unranking functions for $\mathcal{T}(n)$. In order to accomplish this, it suffices to describe bijections between $\mathcal{V}^{n-2}$ and $\{0, \ldots, n^{n-2} - 1\}$, which is straightforward to do. These bijections are presented in Algorithms 3.20 and 3.21.

Algorithm 3.20: PRUFERTORANK (L, r)

$r \leftarrow 0$
$p \leftarrow 1$
for $i \leftarrow n - 2$ **downto** 1
$\quad$ **do** $\begin{cases} r \leftarrow r + p\,(L[i] - 1) \\ p \leftarrow pn \end{cases}$
return (r)

Algorithm 3.21: RANKTOPRUFER (r, L)

for $i \leftarrow n - 2$ **downto** 1
$\quad$ **do** $\begin{cases} L[i] \leftarrow \text{mod}(r, n) + 1 \\ r \leftarrow \left\lfloor \frac{r - L[i] + 1}{n} \right\rfloor \end{cases}$
return (L)

In the example above, we would obtain $\text{rank}(T_8) = 253673$. As another example, the 16 labeled trees on four vertices, and their ranks, are listed in Table 3.5.

TABLE 3.5
The labeled trees on four vertices and their ranks

T	Prüfer(T)	rank(T)	T	Prüfer(T)	rank(T)
	$(1, 1)$	0		$(1, 2)$	1
	$(1, 3)$	2		$(1, 4)$	3
	$(2, 1)$	4		$(2, 2)$	5
	$(2, 3)$	6		$(2, 4)$	7
	$(3, 1)$	8		$(3, 2)$	9
	$(3, 3)$	10		$(3, 4)$	11
	$(4, 1)$	12		$(4, 2)$	13
	$(4, 3)$	14		$(4, 4)$	15

TABLE 3.6
The Catalan families C_n for $1 \leq n \leq 4$

n		C_n				C_n
1	01					1
2	0011	0101				2
3	000111	001011	001101	010011	010101	5
4	00001111	00010111	00011011	00011101	00100111	14
	00101011	00101101	00110011	00110101	01000111	
	01001011	01001101	01010011	01010101		

3.4 Catalan families

The Catalan numbers are a sequence of numbers which arise naturally in many different combinatorial enumeration problems. In this section, we discuss ranking and unranking of objects that are enumerated by Catalan numbers.

We begin by presenting one way — out of many — to define Catalan numbers. Let n be a positive integer, and let $a = [a_1, a_2, \ldots, a_{2n-1}, a_{2n}] \in (\mathbb{Z}_2)^{2n}$. We say that the sequence a is a *totally balanced sequence* if the following two properties are satisfied:

1. a contains n 0s and n 1s, and
2. for any i, $1 \leq i \leq 2n$, it holds that

$$|\{j : 1 \leq j \leq i, a_i = 0\}| \geq |\{j : 1 \leq j \leq i, a_i = 1\}|.$$

Let C_n denote the set of all totally balanced sequences in $(\mathbb{Z}_2)^{2n}$. We will refer to C_n as a *Catalan family* of order n. The *Catalan number* C_n is defined to be $C_n = |C_n|$. To illustrate, we list all the sequences in the Catalan families C_n, for $1 \leq n \leq 4$, in Table 3.6.

The first few Catalan numbers are $C_1 = 1$, $C_2 = 2$, $C_3 = 5$, $C_4 = 14$, $C_5 = 42$, $C_6 = 132$, $C_7 = 429$, $C_8 = 1430$, $C_9 = 4862$ and $C_{10} = 16796$. It turns out that there is a very simple formula for the Catalan numbers, which is stated in the following theorem.

THEOREM 3.17 *For any integer $n \geq 1$, the Catalan number C_n is*

$$C_n = \frac{1}{n+1}\binom{2n}{n}.$$

In the remainder of this section, we will give a nice proof of this result. The approach we use is based on a pictorial interpretation of the sequences in $(\mathbb{Z}_2)^{2n}$.

Let $a \in (\mathbb{Z}_2)^{2n}$. We will describe a graphical representation of a in terms of a path in the (x, y)-plane. We begin at the origin, $(0, 0)$, and process the $2n$

elements $a_1, \ldots, a_{2n}$, constructing a path $P = P(a)$ as we go along. When we are situated at the point (x, y), we will process the term a_x. If $a_x = 0$, then we extend the path P by adding the point $(x + 1, y + 1)$ to P; if $a_x = 1$, then we add the point $(x + 1, y - 1)$ to P. A pseudocode description of this algorithm is presented as Algorithm 3.22. In this algorithm, any binary sequence $a \in (\mathbb{Z}_2)^{2n}$ is converted to a path P beginning at the origin, which is represented as a set of $2n + 1$ points:

$$P = \{(0,0), (1, y_1), \ldots, (2n - 1, y_{2n-1}), (2n, y_{2n-1})\}.$$

Algorithm 3.22: SEQUENCETOPATH $(n, [a_1, \ldots, a_{2n}])$

$P \leftarrow \{(0,0)\}$
$y \leftarrow 0$
for $x \leftarrow 1$ **to** $2n$
$\quad$ **do** $\begin{cases} \textbf{if } a_x = 0 \\ \quad \textbf{then } y \leftarrow y + 1 \\ \quad \textbf{else } y \leftarrow y - 1 \\ P \leftarrow P \cup \{(x, y)\} \end{cases}$
return (P)

The path P thus constructed can be thought of as a *mountain range* of width $2n$. If $a \in C_n$, then the mountain range P begins and ends at sea level (i.e., $y = 0$). The totally balanced condition means that y-coordinates of the points in P are all greater than or equal to zero. Stated a bit more descriptively, the elevation of P never drops below sea level! Conversely, any mountain range of this type corresponds to a unique sequence in C_n.

As an example, the totally balanced sequence $a = 00101101$ corresponds to the following mountain range:

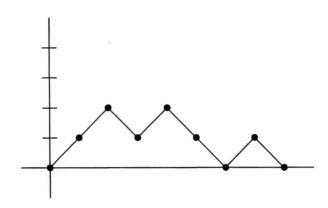

It is easy to see that the total number of mountain ranges from $(0,0)$ to $(2n,0)$ is $\binom{2n}{n}$. This is because any such mountain range corresponds to a sequence of n 0s and n 1s, and there are $\binom{2n}{n}$ binary sequences of length $2n$ that contain exactly n 0s. Thus, we have that

$$C_n = \binom{2n}{n} - \text{the number of mountain ranges that drop below sea level.}$$

Clearly we can compute C_n if we can obtain a formula for the number of mountain ranges that drop below sea level.

Suppose that P is a mountain range from $(0,0)$ to $(2n,0)$ that drops below sea level at some point. Recall that P is comprised of a set of points:

$$P = \{(0,0),(1,y_1),\ldots,(2n-1,y_{2n-1}),(2n,0)\}.$$

Let X be the first point in P with y-coordinate less than zero; then $X = (x_0,-1)$ for some x_0, where $1 \leq x_0 \leq 2n-1$. Now, define a new mountain range P^*, as follows:

$$P^* = \{(x,y) \in P : x_0 \leq x \leq 2n\} \cup \{(-2-x,y) : 1 \leq x \leq x_0-1,(x,y) \in P\}.$$

P^* can be described geometrically by saying that the initial portion of P, from $(0,0)$ to X, is reflected in the line $y = -1$. Therefore P^* is a mountain range from $(0,-2)$ to $(2n,0)$.

As an example, we show the mountain range P corresponding to the sequence $a = 0101100011$, and the reflected mountain range P^*.

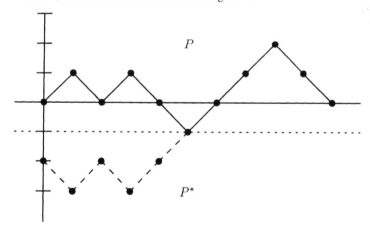

A moment's reflection shows that the mapping $P \mapsto P^*$ defines a bijection from the set of mountain ranges from $(0,0)$ to $(2n,0)$ that drop below sea level to the set of all mountain ranges from $(0,-2)$ to $(2n,0)$. Now, it is easy to show that a mountain range from $(0,-2)$ to $(2n,0)$ corresponds to a binary string of length $2n$ that contains $n+1$ 0s and $n-1$ 1s. (This is because the gain in elevation is equal to 2 and hence we need two more 0s than 1s.) Therefore there are precisely $\binom{2n}{n+1}$ mountain ranges P^*. Thus we have shown that

$$C_n = \binom{2n}{n} - \binom{2n}{n+1} = \frac{1}{n+1}\binom{2n}{n},$$

and Theorem 3.17 is proved.

3.4.1   Ranking and unranking

In this section we provide ranking and unranking algorithms for the lexicographic ordering of the Catalan families. The algorithms are most easily described in terms of the mountain ranges we introduced in the previous section. Let n be a positive integer, and let $(x,y) \in \mathbb{Z} \times \mathbb{Z}$, where $0 \le x \le 2n$. Define $\mathsf{M}_n(x,y)$ to be the set of all mountain ranges from (x,y) to $(2n,0)$ that do not drop below sea level, and denote $M_n(x,y) = |\mathsf{M}_n(x,y)|$.

It is clear that $M_n(0,0) = C_n$, and $M_n(x,y) = 0$ if $y < 0$. As well, $M_n(x,y) = 0$ if $x + y > 2n$, since any mountain range that passes through (x,y) cannot reach sea level before the point $(x+y, 0)$. Finally, it is easy to see that $M_n(x,y) = 0$ if $x + y$ is odd (this is because $x + y$ is even for any point (x,y) on a mountain range that includes the point $(2n,0)$). The following result provides a formula for $M_n(x,y)$ for the remaining (non-trivial) cases. It is proved using the same reflection idea that we used to prove Theorem 3.17.

LEMMA 3.18   *Let n be a positive integer. Suppose that $0 \le x \le 2n$, $y \ge 0$, $x + y$ is even and $x + y \le 2n$. Then*

$$M_n(x,y) = \binom{2n - x}{n - \frac{x+y}{2}} - \binom{2n - x}{n - 1 - \frac{x+y}{2}}.$$

Suppose we are given a sequence $a \in \mathsf{C}_n$, and we want to compute the rank of a. Note that $0 \le \mathrm{rank}(a) \le C_n - 1$. Let $P = P(a)$ be the corresponding mountain range $P \in \mathsf{M}_n(0,0)$, which can be constructed using Algorithm 3.22. We are going to compute $\mathrm{rank}(a)$ by following P from left to right and using a type of binary search technique.

At any given time in the algorithm, we will be looking at a point $(x-1, y) \in P$, and we will have an upper and lower bound on $\mathrm{rank}(a)$, say

$$lo \le \mathrm{rank}(a) \le hi,$$

where

$$hi - lo + 1 = M_n(x - 1, y).$$

(Initially, we set $lo \leftarrow 0$ and $hi \leftarrow C_n - 1$.) Let X be the next point on P. Then

$$X = \begin{cases} (x, y+1) & \text{if } a_x = 0 \\ (x, y-1) & \text{if } a_x = 1. \end{cases}$$

Let P' denote the initial portion of P, from $(0,0)$ to $(x-1,y)$ and let P'' denote the remaining part of P, from X to $(2n,0)$. There are exactly $M_n(x,y+1)$ mountain ranges in $M_n(x,y+1)$, and $M_n(x,y-1)$ mountain ranges in $M_n(x,y-1)$. It is easy to verify the following relation:

$$M_n(x-1,y) = M_n(x,y+1) + M_n(x,y-1). \tag{3.4}$$

Among the mountain ranges in $M_n(0,0)$ with initial portion P', the $M_n(x,y+1)$ lexicographically least of them have $(x,y+1)$ as their next point. This allows us to update either hi or lo. If $a_x = 0$, then $hi \leftarrow hi - M_n(x,y-1)$, and if $a_x = 1$, then $lo \leftarrow lo + M_n(x,y+1)$. Thus the interval $[lo, hi]$, in which $\mathsf{rank}(a)$ must be found, is replaced by one of two possible complementary subintervals:

$$[lo, hi - M_n(x,y-1)] \quad \text{or} \quad [lo + M_n(x,y+1), hi].$$

This is reminiscent of a binary search. Note that

$$lo + M_n(x,y+1) = hi - M_n(x,y-1) + 1$$

since

$$M_n(x,y+1) + M_n(x,y-1) = M_n(x,y-1) = hi - lo + 1.$$

Unranking is almost identical. The only difference is that we use the value of the given rank r to compute the terms a_x and to replace an interval $[lo, hi]$ by an appropriate subinterval. However, it turns out that we do not actually need to keep track of the value hi in the ranking and unranking algorithms. This simplifies the algorithms a bit.

The algorithms are presented as Algorithms 3.23 and 3.24. We will assume the existence of a procedure $\mathsf{M}(n,x,y)$ which computes the function $M_n(x,y)$. The procedure M could either use the formula developed in Lemma 3.18, or precompute all values $M_n(x,y)$ with $x+y \leq 2n$ using the recurrence relation from Equation (3.4).

Algorithm 3.23: CATALANRANK $(n, [a_1, \ldots, a_{2n}])$

external M()

$y \leftarrow 0$
$lo \leftarrow 0$
for $x \leftarrow 1$ **to** $2n - 1$
$\quad$ **do** $\begin{cases} \textbf{if } a_x = 0 \\ \quad \textbf{then } y \leftarrow y + 1 \\ \quad \textbf{else } \begin{cases} lo \leftarrow lo + \mathsf{M}(n, x, y+1) \\ y \leftarrow y - 1 \end{cases} \end{cases}$
return (lo)

y										
5					1					
4				5		1				
3			14		4		1			
2		28		9		3		1		
1		42		14		5		2		1	
0	42		14		5		2		1		1
	$x = 0$	1	2	3	4	5	6	7	8	9	10

FIGURE 3.1
Values of $M_5(x, y)$.

Algorithm 3.24: CATALANUNRANK (n, r)

external M$()$

$y \leftarrow 0$
$lo \leftarrow 0$
for $x \leftarrow 1$ **to** $2n$

$\text{do} \begin{cases} m \leftarrow \text{M}(n, x, y + 1) \\ \textbf{if } r \leq lo + m - 1 \\ \quad \textbf{then } \begin{cases} y \leftarrow y + 1 \\ a_x \leftarrow 0 \end{cases} \\ \quad \textbf{else } \begin{cases} lo \leftarrow lo + m \\ y \leftarrow y - 1 \\ a_x \leftarrow 1 \end{cases} \end{cases}$

return $([a_1, \ldots, a_{2n}])$

To illustrate Algorithms 3.23 and 3.24, we will do an example with $n = 5$. For future reference, we precompute the non-zero values $M_5(x, y)$, which are recorded in Figure 3.1. Note that the diagonals of this figure can be computed in the same fashion as the rows in Pascal's triangle, in view of the similarity of the recurrence in Equation (3.4) to Pascal's identity,

$$\binom{n}{r} = \binom{n-1}{r-1} + \binom{n-1}{r}.$$

Suppose we are given the sequence $a = 0010110101 \in C_5$. Then rank(a)

would be computed as follows.

x	a_x	y	lo
0		0	0
1	0	1	0
2	0	2	0
3	1	1	$lo \leftarrow 0 + M_5(3,3) = 14$
4	0	2	14
5	1	1	$lo \leftarrow 14 + M_5(5,3) = 18$
6	1	0	$lo \leftarrow 18 + M_5(6,2) = 21$
7	0	1	21
8	1	0	$lo \leftarrow 21 + M_5(8,2) = 22$
9	0	1	22
10	1	0	

Therefore $\mathsf{rank}(0010110101) = 22$.

We reverse the process by computing $\mathsf{unrank}(22)$:

m	x	y	lo	a_x
	0	0	0	
$M_5(1,1) = 42$	1	1	0	0
$M_5(2,2) = 28$	2	2	0	0
$M_5(3,3) = 14$	3	1	$lo \leftarrow 0 + 14 = 14$	1
$M_5(4,2) = 9$	4	2	14	0
$M_5(5,3) = 4$	5	1	$lo \leftarrow 14 + 4 = 18$	1
$M_5(6,2) = 3$	6	0	$lo \leftarrow 18 + 3 = 21$	1
$M_5(7,1) = 2$	7	1	21	0
$M_5(8,2) = 1$	8	0	$lo \leftarrow 21 + 1 = 22$	1
$M_5(9,1) = 1$	9	1	22	0
$M_5(10,2) = 0$	10	0	$lo \leftarrow 22 + 0 = 22$	1

Hence, $\mathsf{unrank}(22) = 0010110101$.

3.4.2 Other Catalan families

We defined Catalan families as totally balanced sequences. There are many other families which have the same cardinality, and which therefore could equally well be taken as a definition of Catalan families. These include triangulations of polygons, rooted plane binary trees, rooted plane bushes, and non-crossing handshakes, to name a few examples. If we have an explicit bijection between C_n and any other Catalan family, then our ranking and unranking algorithms can be used to rank and unrank the objects in the given family. We illustrate this with a particular type of Catalan family.

Let n be a positive integer. A 2 by n *standard tableau* is an arrangement of the integers $1, \ldots, 2n$ in a 2 by n array, such that the entries across any row or down any column are increasing. Let $\mathsf{Tab}(n)$ denote the set of all $2 \times n$ standard tableaux. As an example, we present the tableaux in $\mathsf{Tab}(3)$:

1	2	3
4	5	6

1	2	4
3	5	6

1	2	5
3	4	6

1	3	4
2	5	6

1	3	5
2	4	6

It is not difficult to find a bijection $\phi : \mathsf{Tab}(n) \to C_n$. The existence of such a bijection will show that $|\mathsf{Tab}(n)| = C_n$. The definition of the function ϕ is very simple. Suppose that $T \in \mathsf{Tab}(n)$, and denote the entries of T by $T[i,j], i = 1, 2$, $1 \leq j \leq n$. Then define $\phi(T) = [a_1, \ldots, a_{2n}]$, where

$$a_i = \begin{cases} 0 & \text{if } i \text{ is in the first row of } T \\ 1 & \text{if } i \text{ is in the second row of } T. \end{cases}$$

Algorithm 3.25 performs this operation.

Algorithm 3.25: TABLEAUTOSEQUENCE (n, T)

$\mathbf{for}\ i \leftarrow 1\ \mathbf{to}\ 2$

$\quad \mathbf{do} \begin{cases} \mathbf{for}\ j \leftarrow 1\ \mathbf{to}\ n \\ \quad \mathbf{do}\ a_{T[i,j]} \leftarrow i - 1 \end{cases}$

$\mathbf{return}\ ([a_1, \ldots, a_{2n}])$

It is quite easy to show that $\phi(T) \in C_n$ for all $T \in \mathsf{Tab}(n)$. To show that ϕ is a bijection (i.e., that it is one-to-one and onto), it is sufficient to find the inverse function $\phi^{-1} : C_n \to \mathsf{Tab}(n)$. The function ϕ^{-1} can be computed using Algorithm 3.26.

Algorithm 3.26: SEQUENCETOTABLEAU $(n, [a_1, \ldots, a_{2n}])$

$c[1] \leftarrow 0$
$c[2] \leftarrow 0$
$\mathbf{for}\ i \leftarrow 1\ \mathbf{to}\ 2n$

$\quad \mathbf{do} \begin{cases} r \leftarrow a_i + 1 \\ c[r] \leftarrow c[r] + 1 \\ T[r, c[r]] \leftarrow i \end{cases}$

$\mathbf{return}\ (T)$

We leave it to the reader to prove that $\phi^{-1}(\phi(T)) = T$ for all $T \in \mathsf{Tab}(n)$; and $\phi(\phi^{-1}(a)) = a$ for all $a \in C_n$. This will prove that ϕ and ϕ^{-1} are both bijections.

Now, if we want to compute the rank of a tableau $T \in \mathsf{Tab}(n)$, we simply compute $\phi(T)$ using Algorithm 3.25, and then compute the rank of $\phi(T)$ using Algorithm 3.23. If we want to unrank r to a tableau, we first use Algorithm 3.24 to unrank t to a sequence, say a, and then compute $\phi^{-1}(a)$ using Algorithm 3.26.

3.5 Notes

Section 3.1

The study of partitions is one of the oldest and richest parts of combinatorics, and has been a topic of interest to mathematicians for over 300 years. There are many sources of information on partitions, for example, Andrews [2], Goulden and Jackson [37], and the article by Gessel and Stanley [32] in the *Handbook of Combinatorics* [38]. A more algorithmic approach can be found in [90, 101, 115].

Section 3.2

Many of the references for the previous section also contain information on Bell and Stirling numbers. The ranking and unranking algorithms using restricted growth functions are based on the treatment of Stanton and White [101].

Section 3.3

Many textbooks discuss the Prüfer correspondence; two examples are Stanton and White [101] and Williamson [115].

Section 3.4

For an entertaining treatment of Catalan numbers, see Conway and Guy [21]. Stanley [100] presents an extensive list of different Catalan families. The ranking and unranking algorithms are algorithms of our own design.

Exercises

3.1 Prove that
$$P(m,2) = \left\lfloor \frac{m}{2} \right\rfloor$$
and
$$P(m,3) = \left\lfloor \frac{m^2 + 4}{12} \right\rfloor .$$

3.2 Develop a recursive algorithm that directly generates all partitions of m into n parts.

3.3 It is known that the number of partitions of m into odd parts is equal to the number of partitions of m into unequal parts. Develop recursive algorithms to generate all partitions of m of these two types, and run your algorithms for $m = 10$ to generate all such partitions.

3.4 Find the successor and rank of the partition $[8, 6, 6, 4, 3, 1]$.

3.5 Find the partition in $\mathcal{P}(15, 5)$ having rank 18.

3.6 Prove that $S(m, 2) = 2^m - 1$ for all integers $m \geq 1$.

3.7 Prove that $S(m, m - 1) = m(m - 1)/2$ for all integers $m \geq 1$.

3.8 Prove Theorem 3.12.

3.9 Find the rank of the partition $\{\{1, 3, 5\}, \{2\}, \{4\}\}$.

3.10 Find the partition of $\{1, \ldots, 6\}$ that has rank 153.

3.11 Prove Theorem 3.14.

3.12 Use Theorem 3.8 to prove that

$$\sum_{n=0}^{m} S(m, n)\, x(x - 1) \cdots (x - n + 1) = x^m.$$

3.13 Use Theorem 3.14 to prove that

$$\sum_{n=0}^{m} s(m, n)\, x^n = x(x - 1) \cdots (x - m + 1).$$

3.14 Use Exercises 3.12 and 3.13 to prove Theorem 3.15.

3.15 Find the tree on eight vertices having rank 126998.

3.16 Prove Lemma 3.18.

3.17 Prove that

$$C_n = \sum_{i=0}^{n-1} C_i\, C_{n-1-i},$$

where we define $C_0 = 0$.

3.18 Prove that the number of ways of triangulating a polygon with $n + 2$ sides is equal to C_n.

3.19 Compute the mountain range in C_6 having rank 99.

4

Backtracking Algorithms

4.1 Introduction

A *backtracking algorithm* is a recursive method of building up feasible solutions to a combinatorial optimization problem one step at a time (recall that basic terminology relating to optimization problems was introduced in Section 1.3). A backtracking algorithm is an *exhaustive search*; that is, all feasible solutions are considered and it will thus always find the optimal solution. *Pruning* methods can be used to avoid considering some feasible solutions that are not optimal.

To illustrate the basic principles of backtracking, we consider the Knapsack (optimization) problem, which was presented as Problem 1.4. Recall that a problem instance consists of a list of profits, $P = [p_0, \ldots, p_{n-1}]$; a list of weights, $W = [w_0, \ldots, w_{n-1}]$; and a capacity, M. It is required to find the maximum value of $\sum p_i x_i$ subject to $\sum w_i x_i \leq M$ and $x_i \in \{0, 1\}$ for all i. An n-tuple $[x_0, x_1, x_2, \ldots, x_{n-1}]$ of 0s and 1s is a feasible solution if $\sum w_i x_i \leq M$.

One naive way to solve this problem is to try all 2^n possible n-tuples of 0s and 1s. We can build up an n-tuple one coordinate at a time by first choosing a value for x_0, then choosing a value for x_1, etc. Backtracking provides a simple method for generating all possible n-tuples. After each n-tuple is generated it is checked for feasibility. If it is feasible, then its profit is compared to the current best solution found to that point. The current best solution is updated whenever a better feasible solution is found.

We will denote by $X = [x_0, x_1, \ldots, x_{n-1}]$ the current n-tuple being constructed, and $CurP$ will denote its profit. $OptX$ will denote the current optimal solution and $OptP$ is its profit. A recursive backtracking algorithm for the Knapsack (optimization) problem is presented now.

Algorithm 4.1: KNAPSACK1 (ℓ)

global $X, OptP, OptX$
if $\ell = n$

then $\begin{cases} \textbf{if } \displaystyle\sum_{i=0}^{n-1} w_i x_i \le M \\[2em] \quad \textbf{then} \begin{cases} CurP \leftarrow \displaystyle\sum_{i=0}^{n-1} p_i x_i \\[1.5em] \textbf{if } CurP > OptP \\ \quad \textbf{then} \begin{cases} OptP \leftarrow CurP \\ OptX \leftarrow [x_0, \ldots, x_{n-1}] \end{cases} \end{cases} \end{cases}$

else $\begin{cases} x_\ell \leftarrow 1 \\ \text{KNAPSACK1} (\ell + 1) \\ x_\ell \leftarrow 0 \\ \text{KNAPSACK1} (\ell + 1) \end{cases}$

Note that Algorithm 4.1 is invoked initially with $\ell = 0$.

The recursive calls to Algorithm 4.1 produce a binary tree called the *state space tree* for the given problem instance. When $n = 3$, this tree is given in Figure 4.1. A backtrack search performs a depth-first traversal of the state space tree.

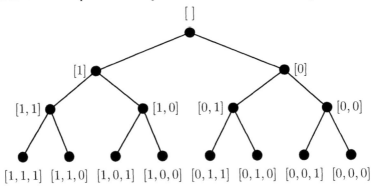

FIGURE 4.1
State space tree when $n = 3$.

Algorithm 4.1 generates the 2^n binary n-tuples in reverse lexicographic order. It takes time $\Theta(n)$ to check each solution, and so the asymptotic running time for this algorithm is $\Theta(n2^n)$. Of course this approach is impractical for $n > 40$, say. Notice that not all n-tuples of 0s and 1s are feasible, and a fairly simple modification to the backtracking algorithm would take this into account. This and other improvements will be considered in the remaining sections of this chapter.

4.2 A general backtrack algorithm

We now present a general backtrack algorithm. For many combinatorial optimization problems of interest, the (optimal) solution can be represented as a list $X = [x_0, x_1, \ldots]$ in which each x_i is chosen from a finite *possibility set*, $\mathcal{P}_i$. The x_is are defined one at a time, in order, as the state space tree is traversed. Hence, the backtrack algorithm considers all members of $\mathcal{P}_0 \times \mathcal{P}_1 \times \cdots \times \mathcal{P}_i$ for each $i = 0, 1, 2, \ldots$. The length of the list X is the same as the depth of the corresponding node in the state space tree.

Given a partial solution $X = [x_0, x_1, \ldots, x_{\ell-1}]$, the constraints for the optimization problem will restrict the possible values for x_ℓ to a subset $\mathcal{C}_\ell \subseteq \mathcal{P}_\ell$ that we call a *choice set*. The computation of the set $\mathcal{C}_\ell$ is referred to as pruning. If $y \in \mathcal{P}_\ell \setminus \mathcal{C}_\ell$, then nodes in the subtree with root node $[x_0, x_1, \ldots, x_{\ell-1}, y]$ will not be considered by the backtracking algorithm. Thus we say that this subtree has been "pruned" from the original state space tree. The general backtracking algorithm with pruning is presented as Algorithm 4.2.

Algorithm 4.2: BACKTRACK (ℓ)

global $X, \mathcal{C}_\ell \quad (\ell = 0, 1, \ldots)$
comment: $X = [x_0, x_1, \ldots]$
if $[x_0, x_1, \ldots, x_{\ell-1}]$ is a feasible solution
   then process it
Compute $\mathcal{C}_\ell$
for each $x \in \mathcal{C}_\ell$
   do $\begin{cases} x_\ell \leftarrow x \\ \text{BACKTRACK}(\ell + 1) \end{cases}$

The first step of the algorithm is to identify if the current partial solution, X, is indeed a feasible solution. The operation "process it" could mean several things, e.g., save X for future use; print it out; or check to see if it is better than the best solution found so far (according to an optimality measure), as was done in Algorithm 4.1.

The second step is to construct the choice set $\mathcal{C}_\ell$ for the current value of X. The third step is to assign every possible value in $\mathcal{C}_\ell$ in turn as the next coordinate, x_ℓ, calling the algorithm recursively each time an assignment is made.

In many problems, it may be the case that no feasible solution can possibly be extended. For example, in the Knapsack (optimization) problem, a feasible solution is an n-tuple, where n is specified in the problem instance. In this situation, the choice set would be defined to be empty, so that no recursive calls would be made at that point. An alternative method, which accomplishes the same thing, would be to use an if-then-else construct, as was done in Algorithm 4.1.

In Algorithm 4.1, no pruning is performed. We could think of this algorithm as being an application of Algorithm 4.2 with $C_\ell = \{1, 0\}$. In order for backtracking to have practical value, we need efficient ways of reducing the size of the choice sets C_ℓ. That is, we would like to eliminate branches of the search tree that cannot lead to solutions, without actually traversing them.

For the Knapsack (optimization) problem, one simple method of pruning is to observe that we must have

$$\sum_{i=0}^{\ell-1} w_i x_i \leq M$$

for any partial solution $[x_0, x_1, \ldots, x_{\ell-1}]$. In other words, we can check partial solutions to see if the feasibility condition is satisfied. Consequently, if $\ell \leq n - 1$ and we set

$$CurW = \sum_{i=0}^{\ell-1} w_i x_i,$$

then we have

$$C_\ell = \begin{cases} \{1, 0\} & \text{if } CurW + w_\ell \leq M; \\ \{0\} & \text{otherwise.} \end{cases}$$

Applying these ideas, using Algorithm 4.2 as a template, we obtain Algorithm 4.3, which is invoked with $\ell = CurW = 0$.

Algorithm 4.3: KNAPSACK2 $(\ell, CurW)$

global $X, OptX, OptP, C_\ell \quad (\ell = 0, 1, \ldots)$

if $\ell = n$

then $\begin{cases} \textbf{if } \sum_{i=0}^{n-1} p_i x_i > OptP \\ \\ \quad \textbf{then } \begin{cases} OptP \leftarrow \sum_{i=0}^{n-1} p_i x_i \\ OptX \leftarrow [x_0, \ldots, x_{n-1}] \end{cases} \end{cases}$

if $\ell = n$

  then $C_\ell \leftarrow \emptyset$

  else $\begin{cases} \textbf{if } CurW + w_\ell \leq M \\ \quad \textbf{then } C_\ell \leftarrow \{1, 0\} \\ \quad \textbf{else } C_\ell \leftarrow \{0\} \end{cases}$

  for each $x \in C_\ell$

  do $\begin{cases} x_\ell \leftarrow x \\ \text{KNAPSACK2}(\ell + 1, CurW + w_\ell x_\ell) \end{cases}$

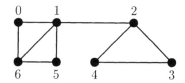

FIGURE 4.2
A graph with four maximal cliques.

4.3 Generating all cliques

Recall from Section 1.2 that a clique in an undirected graph $\mathcal{G} = (\mathcal{V}, \mathcal{E})$ is a subset of vertices $S \subseteq \mathcal{V}$ such that $\{x, y\} \in \mathcal{E}$ for all $x, y \in S$. We consider the empty set to be a clique, and $\{x\}$ is a clique for any $x \in \mathcal{V}$. A clique is a *maximal clique* if it is not a subset of a larger clique. For example, in the graph given in Figure 4.2, the cliques are $\{\}, \{0, 1\}, \{0, 6\}, \{1, 2\}, \{1, 5\}, \{1, 6\}, \{2, 4\}, \{2, 3\},$ $\{3, 4\}, \{0, 1, 6\}, \{1, 5, 6\}$ and $\{1, 3, 4\}$. The maximal cliques are $\{0, 1, 6\}, \{1, 2\},$ $\{1, 5, 6\}$ and $\{1, 3, 4\}$. As another example, the graph given in Figure 4.3 has 90 maximal cliques, one of which is $\{0, 9, 14\}$.

Many combinatorial search problems can be reduced to finding (maximal) cliques in an appropriately chosen graph. Here is an example to illustrate. There are $\binom{6}{2}$ unordered pairs that can be formed from a 6-element set $X = \{0, 1, 2, 3, 4, 5\}$. Suppose we number them in lexicographic order, i.e., $0, 1, \ldots, 14$. The graph given in Figure 4.3 contains an edge xy if and only if the pairs corresponding to the vertices x and y are disjoint. A maximal clique in this graph is just a partition of X into disjoint pairs. (This partition is in fact a perfect matching of the elements of X.)

The problem that we will study in this section is the generation of all the cliques, without repetition, in a given graph. See Problem 4.1.

Problem 4.1:	All Cliques
Instance:	a graph $\mathcal{G} = (\mathcal{V}, \mathcal{E})$
Find:	all the cliques of $\mathcal{G}$ without repetition.

To generate all of the cliques of a graph $\mathcal{G}$ by backtracking, we need to define what a partial solution is and give a method to compute the choice sets C_ℓ. The first part is easy: a sequence $X = [x_0, x_1, \ldots, x_{\ell-1}]$ of vertices is a partial solution if and only if $\{x_0, x_1, \ldots, x_{\ell-1}\}$ is a clique. Now, denote $S_{\ell-1} = \{x_0, \ldots, x_{\ell-1}\}$ and $C_0 = \mathcal{V}$. Then the choice set C_ℓ is given by

$$C_\ell = \{v \in \mathcal{V} \setminus S_{\ell-1} : \{v, x\} \in \mathcal{E} \text{ for each } x \in S_{\ell-1}\}$$

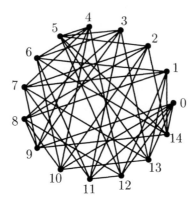

FIGURE 4.3
The graph of non-intersecting pairs from a 6-element set.

$$= \{v \in C_{\ell-1} \setminus \{x_{\ell-1}\} : \{v, x_{\ell-1}\} \in \mathcal{E}\}.$$

Unfortunately, if a clique has size k, then an algorithm based on the above choice function will generate it $k!$ times, once for each possible ordering of its vertices. To avoid this duplication of work, we arbitrarily place a total ordering "$<$" on the vertices $\mathcal{V}$, and list them according to this ordering. That is, we think of $\mathcal{V}$ as an ordered list, i.e., we write $\mathcal{V} = [v_0, v_1, v_2, \ldots., v_{n-1}]$, where $v_0 < v_1 < \cdots < v_{n-1}$.

Then, we redefine C_ℓ as follows:

$$C_\ell = \{v \in C_{\ell-1} : \{v, x_{\ell-1}\} \in \mathcal{E} \text{ and } v > x_{\ell-1}\}.$$

These choice functions can be more efficiently computed if we define, for each vertex $v \in \mathcal{V}$, the auxiliary sets

$$A_v = \{u \in \mathcal{V} : \{u, v\} \in \mathcal{E}\} \tag{4.1}$$

and

$$B_v = \{u \in \mathcal{V} : u > v\}. \tag{4.2}$$

These can be precomputed before the backtracking algorithm begins. Now, we have

$$C_\ell = A_{x_{\ell-1}} \cap B_{x_{\ell-1}} \cap C_{\ell-1}. \tag{4.3}$$

Suppose we also define

$$N_\ell = N_{\ell-1} \cap A_{x_{\ell-1}},$$

where $N_0 = \mathcal{V}$. Then $X = [x_0, \ldots, x_{\ell-1}]$ is a maximal clique if and only if $N_\ell = \emptyset$. The backtrack algorithm, Algorithm 4.4, generates each clique exactly once, and identifies the maximal cliques on the fly.

In Example 4.1, we illustrate the application of Algorithm 4.4 on a small graph. This graph contains five maximal cliques (which are marked with a star), and 21 cliques in total.

Example 4.1   *Finding all the cliques in a graph*

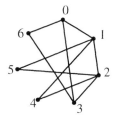

v	A_v	B_v
0	1, 3, 6	1, 2, 3, 4, 5, 6
1	0, 2, 4, 5	2, 3, 4, 5, 6
2	1, 3, 4, 5	3, 4, 5, 6
3	0, 2, 6	4, 5, 6
4	1, 2	5, 6
5	1, 2	6
6	0, 3	

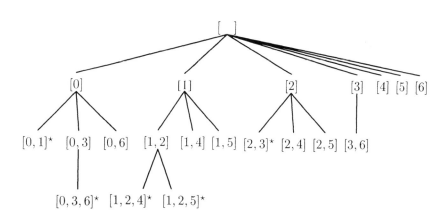

Maximal cliques are indicated with a ⋆.   ⬚

Algorithm 4.4: ALLCLIQUES (ℓ)

global $X, \mathcal{C}_\ell$ ($\ell = 0, 1, \ldots, n - 1$)
comment: every X generated by the algorithm is a clique
if $\ell = 0$
  then output ([ ])
  else output ($[x_0, \ldots, x_{\ell-1}]$)
if $\ell = 0$
  then $N_\ell \leftarrow \mathcal{V}$
  else $N_\ell \leftarrow A_{x_{\ell-1}} \cap N_{\ell-1}$
if $N_\ell = \emptyset$
  then $\{x_0, \ldots, x_{\ell-1}\}$ is a maximal clique
if $\ell = 0$
  then $\mathcal{C}_\ell \leftarrow \mathcal{V}$
  else $\mathcal{C}_\ell \leftarrow A_{x_{\ell-1}} \cap B_{x_{\ell-1}} \cap \mathcal{C}_{\ell-1}$
for each $x \in \mathcal{C}_\ell$
  do $\begin{cases} x_\ell \leftarrow x \\ \text{ALLCLIQUES}(\ell + 1) \end{cases}$

4.3.1 Average-case analysis

In this section, we determine the average-case complexity of Algorithm 4.4.

Let n be a positive integer, and let $\mathcal{G}(n)$ denote the set of all graphs on vertex set $\mathcal{V} = \{0, \ldots, n - 1\}$. There are $2^{\binom{n}{2}}$ graphs in $\mathcal{G}(n)$, because any unordered pair $\{x, y\} \subseteq \mathcal{V}$ can either be included as an edge or left out.

For a graph $\mathcal{G} \in \mathcal{G}(n)$, define $c(\mathcal{G})$ to be the number of cliques in the graph $\mathcal{G}$. When Algorithm 4.4 is run on the graph $\mathcal{G}$, the number of nodes in the state space tree is precisely $c(\mathcal{G})$. The running time of Algorithm 4.4 is easily seen to be $O(n\, c(\mathcal{G}))$. Of course, this running time depends on the particular graph $\mathcal{G}$, and so can vary greatly. For example, if $\mathcal{G}$ is the graph containing no edges, then $c(\mathcal{G}) = n + 1$, because the only cliques have size zero or one. At the other extreme, $c(K_n) = 2^n$ because any subset of vertices in a complete graph is a clique. It is because of this variation that we will do an average-case analysis.

We will be looking at the average value of $c(\mathcal{G})$, where the average is computed over all graphs in $\mathcal{G}(n)$. Thus we define

$$\bar{c}(n) = \frac{1}{2^{\binom{n}{2}}} \sum_{\mathcal{G} \in \mathcal{G}(n)} c(\mathcal{G}).$$

We now obtain an explicit formula for $\bar{c}(n)$. Suppose $\mathcal{G} \in \mathcal{G}(n)$ and $\mathcal{W} \subseteq \mathcal{V}$. Define the indicator function

$$\chi(\mathcal{G}, \mathcal{W}) = \begin{cases} 1 & \text{if } \mathcal{W} \text{ is a clique in } \mathcal{G} \\ 0 & \text{otherwise.} \end{cases}$$

Then we have that

$$c(\mathcal{G}) = \sum_{\mathcal{W} \subseteq \mathcal{V}} \chi(\mathcal{G}, \mathcal{W}),$$

and hence

$$\bar{c}(n) = \frac{1}{2^{\binom{n}{2}}} \sum_{\mathcal{G} \in \mathcal{G}(n)} \sum_{\mathcal{W} \subseteq \mathcal{V}} \chi(\mathcal{G}, \mathcal{W})$$

$$= \frac{1}{2^{\binom{n}{2}}} \sum_{\mathcal{W} \subseteq \mathcal{V}} \sum_{\mathcal{G} \in \mathcal{G}(n)} \chi(\mathcal{G}, \mathcal{W}).$$

For a given subset of vertices $\mathcal{W} \subseteq \mathcal{V}$, $\mathcal{W}$ is a clique of a graph $\mathcal{G}$ if and only if the $\binom{|\mathcal{W}|}{2}$ pairs of vertices in $\mathcal{W}$ are all edges of $\mathcal{G}$. There are $\binom{n}{2} - \binom{|\mathcal{W}|}{2}$ remaining possible edges, and so it follows that there are exactly

$$2^{\binom{n}{2} - \binom{|\mathcal{W}|}{2}}$$

graphs $\mathcal{G} \in \mathcal{G}(n)$ in which $\mathcal{W}$ is a clique. Hence we have that

$$\sum_{\mathcal{G} \in \mathcal{G}(n)} \chi(\mathcal{G}, \mathcal{W}) = 2^{\binom{n}{2} - \binom{|\mathcal{W}|}{2}}.$$

Thus, we obtain the following expression for $\bar{c}(n)$:

$$\bar{c}(n) = \frac{1}{2^{\binom{n}{2}}} \sum_{\mathcal{W} \subseteq \mathcal{V}} 2^{\binom{n}{2} - \binom{|\mathcal{W}|}{2}}.$$

Now, for any integer k such that $0 \le k \le n$, there are precisely $\binom{n}{k}$ subsets of vertices $\mathcal{W} \subseteq \mathcal{V}$ with $|\mathcal{W}| = k$. Hence, we have the following formula.

$$\bar{c}(n) = \frac{1}{2^{\binom{n}{2}}} \sum_{k=0}^{n} \binom{n}{k} 2^{\binom{n}{2} - \binom{k}{2}}$$

$$= \sum_{k=0}^{n} \binom{n}{k} 2^{-\binom{k}{2}}.$$

We tabulate some values of $\bar{c}(n)$ in Table 4.1.

Having derived an explicit formula for $\bar{c}(n)$, it is natural to ask how $\bar{c}(n)$ behaves as a function of n. Since $\bar{c}(n)$ is expressed as a sum, this will require some further analysis.

Define

$$t_k = \binom{n}{k} 2^{-\binom{k}{2}};$$

then

$$\bar{c}(n) = \sum_{k=0}^{n} t_k.$$

TABLE 4.1
Average number of nodes in the state space tree for Algorithm 4.4

n	$\bar{c}(n)$	n	$\bar{c}(n)$	n	$\bar{c}(n)$
1	2.0	10	52	110	321948
2	3.5	20	351	120	496385
3	5.6	30	1342	130	744800
4	8.5	40	3863	140	1091392
5	12.3	50	9316	150	1566330
6	17.2	60	19898	160	2206835
7	23.4	70	38876	170	3058400
8	31.1	80	70916	180	4176150
9	40.6	90	122485	190	5626373
		100	202314	200	7488221

Consider the ratio of successive terms in the sequence $t_0, t_1, \ldots, t_n$. Straightforward (but messy!) algebra shows that

$$\frac{t_k}{t_{k-1}} = \frac{n - k + 1}{k \, 2^{k-1}}.$$

Hence, we see that $t_k \geq t_{k-1}$ if and only if $n \geq k - 1 + k \, 2^{k-1}$. Since the function $f(k) = k - 1 + k \, 2^{k-1}$ is a strictly increasing function of k, it follows that the sequence $[t_0, \ldots, t_n]$ is *unimodular*, i.e., there exists an index ℓ such that

$$t_0 \leq \ldots \leq t_{\ell-1} \leq t_\ell \geq t_{\ell+1} \geq \ldots \geq t_n.$$

Then it is obvious that

$$\bar{c}(n) \leq (n + 1)t_\ell.$$

Now, we obtain an upper bound on the value ℓ. First, observe that

$$f(\log_2 n) = \log_2 n - 1 + \frac{n \log_2 n}{2}.$$

When $n \geq 4$, we have that $\log_2 n \geq 2$, and hence

$$f(\log_2 n) \geq n + 1 > n.$$

The inequality $f(\log_2 n) > n$ implies that $t_k < t_{k-1}$ when $k \geq \log_2 n$. This establishes that $\ell \leq \log_2 n$ if $n \geq 4$.

Now, we consider the term t_ℓ:

$$t_\ell = \binom{n}{\ell} 2^{-\binom{\ell}{2}} < \binom{n}{\ell} < n^\ell \leq n^{\log_2 n}.$$

Thus we have shown that $\bar{c}(n) \leq (n + 1)n^{\log_2 n}$ for $n \geq 4$. From this, it follows that $\bar{c}(n)$ is $O(n^{\log_2 n + 1})$, and the average-case running time of Algorithm 4.4 is $O(n^{\log_2 n + 2})$.

4.4 Estimating the size of a backtrack tree

In this section, we present an algorithmic method to estimate the number of nodes in a state space tree T for a backtracking algorithm. The algorithm will provide a quick way of getting a rough estimate on the number $|T|$ of nodes in T without running the entire backtracking algorithm. It is therefore a useful method to predict how long a big backtrack search might take to finish.

To motivate the algorithm, we first consider a special case. Suppose n is a positive integer, and $c_0, c_1, \ldots, c_{n-1}$ are also positive integers. Suppose that T is a tree of depth n in which every node of depth i has c_i children, for $0 \leq i \leq n - 1$. Equivalently, for $0 \leq i \leq n - 1$ and for any partial solution $X = [x_0, \ldots, x_{i-1}]$, there are c_i choices for x_i. The leaf nodes in T would correspond to solutions $[x_0, \ldots, x_{n-1}]$. In this particular tree, it is easy to see that, for each $i = 0, 1, \ldots, n$, the number nodes at depth i is $c_0 c_1 \cdots c_{i-1}$. Thus, the number of nodes in T is given by the following equation:

$$|T| = 1 + c_0 + c_0 c_1 + c_0 c_1 c_2 + \cdots + c_0 c_1 c_2 \cdots c_{n-1}. \quad (4.4)$$

In general, a state space tree T will not have such a regular structure. We will obtain an estimate of the number of nodes in T by probing a random path through T from the root node to a leaf node. As we follow this path, we compute a quantity analogous to Equation (4.4), in which the c_is are replaced by the number of choices available at the nodes in the given path. The algorithm is presented as Algorithm 4.5.

For any given state space tree T, Algorithm 4.5 will return a value $N = N(P)$ which depends on the path P that is probed. The value $N(P)$ is an estimate of $|T|$. It may be larger or smaller than $|T|$. (We can increase the accuracy of the estimate by running the algorithm several times, and computing the average of the values $N(P)$ over the various runs. In this way, we would expect to obtain a more accurate estimate of $|T|$.) We will now show that the expected value of N is in fact equal to $|T|$.

Define the following function on the nodes of T:

$$S([x_0, \ldots, x_{\ell-1}]) = \begin{cases} 1 & \text{if } \ell = 0 \\ |C_{\ell-1}([x_0, \ldots, x_{\ell-2}])| \cdot S([x_0, \ldots, x_{\ell-2}]) & \text{if } \ell \geq 1. \end{cases}$$

Thus, if X is the root node, then $S(X) = 1$. Further, if there are c choices available at a given node X, then $S(Y) = c S(X)$ for all children Y of X. Observe that, if P is the path probed in Algorithm 4.5, then

$$N(P) = \sum_{Y \in P} S(Y).$$

Algorithm 4.5: ESTIMATEBACKTRACK ( )

global N, m, C_ℓ $(\ell = 0, 1, \ldots)$
procedure PROBE(ℓ)
  Compute C_ℓ for the node $[x_0, \ldots, x_{\ell-1}]$
  $c \leftarrow |C_\ell|$
  if $c \neq 0$

  then $\begin{cases} m \leftarrow m\,c \\ N \leftarrow N + m \\ x_\ell \leftarrow \text{a random element of } C_\ell \\ \text{PROBE}(\ell+1) \end{cases}$

main
  $N \leftarrow 1$
  $m \leftarrow 1$
  PROBE(0)
  return (N)

In Example 4.2, we present a small tree and the S-values at each node. In this tree, there are six leaf nodes. The corresponding six paths from the root node to a leaf node have values $N(P) = 7, 15, 15, 9, 9$ and 9, and the paths are chosen with probabilities $1/6, 1/8, 1/8, 1/6, 1/6$ and $1/6$, respectively. Example 4.2 shows that the expected value of $N(P)$ is 10, which is equal to the number of nodes in the tree.

Example 4.2   *A state space tree*

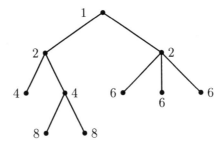

$$\overline{N} = \frac{1}{4} \cdot 7 + \frac{1}{8} \cdot 15 + \frac{1}{8} \cdot 15 + \frac{1}{6} \cdot 9 + \frac{1}{6} \cdot 9 + \frac{1}{6} \cdot 9 = 10$$

$\Box$

We now state and prove the main theorem of this section.

THEOREM 4.1 *For any given state space tree T, let P be the path probed by Algorithm 4.5. If $N = N(P)$ is the value returned by Algorithm 4.5, then the expected value of N is $|T|$.*

PROOF For any leaf node $X = [x_0, \ldots, x_{\ell-1}] \in T$, there is a unique path, say $P(X)$, from the root node to X. We will denote the nodes in the path $P(X)$ by $X_0 = [\,]$ (the root node), $X_1 = [x_0]$, ..., $X_\ell = X = [x_0, \ldots, x_{\ell-1}]$. The probability that X_1 is chosen in Algorithm 4.5 when $\ell = 0$ is $1/|\mathcal{C}_0(X_0)|$. Then, the probability that X_2 is chosen when $\ell = 1$ is $1/|\mathcal{C}_1(X_1)|$. In general, the probability that X_i is chosen, given that $X_0, \ldots, X_{i-1}$ have already been chosen, is $1/|\mathcal{C}_{i-1}(X_{i-1})|$. Therefore, the probability that $P(X)$ is the path chosen in Algorithm 4.5 is

$$\frac{1}{|\mathcal{C}_0(X_0)|} \cdot \frac{1}{|\mathcal{C}_1(X_1)|} \cdots \frac{1}{|\mathcal{C}_{\ell-1}(X_{\ell-1})|} = \frac{1}{S(X)}.$$

Now we can proceed to compute the expected value of N. An expected value is actually a weighted average. In this case, we are computing the weighted average of the values of $N(P(X))$, over all leaf nodes X, where the path $P(X)$ is chosen with probability $1/S(X)$. Let $\mathcal{L}(T)$ denote the set of leaf nodes in the tree T. The desired weighted average, which we denote by $\overline{N}$, is computed as follows:

$$\overline{N} = \sum_{X \in \mathcal{L}(T)} (\text{prob}(P(X)) \cdot N(P(X))) = \sum_{X \in \mathcal{L}(T)} \frac{1}{S(X)} \sum_{Y \in P(X)} S(Y).$$

We can interchange the order of the two summations in the above, obtaining

$$\overline{N} = \sum_{Y \in T} \left(\sum_{\{X \in \mathcal{L}(T): Y \in P(X)\}} \frac{S(Y)}{S(X)} \right).$$

Now, for any $Y \in T$, let's evaluate the inner sum,

$$\sum_{\{X \in \mathcal{L}(T): Y \in P(X)\}} \frac{S(Y)}{S(X)}. \tag{4.5}$$

For any non-leaf node Y in the tree, it is clear from the definition of the function S that

$$\frac{1}{S(Y)} = \sum_{\{Z: Z \text{ is a child of } Y\}} \frac{1}{S(Z)}.$$

Iterating this equation, we see that

$$\frac{1}{S(Y)} = \sum_{\{X: X \text{ is a leaf node that is a descendant of } Y\}} \frac{1}{S(X)}.$$

This is equivalent to saying that

$$\sum_{\{X \in \mathcal{L}(T): Y \in P(X)\}} \frac{1}{S(X)} = \frac{1}{S(Y)}.$$

Note that this equation also holds if Y is a leaf node. Hence, for any node Y in the tree, the sum in Equation (4.5) has the value 1. Thus

$$\overline{N} = \sum_{Y \in T} 1 = |T|,$$

which is the result we wanted to prove.   ∎

4.5   Exact cover

In this section, we study a problem similar to All Cliques that is called Exact Cover.

Problem 4.2:   Exact Cover

Instance: a collection S of subsets of the set $\mathcal{R} = \{0, \ldots, n-1\}$

Question: Does S contains an *exact cover* of $\mathcal{R}$? (In other words, does there exist a subcollection $S' = \{S_{x_0}, S_{x_1}, \ldots, S_{x_{\ell-1}}\} \subseteq S$ such that every element of $\mathcal{R}$ is contained in exactly one member of S'?)

Let $S = \{S_0, S_1, S_2, \ldots, S_{m-1}\}$ be the collection of subsets of $\mathcal{R}$ in an instance of Problem 4.2. Observe that a solution to this problem is a list of subsets from S whose members partition $\mathcal{R}$. Instead of storing the chosen subsets, we will keep track of them using an array $X = [x_0, x_1, \ldots, x_{\ell-1}]$ of their indices. Thus, for example, $x_3 = 22$ will mean that the set $S_{22} \in S$ was chosen as the third set in the (partial) solution represented by the array X.

Let $\mathcal{G}$ be the graph with vertex set $V = \{0, 1, 2, \ldots, m-1\}$ in which two vertices i and j are adjacent if and only if $S_i \cap S_j = \emptyset$. Then the partial solutions for Exact Cover are precisely the cliques in the graph $\mathcal{G}$. We can use Algorithm 4.4 to generate all the cliques in $\mathcal{G}$. Furthermore, we can check each maximal clique to see if it corresponds to a partition of $\mathcal{R}$, and thus solve Problem 4.2.

In order to use Algorithm 4.4, we first impose an ordering on the members of S. Although any ordering can be used, it is beneficial to order the subsets of $\mathcal{R}$ in decreasing lexicographic order, which we denote by $\overset{\text{lex}}{>}$. This ordering is chosen because it is useful for pruning.

We now describe how our choice sets can be determined. First, we define

$$C'_0 = \mathcal{V}$$

and

$$C'_\ell = A_{x_{\ell-1}} \cap B_{x_{\ell-1}} \cap C'_{\ell-1},$$

where

$$A_x = \{y \in \mathcal{V} : S_y \cap S_x = \emptyset\},$$

and

$$B_x = \{y \in \mathcal{V} : S_y \overset{\text{lex}}{>} S_x\},$$

for $x = 0, 1, 2, \ldots, m - 1$. If the sets C'_ℓ are used as choice sets, then the resulting backtracking algorithm is the same as Algorithm 4.4 for solving **All Cliques** (see Equations 4.1, 4.2 and 4.3). However, we can take advantage of the structure of the set system $(\mathcal{R}, \mathcal{S})$ to further prune the search, as follows. Recall that we assumed that the sets in $\mathcal{S}$ are sorted in decreasing lexicographic order, so

$$S_0 \overset{\text{lex}}{>} S_1 \overset{\text{lex}}{>} \cdots \overset{\text{lex}}{>} S_{m-1}.$$

Then the largest sets with respect to this ordering will all contain 0. The next bunch will not contain 0 but will contain 1; the next bunch will contain 2 but neither 1 nor 0, and so on. Thus, the ordering of these sets defines an ordered partition

$$\mathcal{H} = [H_0, H_1, \ldots, H_{n-1}]$$

of $\{0, 1, 2, \ldots, m - 1\}$, in which we have

$$H_i = \{x \in \mathcal{V} : S_x \cap \{0, \ldots, i\} = \{i\}\},$$

for $i = 0, 1, \ldots, n - 1$.

Now, suppose $X = [x_0, x_1, \ldots, x_{\ell-1}]$ is a partial solution that we wish to extend, if possible. If X is not itself a solution, then the set

$$\mathcal{U}_\ell = \mathcal{R} \setminus \left(\bigcup_{i=0}^{\ell-1} S_{x_i} \right)$$

is not the empty set. In this case, let r be the smallest integer in $\mathcal{U}_\ell$. If it is possible to extend X to a solution, say $[x_0, x_1, \ldots, x_{\ell-1}, x_\ell, \ldots]$, then it must be the case that $x_\ell \in H_r$. We can use this observation to facilitate further pruning, by defining the (improved) choice set C_ℓ as follows:

$$C_\ell = C'_\ell \cap H_r.$$

In order to implement these modifications, we proceed as follows. First, before running the backtrack algorithm, we sort S as described above. This can be done in $O(m \log m)$ set operations. Then the sets A_x and B_x, and the partition $\mathcal{H}$, can be constructed easily with one pass through S. The time to construct them is $O(m^2)$ set operations. The resulting algorithm is presented as Algorithm 4.6. An example, including the state space tree that results when this algorithm is run, is given in Example 4.3.

Algorithm 4.6: EXACTCOVER (n, S)

global $X, \mathcal{C}_\ell, \mathcal{C}'_\ell$ $(\ell = 0, 1, \ldots)$
procedure EXACTCOVERBT(ℓ, r')
if $\ell = 0$
   then $\begin{cases} \mathcal{U}_0 \leftarrow \{0, \ldots, n-1\} \\ r \leftarrow 0 \end{cases}$
   else $\begin{cases} \mathcal{U}_\ell \leftarrow \mathcal{U}_{\ell-1} \setminus S_{x_{\ell-1}} \\ r \leftarrow r' \\ \textbf{while } r \notin \mathcal{U}_\ell \textbf{ and } r < n \\ \quad \textbf{do } r \leftarrow r+1 \end{cases}$
if $r = n$
   then output $([x_0, \ldots, x_{\ell-1}])$
if $\ell = 0$
   then $\mathcal{C}'_0 \leftarrow \{0, 1, \ldots, m-1\}$
   else $\mathcal{C}'_\ell \leftarrow A_{x_{\ell-1}} \cap B_{x_{\ell-1}} \cap \mathcal{C}'_{\ell-1}$
$\mathcal{C}_\ell \leftarrow \mathcal{C}'_\ell \cap H_r$
for each $x \in \mathcal{C}_\ell$
   do $\begin{cases} x_\ell \leftarrow x \\ \text{EXACTCOVERBT}(\ell+1, r) \end{cases}$

main
$m \leftarrow |S|$
sort S in decreasing lexicographic order
for $i \leftarrow 0$ **to** $m-1$
   do $A_i \leftarrow \{j : S_i \cap S_j = \emptyset\}$
for $i \leftarrow 0$ **to** $m-1$
   do $B_i \leftarrow \{i+1, i+2, \ldots, m-1\}$
for $i \leftarrow 0$ **to** $n-1$
   do $H_i \leftarrow \{j : S_j \cap \{0, \ldots, i\} = \{i\}\}$
$H_n \leftarrow \emptyset$
EXACTCOVERBT$(0, 0)$

Example 4.3 *An instance of* Exact Cover

j	S_j	rank(S_j)	$A_j \cap B_j$
0	$\{0,1,3\}$	104	$\{10,11,12\}$
1	$\{0,1,5\}$	98	$\{12\}$
2	$\{0,2,4\}$	84	$\{7,8,9,10,11,12\}$
3	$\{0,2,5\}$	82	$\{8,9,10,11,12\}$
4	$\{0,3,6\}$	73	$\{5,6,7,8,9,10,11,12\}$
5	$\{1,2,4\}$	52	$\emptyset$
6	$\{1,2,6\}$	49	$\{11,12\}$
7	$\{1,3,5\}$	49	$\emptyset$
8	$\{1,4,6\}$	37	$\emptyset$
9	$\{1\}$	32	$\{10,11,12\}$
10	$\{2,5,6\}$	19	$\emptyset$
11	$\{3,4,5\}$	14	$\emptyset$
12	$\{3,4,6\}$	13	$\emptyset$

i	0	1	2	3	4	5	6
H_i	$\{0,1,2,3,4\}$	$\{5,6,7,8,9\}$	$\{10\}$	$\{11,12\}$	$\emptyset$	$\emptyset$	$\emptyset$

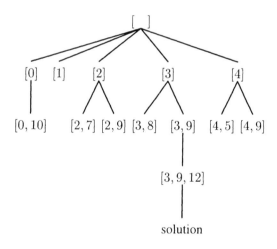

solution

4.6   Bounding functions

A more sophisticated method of pruning is to use a bounding function. We require a few preliminary definitions, which will apply to any backtracking algorithm for a maximization problem. Let $\text{profit}(X)$ denote the profit for any feasible solution X. For a partial feasible solution, say $X = [x_0, x_1, \ldots, x_{\ell-1}]$, define $P(X)$ to be the maximum profit of any feasible solution which is a descendant of X in the state space tree. In other words, $P(X)$ is the maximum value of $\text{profit}(X')$ taken over all feasible solutions $X' = [x'_0, x'_1, \ldots, x'_{n-1}]$ such that $x_i = x'_i$ for $0 \le i \le \ell - 1$. It follows from the definition that, if $X = [\ ]$, then $P(X)$ is the optimal profit of the given problem instance.

In general, $P(X)$ can be computed exactly only by traversing the subtree with root node X. In order to avoid doing this, if possible, we will employ a bounding function. A *bounding function* is a real-valued function B, defined on the set of nodes in the state space tree, satisfying the following condition:

For any feasible partial solution X, $\text{B}(X) \ge \text{P}(X)$.

This property says that $\text{B}(X)$ is an upper bound on the profit of any feasible solution that is a descendant of X in the state space tree. (For a minimization problem, the definition is the same, except that the inequality is reversed.)

Once a bounding function $\text{B}(X)$ has been specified, it can be used to prune the state tree, as follows. Suppose that at some stage of the backtracking algorithm, we have a current partial solution $X = [x_0, x_1, \ldots, x_{\ell-1}]$, and $OptP$ is the current optimal profit. If it happens that $\text{B}(X) \le OptP$, then we have that

$$\text{P}(X) \le \text{B}(X) \le OptP.$$

This means that no descendants of X in the state space tree can improve the current optimal profit. Hence we can prune this entire subtree, i.e., we can define C_ℓ to be the empty set.

It is helpful to think of $\text{B}(X)$ as an approximation to $\text{P}(X)$. We want a bounding function to be:

1. easy to compute, and
2. close to $\text{P}(X)$.

These two properties work against each other. For example, $\text{P}(X)$ is itself a bounding function, but is too difficult to compute. Finding bounding functions which satisfy both of the above properties can be a challenging task, and we will describe some nice examples of useful bounding functions later in this chapter.

A general backtracking algorithm incorporating a bounding function is presented as Algorithm 4.7. In this algorithm, B is any specified bounding function, and the function profit computes the profit for a feasible solution X.

Algorithm 4.7: BOUNDING (ℓ)

external $P(), B()$
global $X, OptP, OptX, C_\ell$ ($\ell = 0, 1, \ldots$)
if $[x_0, \ldots, x_{\ell-1}]$ is a feasible solution

then $\begin{cases} P \leftarrow \mathsf{profit}([x_0, \ldots, x_{\ell-1}]) \\ \textbf{if } P > OptP \\ \quad \textbf{then } \begin{cases} OptP \leftarrow P \\ OptX \leftarrow [x_0, \ldots, x_{\ell-1}] \end{cases} \end{cases}$

Compute C_ℓ
$B \leftarrow \mathsf{B}([x_0, \ldots, x_{\ell-1}])$
for each $x \in C_\ell$

do $\begin{cases} \textbf{if } B \leq OptP \textbf{ then return} \\ x_\ell \leftarrow x \\ \text{BOUNDING}(\ell + 1) \end{cases}$

It is very important to note that we check to see if the pruning condition, $B \leq OptP$, is true every time we consider another element $x \in C_\ell$. This is because the value of $OptP$ can increase as the algorithm progresses, and so we check to see if we can prune every time we are preparing to choose a new value for x_ℓ.

4.6.1 The knapsack problem

We now show how to define a useful bounding function for the Knapsack (optimization) problem. First, we consider a related problem called the Rational Knapsack problem, which we present as Problem 4.3.

Problem 4.3: Rational Knapsack

Instance: *profits* $p_0, p_1, p_2, \ldots, p_{n-1}$;
weights $w_0, w_1, w_2, \ldots, w_{n-1}$; and
capacity M

Find: the maximum value of

$$\sum_{i=0}^{n-1} p_i x_i$$

subject to

$$\sum_{i=0}^{n-1} w_i x_i \leq M$$

where x_i is rational and $0 \leq x_i \leq 1$ for all i.

It is straightforward to see that Algorithm 4.8, which uses a greedy strategy, returns the optimal profit for Rational Knapsack .

Algorithm 4.8: RKNAP $(p_0, p_1, \ldots, p_{n-1}, w_0, w_1, \ldots, w_{n-1}, M)$

permute the indices so that $p_0/w_0 \geq p_1/w_1 \geq p_{n-1}/w_{n-1}$
$i \leftarrow 0$
$P \leftarrow 0$
$W \leftarrow 0$
for $j \leftarrow 0$ **to** $n - 1$
  do $x_j \leftarrow 0$
while $W < M$ **and** $i < n$

$\text{do} \begin{cases} \textbf{if } W + w_i \leq M \\ \quad \textbf{then} \begin{cases} x_i \leftarrow 1 \\ W \leftarrow W + w_i \\ P \leftarrow P + p_i \\ i \leftarrow i + 1 \end{cases} \\ \quad \textbf{else} \begin{cases} x_i \leftarrow (M - W)/w_i \\ W \leftarrow M \\ P \leftarrow P + x_i\, p_i \\ i \leftarrow i + 1 \end{cases} \end{cases}$

return (P)

We use RKNAP to define a bounding function for the Knapsack (optimization) problem as follows. Given a (feasible) partial solution $X = [x_0, x_1, \ldots, x_{\ell-1}]$, define

$$B(X) = \sum_{i=0}^{\ell-1} p_i x_i + \text{RKNAP}(p_\ell, \ldots, p_n, w_\ell, \ldots, w_n, M - \sum_{i=0}^{\ell-1} w_i x_i)$$

$$= \sum_{i=0}^{\ell-1} p_i x_i + \text{RKNAP}(p_\ell, \ldots, p_n, w_\ell, \ldots, w_n, M - CurW)$$

Thus, $B(X)$ is equal to the sum of:

1. the profit obtained from objects $0, 1, \ldots, \ell - 1$,

plus

2. the profit from the remaining objects, using the remaining capacity

$$M - CurW,$$

but allowing rational x_is.

If we restricted each x_i to be 0 or 1 in part 2 above, then we would obtain $P(X)$. Allowing the x_is with $\ell \leq i \leq n$ to be rational may yield a higher profit; so $B(X) \geq P(X)$ and B is indeed a bounding function. It is also easy to compute, and thus may be useful for pruning.

Suppose we want to solve an instance of the Knapsack (optimization) problem. It will be useful to sort the objects in non-decreasing order of profit/weight ahead of time, before we begin the backtracking algorithm. Then, when we wish to evaluate our bounding function, the first step of Algorithm 4.8 will be unnecessary, and consequently RKNAP will run faster. Thus, we will assume that

$$\frac{p_0}{w_0} \geq \frac{p_1}{w_1} \geq \cdots \geq \frac{p_{n-1}}{w_{n-1}}.$$

The improved algorithm is given as Algorithm 4.9.

Algorithm 4.9: KNAPSACK3 $(\ell, CurW)$

external RKNAP()
global $X, OptX, OptP, C_\ell$ $(\ell = 0, 1, \ldots)$
if $\ell = n$

then $\begin{cases} \textbf{if } \sum_{i=0}^{n-1} p_i x_i > OptP \\ \qquad \textbf{then } \begin{cases} OptP \leftarrow \sum_{i=0}^{n-1} p_i x_i \\ OptX \leftarrow [x_0, \ldots, x_{n-1}] \end{cases} \end{cases}$

if $\ell = n$
  then $C_\ell \leftarrow \emptyset$
  else $\begin{cases} \textbf{if } CurW + w_\ell \leq M \\ \quad \textbf{then } C_\ell \leftarrow \{1, 0\} \\ \quad \textbf{else } C_\ell \leftarrow \{0\} \end{cases}$

$B \leftarrow \sum_{i=0}^{\ell-1} p_i x_i + \text{RKNAP}(p_\ell, \ldots, p_n, w_\ell, \ldots, w_n, M - CurW)$
for each $x \in C_\ell$
  do $\begin{cases} \textbf{if } B \leq OptP \textbf{ then return} \\ x_\ell \leftarrow x \\ \text{KNAPSACK3}(\ell + 1, CurW + w_\ell x_\ell) \end{cases}$

Example 4.4 *An instance of the* Knapsack (Optimization) *problem*
Suppose we have five objects, having weights 11, 12, 8, 7, 9; profits 23, 24, 15, 13 and 16 (respectively); and capacity $M = 26$. Note that the objects are already arranged in decreasing order of profit/weight. We draw in Figure 4.4 the space tree traversed in the course of the backtrack algorithm KNAPSACK3 . At each node, we record the current values of $X, B(X)$ and $CurW$.   □

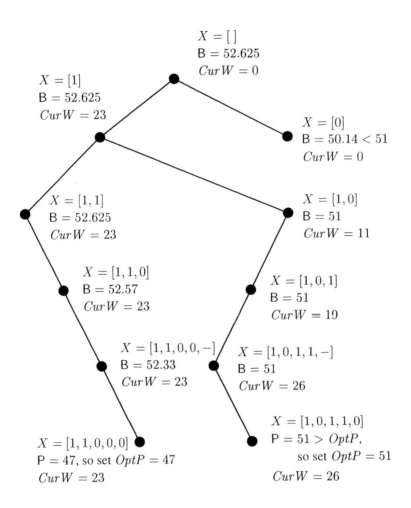

FIGURE 4.4
The state space tree traversed by KNAPSACK3 .

TABLE 4.2
Size of state space trees for Algorithms 4.1, 4.3, and 4.9, on random instances with n weights

n	Algorithm 4.1	Algorithm 4.3	Algorithm 4.9
8	511	332	52
8	511	312	78
8	511	333	72
8	511	321	74
8	511	313	57
12	8191	4598	109
12	8191	4737	93
12	8191	5079	164
12	8191	4988	195
12	8191	4620	87
16	131071	73639	192
16	131071	72302	58
16	131071	76512	168
16	131071	78716	601
16	131071	78510	392
20	2097151	1173522	299
20	2097151	1164523	104
20	2097151	1257745	416
20	2097151	1152046	118
20	2097151	1166086	480
24	33554431	19491410	693
24	33554431	18953093	180
24	33554431	17853054	278
24	33554431	19814875	559
24	33554431	18705548	755

To compare the amount of pruning provided by Algorithms 4.1, 4.3, and 4.9, we give experimental data in Table 4.2. For each $n = 8, 12, 16, 20$, and 24, five instances of Problem 1.4 were generated by randomly selecting n integer weights $w_0, w_1, \ldots, w_{n-1}$ between 0 and 1000000. In an attempt to generate "hard" instances of Problem 1.4, we defined, for each $i = 0, 1, \ldots, n - 1$, the profit $p_i = 2 w_i \epsilon$, where ϵ was chosen at random in the interval $(.9, 1.1)$. Hence the profit of each object is within 10% of twice its weight. The capacity M was chosen to be half the sum of the weights.

Certainly the bounding function has a dramatic effect on the running time of the algorithm in these random problem instances.

4.6.2   The traveling salesman problem

In the traveling salesman problem, a salesman must visit n cities and return home, doing so in such a way that the cost of the trip is minimized. More precisely, the Traveling Salesman problem is defined as follows.

Problem 4.4:   Traveling Salesman

Instance: a complete graph on n vertices, $\mathcal{G} = (\mathcal{V}, E)$;

                a cost function, $\mathrm{cost} : E \to \mathbb{Z}^+$

Find: a Hamiltonian circuit X of $\mathcal{G}$ such that

$$\mathrm{cost}(X) = \sum_{e \in E(X)} \mathrm{cost}(e)$$

                is minimized. (Recall that a Hamiltonian circuit in a graph $\mathcal{G}$ is a circuit that passes through each vertex of $\mathcal{G}$ exactly once.)

Let $\mathcal{V} = \{0, 1, \ldots, n-1\}$ be the vertices of the graph $\mathcal{G}$. For convenience, we will define $\mathrm{cost}(x, y) = \mathrm{cost}(\{x, y\})$ if $x \neq y$, and $\mathrm{cost}(x, y) = \infty$ if $x = y$.

Any Hamiltonian circuit X can be represented as a permutation of $\mathcal{V}$, say $X = [x_0, \ldots, x_{n-1}]$. Without loss of generality, we can regard X as starting and ending at vertex 0; so, we can define $x_0 = 0$. For example, the circuit

$$2\ 5\ 1\ 0\ 3\ 4\ 6\ 2$$

would be represented by the 7-tuple

$$[0, 3, 4, 6, 2, 5, 1] \quad \text{or by} \quad [0, 1, 5, 2, 6, 3, 4].$$

Algorithm 4.10 is a basic backtrack algorithm for the Traveling Salesman problem. To speed up Algorithm 4.10, we will construct some bounding functions. The Traveling Salesman problem is a minimization problem; so, a bounding function, $\mathrm{Bound}(X)$, must provide a lower bound on the cost of any Hamiltonian circuit that is an extension of the partial solution X. Suppose $X = [x_0, x_1, \ldots, x_{\ell-1}]$ is a partial solution. Then $\ell \leq n-1$, and X represents the path $x_0\, x_1\, \cdots\, x_{\ell-1}$ of length $\ell - 1$. Define

$$\mathcal{Y} = \mathcal{V} \setminus \{x_0, x_1, \ldots, x_{\ell-1}\}.$$

Observe that, if $[x_0, \ldots, x_{n-1}]$ is a feasible solution that is an extension of X, then

$$\mathcal{Y} = \{x_\ell, \ldots, x_{n-1}\}.$$

Algorithm 4.10: TSP1 (ℓ)

  global C_ℓ ($\ell = 0, 1, \ldots, n - 1$)
  if $\ell = n$
    then $\begin{cases} C \leftarrow \text{cost}([x_0, \ldots, x_{n-1}]) \\ \textbf{if } C < OptC \\ \quad \textbf{then } \begin{cases} OptC \leftarrow C \\ OptX \leftarrow [x_0, \ldots, x_{n-1}] \end{cases} \end{cases}$
  if $\ell = 0$
    then $C_\ell \leftarrow \{0\}$
  else $\begin{cases} \textbf{if } \ell = 1 \\ \quad \textbf{then } C_\ell \leftarrow \{1, \ldots, n - 1\} \\ \quad \textbf{else } C_\ell \leftarrow C_{\ell-1} \setminus \{x_{\ell-1}\} \end{cases}$
  for each $x \in C_\ell$
    do $\begin{cases} x_\ell \leftarrow x \\ \text{TSP1}(\ell + 1) \end{cases}$

For $x \in V$ and $W \subseteq V$ (where $W \neq \emptyset$), define

$$\mathsf{b}(x, W) = \min\{\text{cost}(x, y) : y \in W\}.$$

We now prove an inequality that will lead to a bounding function.

THEOREM 4.2   *Let* $X' = [x_0, \ldots, x_{n-1}]$ *be the minimum cost Hamiltonian circuit that extends* $[x_0, x_1, \ldots, x_{\ell-1}]$, *where* $\ell \leq n - 1$. *Then it holds that*

$$\text{cost}(X') \geq \sum_{i=0}^{\ell-1} \text{cost}(x_i, x_{i+1}) + \mathsf{b}(x_{\ell-1}, \mathcal{Y}) + \sum_{y \in \mathcal{Y}} \mathsf{b}(y, \mathcal{Y} \cup \{x_0\}).$$

PROOF   Define $x_n = x_0$ for convenience; then we have

$$\text{cost}(X') = \sum_{i=0}^{n-1} \text{cost}(x_i, x_{i+1}).$$

First, the sum

$$\sum_{i=0}^{\ell-1} \text{cost}(x_i, x_{i+1})$$

represents the sum of the costs of the edges already chosen in X. Next, we have

$$\text{cost}(x_{\ell-1}, x_\ell) \geq \mathsf{b}(x_{\ell-1}, \mathcal{Y}),$$

because $x_\ell \in \mathcal{Y}$. Finally, for $\ell \leq i \leq n - 1$, we have

$$\text{cost}(x_i, x_{i+1}) \geq \mathsf{b}(x_i, \mathcal{Y} \cup \{x_0\}),$$

because $x_{i+1} \in \mathcal{Y} \cup \{x_0\}$ for $\ell \leq i \leq n - 1$. Since $\mathcal{Y} = \{x_\ell, \ldots x_{n-1}\}$, the result follows. ∎

Let $X = [x_0, x_1, \ldots, x_{\ell-1}]$. If $\ell \leq n - 1$, then define

$$\text{MINCOSTBOUND}(X) = \sum_{i=0}^{\ell-1} \text{cost}(x_i, x_{i+1}) + \text{b}(x_{\ell-1}, \mathcal{Y}) + \sum_{y \in \mathcal{Y}} \text{b}(y, \mathcal{Y} \cup \{x_0\}),$$

whereas if $\ell = n$, then define

$$\text{MINCOSTBOUND}(X) = \sum_{i=0}^{\ell-1} \text{cost}(x_i, x_{i+1}) + \text{cost}(x_{n-1}, x_0).$$

Theorem 4.2 establishes that MINCOSTBOUND is a bounding function. It is straightforward to describe an algorithm to compute MINCOSTBOUND in $O(n^2)$ time.

Another bounding function can be achieved using *reduced matrices*, which we will now study. A matrix M of integers is said to be *reduced* if the following three properties are satisfied:

1. all entries of M are non-negative;
2. every row of M contains at least one entry equal to 0;
3. every column of M contains at least one entry equal to 0.

Suppose M is an m by m matrix in which all the entries are non-negative. Algorithm 4.11 transforms M into a reduced matrix, and computes a quantity $val = \text{REDUCE}(M)$ which we call the *value* of the matrix M.

The following result shows the relevance of reduced matrices to the Traveling Salesman problem.

THEOREM 4.3   *Suppose* cost *is a cost function for the complete graph $\mathcal{G}$ on m vertices. Define M to be the m by m matrix in which $M[i, j] = \text{cost}(i, j)$. Then any Hamiltonian circuit in $\mathcal{G}$ has cost at least* $\text{REDUCE}(M)$.

Algorithm 4.11: REDUCE (M)

comment: M is an m by m matrix
$val \leftarrow 0$
for $i \leftarrow 0$ **to** $m - 1$
$\quad$ **do** $\begin{cases} min \leftarrow M[i, 0] \\ \textbf{for } j \leftarrow 1 \textbf{ to } m - 1 \\ \quad \textbf{do} \begin{cases} \textbf{if } M[i, j] < min \\ \quad \textbf{then } min \leftarrow M[i, j] \end{cases} \\ \textbf{for } j \leftarrow 0 \textbf{ to } m - 1 \\ \quad \textbf{do } M[i, j] \leftarrow M[i, j] - min \\ val \leftarrow val + min \end{cases}$
for $j \leftarrow 0$ **to** $m - 1$
$\quad$ **do** $\begin{cases} min \leftarrow M[0, j] \\ \textbf{for } i \leftarrow 1 \textbf{ to } m - 1 \\ \quad \textbf{do} \begin{cases} \textbf{if } M[i, j] < min \\ \quad \textbf{then } min \leftarrow M[i, j] \end{cases} \\ \textbf{for } i \leftarrow 0 \textbf{ to } m - 1 \\ \quad \textbf{do } M[i, j] \leftarrow M[i, j] - min \\ val \leftarrow val + min \end{cases}$
return (val)

Before giving a proof of this theorem, we present an example to illustrate it.

Example 4.5 *An illustration of Theorem 4.3*
Suppose

$$M = \begin{bmatrix} \infty & 3 & 5 & 8 \\ 3 & \infty & 2 & 7 \\ 5 & 2 & \infty & 6 \\ 8 & 7 & 6 & \infty \end{bmatrix}$$

is the cost matrix for a graph $\mathcal{G}$. Reducing M, using Algorithm 4.11, we see that REDUCE$(M) = 18$. On the other hand, it is not difficult to check the costs of the three possible Hamiltonian circuits:

the circuit $0\,1\,2\,3$ has cost $3 + 2 + 6 + 8 = 19$
the circuit $0\,1\,3\,2$ has cost $3 + 7 + 6 + 5 = 21$
the circuit $0\,2\,1\,3$ has cost $5 + 2 + 7 + 8 = 22$

Hence the minimum-cost Hamiltonian circuit has cost 19, and indeed,

$$19 \geq \text{REDUCE}(M) = 18.$$

This example shows that $\text{REDUCE}(M)$ is indeed only a lower bound on the cost of any Hamiltonian circuit; it need not give the exact value of the minimum-cost circuit.   ⬜

PROOF   (of Theorem 4.3) Let $X = [x_0, x_1, \ldots, x_{n-1}]$ be any Hamiltonian circuit of $\mathcal{G}$. Define $x_n = x_0$. Then

$$\text{cost}(X) = M[x_0, x_1] + M[x_1, x_2] + \cdots + M[x_{n-1}, x_n].$$

This sum uses exactly one cell from each row and column of M. Define

$$r_i = \min\{M[i, j] : 0 \le j \le n - 1\}$$

and

$$c_j = \min\{M[i, j] - r_i : 0 \le i \le n - 1\},$$

for $0 \le i \le n - 1$ and $0 \le j \le n - 1$. Observe that

$$\text{REDUCE}(M) = \sum_{i=0}^{n-1} r_i + \sum_{j=0}^{n-1} c_j.$$

Clearly, we have

$$r_{x_i} + c_{x_{i+1}} \le M[x_i, x_{i+1}],$$

for all i, $0 \le i \le n - 1$. Hence, summing over i, we see that

$$\text{REDUCE}(M) \le \text{cost}(X),$$

and the result follows.   ∎

We now use the idea of reduced matrices to define a bounding function. We need a way to determine a lower bound on the cost of any Hamiltonian circuit that is the completion of a given partial solution. Suppose we have a partial solution

$$X = [x_0, x_1, \ldots, x_{\ell-1}],$$

where $\ell \le n - 1$. As mentioned previously, X represents the path $x_0\, x_1 \cdots x_{\ell-1}$. Perform the following operations on the cost matrix M:

1. **if** $\ell < n$
      then $M[x_{\ell-1}, 0] = \infty$;
2. **delete** rows $x_0, x_1, \ldots, x_{\ell-2}$ of M; and
3. **delete** columns $x_1, \ldots, x_{\ell-1}$ of M.

Call the resulting matrix $M'(X)$. Observe that $M'(X)$ is an $(n - \ell + 1)$ by $(n - \ell + 1)$ matrix. Now, we define our bounding function by the following formula:

$$\text{REDUCEBOUND}(X) = \text{REDUCE}(M'(X)) + M[x_0, x_1] + \cdots + M[x_{\ell-2}, x_{\ell-1}].$$

That is, $\text{REDUCEBOUND}(X)$ is the sum of the costs of the edges in the partial solution X, and the value of the matrix $M'(X)$.

To show that this formula is indeed a bounding function, we need to prove that $\text{REDUCE}(M'(X))$ is a lower bound on the sum of the costs of any completion of X to a Hamiltonian circuit. We argue informally that this is the case.

Consider the effects of operations 1, 2, and 3. Operation 1 rules out using the edge $\{x_{\ell-1}, x_0\}$ if $\ell < n$, because using this edge would close the circuit prematurely. Operation 2 rules out using edges leaving vertices $x_0, x_1, \dots, x_{\ell-2}$ and operation 3 rules out using edges entering vertices $x_1, \dots, x_{\ell-1}$. Any $n - \ell$ edges which provide a completion of X to a Hamiltonian circuit will thus hit $M'(X)$ exactly once in every row and column. It can be shown that the sum of the costs of these edges must be no less than the value of $M'(X)$, in a manner similar to the proof of Theorem 4.3.

We now give a detailed description of the resulting bounding function.

Algorithm 4.12: REDUCEBOUND (X)

external cost(), REDUCE()

global M, V

comment: $X = [x_0, \dots, x_{m-1}]$

if $m = n$ **then return** $(\text{cost}(X))$

$M'[0, 0] \leftarrow \infty$

$j \leftarrow 1$

for each $y \in V \setminus \{x_0, x_1, \dots, x_{m-1}\}$ **do** $\begin{cases} M'[0, j] \leftarrow M[x_{m-1}, y] \\ j \leftarrow j + 1 \end{cases}$

$i \leftarrow 1$

for each $x \in V \setminus \{x_0, x_1, \dots, x_{m-1}\}$ **do** $\begin{cases} M'[i, 0] \leftarrow M[x, x_0] \\ i \leftarrow i + 1 \end{cases}$

$i \leftarrow 1$

for each $x \in V \setminus \{x_0, x_1, \dots, x_{m-1}\}$

$\text{\textbf{do}} \begin{cases} j \leftarrow 1 \\ \text{\textbf{for each} } y \in V \setminus \{x_0, x_1, \dots, x_{m-1}\} \text{ \textbf{do} } \begin{cases} M'[i, j] \leftarrow M[x, y] \\ j \leftarrow j + 1 \end{cases} \\ i \leftarrow i + 1 \end{cases}$

$ans \leftarrow \text{REDUCE}(M')$

for $i \leftarrow 1$ **to** $m - 1$

do $ans \leftarrow ans + M[x_{i-1}, x_i]$

return (ans)

TABLE 4.3
Size of state space trees for Algorithms 4.10 and 4.13, on random instances of the Traveling Salesman problem with n vertices. Algorithm 4.13 is applied with bounding functions MINCOSTBOUND and REDUCEBOUND

n	Optimal Cost	Algorithm 4.10	Algorithm 4.13 MINCOSTBOUND	REDUCEBOUND
5	137	65	45	18
10	160	986,410	5,199	1,287
15	234	236,975,164,805	1,538,773	53,486
20	173	$\approx 3.3 \cdot 10^{17}$	64,259,127	1,326,640

The value of an n by n matrix is computed in time $O(n^2)$ by Algorithm 4.11. Hence REDUCEBOUND(X) can also be computed in time $O(n^2)$. Algorithm 4.13 is a backtracking algorithm for the Traveling Salesman problem that incorporates an arbitrary bounding function.

To compare the effect of the two bounding functions MINCOSTBOUND and REDUCEBOUND described in this section, we generated random instances of the Traveling Salesman problem on 5, 10, 15, and 20 vertices. The edge costs were randomly chosen integers between 0 and 100. In Table 4.3 we report the number of nodes in the state space trees when Algorithm 4.13 is used with these two bounding functions.

Algorithm 4.13: TSP2 (ℓ)

external B()
global C_ℓ $(\ell = 0, 1, \ldots, n - 1)$
if $\ell = n$
then $\begin{cases} C \leftarrow \text{cost}([x_0, \ldots, x_{n-1}]) \\ \textbf{if } C < OptC \\ \quad \textbf{then } \begin{cases} OptC \leftarrow C \\ OptX \leftarrow [x_0, \ldots, x_{n-1}] \end{cases} \end{cases}$
if $\ell = 0$ **then** $C_\ell \leftarrow \{0\}$
  else if $\ell = 1$ **then** $C_\ell \leftarrow \{1, \ldots, n - 1\}$
  else $C_\ell \leftarrow C_{\ell-1} \setminus \{x_{\ell-1}\}$
$B \leftarrow B([x_0, \ldots, x_{\ell-1}])$
for each $x \in C_\ell$
do $\begin{cases} \textbf{if } B \geq OptC \\ \quad \textbf{then return} \\ x_\ell \leftarrow x \\ \text{TSP2}(\ell + 1) \end{cases}$

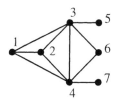

FIGURE 4.5
A graph with maximum clique $\{1, 2, 3, 4\}$.

4.6.3 The maximum clique problem

Recall that a maximum clique in a graph $\mathcal{G}$ is a clique of largest cardinality. For example, the maximal cliques in the graph in Figure 4.5 are $\{1, 2, 3, 4\}$, $\{3, 4, 6\}$, $\{3, 5\}$, and $\{4, 7\}$. The clique $\{1, 2, 3, 4\}$ is the only maximum clique. In general, a graph may have more than one maximum clique. The problem of finding a maximum clique in a graph $\mathcal{G}$ is known as the Maximum Clique problem. A decision version of this problem was introduced in Section 1.6 as Problem 1.7. The optimization version of the problem is defined as follows.

Problem 4.5:	Maximum Clique
Instance:	A graph $\mathcal{G} = (\mathcal{V}, \mathcal{E})$
Find:	a maximum clique of $\mathcal{G}$.

This problem has been shown to be NP-complete, but in spite of its inherent difficulty, many algorithms have been developed that perform well in practice.

In Section 4.3 we developed Algorithm 4.4 for generating all the cliques in a graph $\mathcal{G} = (\mathcal{V}, \mathcal{E})$. This algorithm can easily be modified to find a maximum clique; see Algorithm 4.14. Note that we no longer need to maintain the sets N_ℓ; we simply check to see if each clique constructed is larger than any previously constructed clique.

We now turn to the development of bounding functions for this problem. First we require a definition. Suppose $\mathcal{G} = (\mathcal{V}, \mathcal{E})$ is a graph, and $\mathcal{W} \subseteq \mathcal{V}$. The *induced subgraph* $\mathcal{G}[\mathcal{W}]$ has vertex set $\mathcal{W}$, and edge set

$$\{\{u, v\} \in \mathcal{E} : \{u, v\} \subseteq \mathcal{W}\}.$$

Now, at a typical point in Algorithm 4.14, we have the partial solution (i.e., clique) $X = [x_0, x_1, \ldots, x_{\ell-1}]$. Suppose $X' = [x_0, x_1, x_2, \ldots, x_j]$ is a clique which extends the partial solution X, where $j \geq \ell - 1$. Then $\{x_\ell, \ldots, x_j\}$ must be a clique in the induced subgraph $\mathcal{G}[\mathcal{C}_\ell]$. Thus, we can obtain a bounding function by placing an upper bound on the size of a maximum clique in $\mathcal{G}[\mathcal{C}_\ell]$. If the size of a maximum clique in $\mathcal{G}[\mathcal{C}_\ell]$ is denoted by $mc(\ell)$, and $mc(\ell) \leq ub(\ell)$, then

$$B(X) = \ell + ub(\ell)$$

is a bounding function.

Algorithm 4.14: MAXCLIQUE1 (ℓ)

global A_ℓ, B_ℓ, C_ℓ $(\ell = 0, 1, \ldots, n-1)$
if $\ell > OptSize$
  then $\begin{cases} OptSize \leftarrow \ell + 1 \\ OptClique \leftarrow [x_0, \ldots, x_{\ell-1}] \end{cases}$
if $\ell = 0$
  then $C_\ell \leftarrow \mathcal{V}$
  else $C_\ell \leftarrow A_{x_{\ell-1}} \cap B_{x_{\ell-1}} \cap C_{\ell-1}$
for each $x \in C_\ell$
  do $\begin{cases} x_\ell \leftarrow x \\ \text{MAXCLIQUE1}(\ell + 1) \end{cases}$
main
  $OptSize \leftarrow 0$
  MAXCLIQUE1 (0)
  output $(OptClique)$

We can use this idea to obtain several different bounding functions. The simplest of them is to observe that

$$mc(\ell) \leq |C_\ell|.$$

This gives rise to the bounding function presented in Algorithm 4.15, which we call the *size bound*.

Algorithm 4.15: SIZEBOUND (X)

global C_ℓ
comment: $X = [x_0, \ldots, x_{\ell-1}]$
return $(\ell + |C_\ell|)$

Other, more sophisticated, methods of obtaining bounding functions use the idea of *vertex coloring* (see Problem 1.5). Recall that, if $\mathcal{G} = (\mathcal{V}, \mathcal{E})$ is a graph and k is a positive integer, then a (vertex) k-coloring of $\mathcal{G}$ is a function

$$color : \mathcal{V} \rightarrow \{0, \ldots, k-1\}$$

such that $color(x) \neq color(y)$ for all $\{x, y\} \in \mathcal{E}$. The relevance of vertex coloring to the Maximum Clique problem is stated in the following simple lemma.

LEMMA 4.4 *Let G be a graph, and suppose that G has a vertex k-coloring. Then the maximum clique in G has size at most k.*

PROOF If vertices x and y receive the same color, then they cannot both be in the same clique. ∎

Even though finding a vertex k-coloring in which k is minimized is NP-hard, it is not difficult to find k-colorings for values of k that are larger than the minimum. One easy way to do this is to color the vertices by a greedy strategy (recall that greedy algorithms were introduced in Section 1.8.1). In a greedy algorithm, the vertices are processed in order, each vertex receiving the first available color. Algorithm 4.16 presents such an algorithm. In Algorithm 4.16, we assume that the vertex set is written as $V = \{0, \ldots, n-1\}$. The algorithm constructs a k-coloring for some positive integer k, and returns that value of k. The actual k-coloring is stored as a (global) array, *color*. In the process of constructing this coloring, the algorithm constructs an array of sets called *ColorClass*, which is defined as follows:

$$ColorClass[h] = \{i \in V : color[i] = h\}$$

for $0 \le h \le k - 1$.

Algorithm 4.16: GREEDYCOLOR $(G = (V, \mathcal{E}))$

global *color*
comment: $V = \{0, \ldots, n-1\}$
$k \leftarrow 0$
for $i \leftarrow 0$ **to** $n - 1$

$\quad$ **do** $\begin{cases} h \leftarrow 0 \\ \textbf{while } h < k \textbf{ and } A_i \cap ColorClass[h] \neq \emptyset \textbf{ do } h \leftarrow h + 1 \\ \textbf{if } h = k \textbf{ then } \begin{cases} k \leftarrow k + 1 \\ ColorClass[h] \leftarrow \emptyset \end{cases} \\ ColorClass[h] \leftarrow ColorClass[h] \cup \{i\} \\ color[i] \leftarrow h \end{cases}$

return (k)

There are several ways in which Algorithm 4.16 can be incorporated into a bounding function. One way is to find an initial greedy coloring of the graph before the backtracking algorithm begins. Suppose that this coloring is denoted *color* and it uses k colors. For each induced subgraph $G[\mathcal{C}_\ell]$, the function *color*, restricted to the vertices in $\mathcal{C}_\ell$, defines a coloring of $G[\mathcal{C}_\ell]$ which may use fewer than k colors. The number of colors in this induced coloring yields an upper bound on the size of a maximum clique in $G[\mathcal{C}_\ell]$. The resulting bounding function, which we call the *sampling bound*, is presented in Algorithm 4.17.

Algorithm 4.17: SAMPLINGBOUND (X)

global C_ℓ, *color*
comment: $X = [x_0, \ldots, x_{\ell-1}]$
return $(\ell + |\{color[x] : x \in C_\ell\}|)$

Another way to use the greedy coloring algorithm in a bounding function is to apply Algorithm 4.16 to the induced subgraph $\mathcal{G}[C_\ell]$ every time we want to compute the bounding function. The resulting bounding function is called the *greedy bound* and it is presented in Algorithm 4.18.

Algorithm 4.18: GREEDYBOUND (X)

external GREEDYCOLOR()
global C_ℓ
comment: $X = [x_0, \ldots, x_{\ell-1}]$
$k \leftarrow$ GREEDYCOLOR($\mathcal{G}[C_\ell]$)
return $(\ell + k)$

Any of the three bounding functions discussed above (or any other bounding function, for that matter) can be incorporated into our backtracking algorithm as the function B(X). Algorithm 4.19 is the result.

As was done in other algorithms incorporating bounding functions, we check to see if the condition $M \leq OptSize$ is true in every iteration of the loop. This is because the value of *OptSize* can increase as the algorithm progresses, and so we check to see if we can prune every time we are preparing to add a new node to the clique being considered.

In Table 4.4 we list the number of nodes in the state space tree, for graphs of various sizes, when Algorithm 4.19 is run using the different bounding functions we have discussed. We also list the number of edges, and the size of the maximum cliques in these graphs. The graphs we used were generated at random from the class $\mathcal{G}(n)$ defined in Section 4.3.1. There are several ways to do this. One nice method uses ranking and unranking algorithms we developed in Chapter 2. Note that the function RandomInteger(a, b) generates a random integer in the interval $[a, b]$. Algorithm 4.20 constructs a random graph in the class $\mathcal{G}(n)$.

Algorithm 4.19: MAXCLIQUE2 (ℓ)

external B()
global $A_\ell, B_\ell, C_\ell \quad (\ell = 0, 1, \ldots, n - 1)$
if $\ell > OptSize$
  then $\begin{cases} OptSize \leftarrow \ell \\ OptClique \leftarrow [x_0, \ldots, x_{\ell-1}] \end{cases}$
if $\ell = 0$
  then $C_\ell \leftarrow \mathcal{V}$
  else $C_\ell \leftarrow A_{x_{\ell-1}} \cap B_{x_{\ell-1}} \cap C_{\ell-1}$
$M \leftarrow B([x_0, \ldots, x_{\ell-1}])$
for each $x \in C_\ell$
  do $\begin{cases} \textbf{if } M \leq OptSize \\ \quad \textbf{then return} \\ x_\ell \leftarrow x \\ \text{MAXCLIQUE2}(\ell + 1) \end{cases}$
main
  $OptSize \leftarrow 0$
  MAXCLIQUE2(0)
  output ($OptClique$)

Algorithm 4.20: GENERATERANDOMGRAPH (n)

external $\begin{cases} \text{RandomInteger}() \\ \text{SUBSETLEXUNRANK}() \\ \text{kSUBSETLEXUNRANK}() \end{cases}$
$r \leftarrow \text{RandomInteger}(0, 2^{\binom{n}{2}} - 1)$
$T \leftarrow \text{SUBSETLEXUNRANK}(\binom{n}{2}, r)$
$\mathcal{E} \leftarrow \emptyset$
for each $j \in T$
  do $\begin{cases} \{x, y\} \leftarrow \text{kSUBSETLEXUNRANK}(j, 2, n) \\ \mathcal{E} \leftarrow \mathcal{E} \cup \{x - 1, y - 1\} \end{cases}$
return ($\mathcal{G} = (\{0, \ldots, n - 1\}, \mathcal{E})$)

The only aspect of Algorithm 4.20 that might require explanation is the last line, where we add the edge $\{x - 1, y - 1\}$ to $\mathcal{E}$. This is because the algorithm kSUBSETLEXUNRANK returns a 2-subset of $\{1, \ldots, n\}$, whereas we want a 2-subset of $\{0, \ldots, n - 1\}$. Thus we subtract one from x and y to create the edge to be included in $\mathcal{E}$.

Notice that the expected (i.e., average) sizes of state space trees when no pruning was done were denoted in Section 4.3.1 by $\overline{c}(n)$, and some values of $\overline{c}(n)$

TABLE 4.4
Size of state space trees for Algorithm 4.19 on random graphs with edge density .5

number of vertices	50	100	150	200	250
number of edges	607	2535	5602	9925	15566
size of maximum clique	7	9	10	11	11
bounding function					
none	8687	257145	1659016	7588328	26182672
size bound	3204	57225	350310	1434006	5008767
sampling bound	2268	44072	266246	1182514	4093535
greedy bound	430	5734	22599	91671	290788

TABLE 4.5
Size of state space trees for Algorithm 4.19 on random graphs with edge density .75

number of vertices	25	50	75	100	125
number of edges	236	959	2045	3720	5780
size of maximum clique	11	14	15	17	18
bounding function					
none	25570	2083770	12385596	186543706	1414266577
size bound	1840	91663	426279	5370268	35108264
sampling bound	794	37218	195567	2225982	15615755
greedy bound	91	2843	10476	70404	413421

were presented in Table 4.1. It is interesting to compare these values to the experimental results obtained in Table 4.4.

The edge density of a graph is the ratio of the number of its edges to $\binom{n}{2}$ (which is the total possible number of edges). The random graphs generated by Algorithm 4.20 will have edge density approximately .5. To obtain a random graph with a given edge density δ, $0 \leq \delta \leq 1$, Algorithm 4.21 can be used. In this algorithm the function $\text{Random}(a, b)$ generates a random real number in the interval $[a, b]$. Table 4.5 presents data similar to Table 4.4, but for randomly generated graphs with edge density approximately .75.

Algorithm 4.21: GENERATERANDOMGRAPH2 (n, δ)

external Random()
for $x \leftarrow 0$ **to** $n - 2$
$\quad \textbf{do} \begin{cases} \textbf{for } y \leftarrow x + 1 \textbf{ to } n - 1 \\ \quad \textbf{do} \begin{cases} r \leftarrow \text{Random}(0, 1) \\ \textbf{if } r \geq 1 - \delta \\ \quad \textbf{then } \mathcal{E} \leftarrow \mathcal{E} \cup \{x - 1, y - 1\} \end{cases} \end{cases}$
return $(\mathcal{G} = (\{0, \ldots, n - 1\}, \mathcal{E}))$

Tables 4.4 and 4.5 show that the size of the state space tree decreases significantly as better bounding functions are employed. Of course, the optimal choice for a bounding function depends on both the time required to compute the bounding function, and on the amount by which the size of the state space tree is reduced. The relative computation times for the different bounding functions can depend heavily on the implementation. However, we can make a couple of observations on the complexity of these computations. First, when given as input a graph having n vertices, the greedy coloring algorithm takes time $O(n^2)$. Therefore the greedy bound is computed in time $O(|\mathcal{C}_\ell|^2)$. The size bound and sampling bound, on the other hand, can be computed in time $O(|\mathcal{C}_\ell|)$ using standard algorithms. Hence, there is a tradeoff, because the more effective greedy bound has a slower computation time. In general, the greedy bound will result in a faster algorithm for "large enough" graphs. The crossover point, however, will depend on the implementation and is best determined by experimentation.

4.7 Branch and bound

Another way in which we can take advantage of a bounding function is a method called *branch and bound*. The usual implementation of backtracking is to examine each of the choices $x_\ell \in \mathcal{C}_\ell$ in some predetermined order, calling the algorithm recursively for each choice. A better strategy is to use a bounding function to determine the order in which the recursive calls are made. A branch and bound algorithm for a general maximization problem is presented as Algorithm 4.22.

We illustrate the branch and bound technique using the Traveling Salesman problem. Suppose $X = [x_0, x_1, x_2, \ldots, x_{\ell-1}]$ is a partial solution for an instance of the Traveling Salesman problem, and $\ell \leq n - 1$. Then there are $(n - 1) - (\ell - 1) = n - \ell$ choices for x_ℓ. Consider the node in the state space tree corresponding to the partial solution X. Algorithm 4.13 would look at the $n - \ell$ children of X in increasing order of x_ℓ. There is no particular reason to proceed in this order.

Algorithm 4.22: BRANCHANDBOUND (ℓ)

external B(), profit()
global C_ℓ ($\ell = 0, 1, \ldots$)
if $[x_0, \ldots, x_{\ell-1}]$ is a solution
then $\begin{cases} P \leftarrow \text{profit}([x_0, \ldots, x_{\ell-1}]) \\ \textbf{if } P > OptP \\ \quad \textbf{then } \begin{cases} OptP \leftarrow P \\ OptX \leftarrow [x_0, \ldots, x_{\ell-1}] \end{cases} \end{cases}$
Compute C_ℓ
$count \leftarrow 0$
for each $x \in C_\ell$
do $\begin{cases} x_\ell \leftarrow x \\ nextchoice[count] \leftarrow x \\ nextbound[count] \leftarrow \text{B}([x_0, \ldots, x_{\ell-1}, x]) \\ count \leftarrow count + 1 \end{cases}$
Sort $nextchoice$ and $nextbound$
  so that $nextbound$ is in decreasing order
for $i \leftarrow 0$ **to** $count$ 1
do $\begin{cases} \textbf{if } nextbound[i] \leq OptP \\ \quad \textbf{then return} \\ x_\ell \leftarrow nextchoice[i] \\ \text{BRANCHANDBOUND}(\ell + 1) \end{cases}$

In a branch and bound algorithm, for the Traveling Salesman problem we will calculate B(X') for each of the $n - \ell$ children X' of X before we make any recursive calls from this node. Then, we will make recursive calls in increasing order of the $n - \ell$ values of B(X') that we computed (since it is a minimization problem). We hope that an optimal solution is most likely to be found in the branch of the state space tree where the bounding function is smallest. The remaining branches of the state space tree can then possibly be pruned without having to traverse them.

The resulting algorithm is presented as Algorithm 4.23. Other than the modification to the order in which recursive calls are made, Algorithm 4.23 is unchanged from the previous algorithm, Algorithm 4.13.

TABLE 4.6
Size of state space trees for Algorithms 4.13 and 4.23, on random instances with n vertices. Algorithms 4.13 and 4.23 are shown with both bounding functions MINCOSTBOUND and REDUCEBOUND

n	Algorithm 4.13		Algorithm 4.23	
	MINCOSTBOUND	REDUCEBOUND	MINCOSTBOUND	REDUCEBOUND
5	45	18	25	9
10	5,199	1,287	490	102
15	1,538,773	53,486	128,167	5,078
20	64,259,127	1,326,640	6,105,089	39,035

Algorithm 4.23: TSP3 (ℓ)

external Sort(), cost()

global $C_\ell \quad (\ell = 0, 1, \ldots, n-1)$

if $\ell = n$

then $\begin{cases} C \leftarrow \text{cost}([x_0, \ldots, x_{n-1}]) \\ \textbf{if } C < OptC \\ \qquad \textbf{then} \begin{cases} OptC \leftarrow C \\ OptX \leftarrow [x_0, \ldots, x_{n-1}] \end{cases} \end{cases}$

if $\ell = 0$ **then** $C_\ell \leftarrow \{0\}$

    else if $\ell = 1$ **then** $C_\ell \leftarrow \{1, \ldots, n-1\}$

    else $C_\ell \leftarrow C_{\ell-1} \setminus \{x_{\ell-1}\}$

$count \leftarrow 0$

for each $x \in C_\ell$

do $\begin{cases} x_\ell \leftarrow x \\ nextchoice[count] \leftarrow x \\ nextbound[count] \leftarrow \text{B}([x_0, \ldots, x_{\ell-1}, x]) \\ count \leftarrow count + 1 \end{cases}$

Sort $nextchoice$ and $nextbound$

    so that $nextbound$ is in increasing order

for $i \leftarrow 0$ **to** $count - 1$

do $\begin{cases} \textbf{if } nextbound[i] \geq OptP \\ \qquad \textbf{then return} \\ x_\ell \leftarrow nextchoice[i] \\ \text{TSP3}(\ell + 1) \end{cases}$

To evaluate the effect of the two bounding functions in a branch and bound algorithm, we used the same random instances as we did in Section 4.6.2. In Table 4.6 we report the number of nodes in the state space trees when Algorithms 4.13 and 4.23 are used, for both of the bounding functions.

4.8 Notes

Section 4.1

Backtracking algorithms are described in several textbooks and monographs, for example Brassard and Bratley [9], Goldberg [36], Horowitz and Sahni [43], Purdom and Brown [84], Reingold, Nievergelt and Deo [90] and Stinson [103].

Section 4.3.1

The average-case analysis of Algorithm 4.4 is due to Wilf; see [113, Section 5.6]. An example of an average-case analysis of a backtracking algorithm for a different problem (the Satisfiability problem) can be found in [84, Section 4.3].

Section 4.4

For a thorough treatment of the estimation of backtrack trees, see Purdom and Brown [84, Section 11.10].

Section 4.5

An earlier version of Algorithm 4.6 was developed and used by Frenz and Kreher in [30] to enumerate inequivalent cyclic Steiner triple systems.

Wells discusses the Exact Cover problem in [111, Section 6.4]. He develops an algorithm which he then uses to construct Steiner triple systems. This is followed by a couple of additional refinements to the algorithm. If the subsets in S are restricted to each having exactly 3 elements, then the problem is known as Exact Cover by 3-sets. Exact Cover by 3-sets was shown to be NP-complete by Karp in [51] (see also [31]).

Section 4.6.1

The bounding function we use for the Knapsack (optimization) problem is described in Horowitz and Sahni [43, Section 8.2]. The instances that are referred to in Table 4.2 can be found in the web pages at the following URL:

```
http://www.math.mtu.edu/~kreher/cages/Data.html
```

Section 4.6.2

An overview of bounding functions for the Traveling Salesman problem can be found in [63, Chapter 10]. The method of reduced matrices is described in several places, for example, in Horowitz and Sahni [43, Section 8.3]. The instances that are referred to in Tables 4.3 and 4.6 can be found in the web pages at the following URL:

```
http://www.math.mtu.edu/~kreher/cages/Data.html
```

Section 4.6.3

The monograph [49] edited by Johnson and Trick is a recent work devoted to three fundamental NP-hard problems, namely, the Maximum Clique problem, the Vertex Coloring problem, and the Satisfiability problem.

The 1973 branch and bound technique of Bron and Kerbosch [13], used in their algorithm CACM457, is the basis for most of the recent maximum clique algorithms. Among these, the most notable are the algorithms by Balas and Yu [5] and Babel [4]. The 1986 Balas-Yu Algorithm uses a greedy coloring and maximally triangulated induced subgraphs to achieve tighter bounds on the maximum clique size. This algorithm was one of the fastest until 1990, when Babel [4] introduced an algorithm that uses the DSATUR coloring method of Brelaz [10]. In 1998, Myrvold, Prsa and Walker [78] developed a promising method for testing maximum clique algorithms when the number of vertices becomes prohibitively large.

The instances of the Maximum Clique problem that are referred to in Tables 4.4 and 4.5 can be found in the web pages at the following URL:

`http://www.math.mtu.edu/~kreher/cages/Data.html`

Exercises

4.1 Define choice sets and describe backtracking algorithms for the following problems:
  (a) Find all ways of placing n mutually non-attacking queens on an n by n chess board.
  (b) Find all self-avoiding walks of length n. (A *self-avoiding walk* is described by a sequence of edges in the Euclidean plane, beginning at the origin, such that each of the edges is a vertical or horizontal line segment of length one, and such that no point in the plane is visited more than once. There are precisely three such walks of length one, 12 walks of length two, and 36 walks of length three.)
  (c) Find all k-vertex colorings of a graph $\mathcal{G}$.

4.2 Find a formula for the number of nodes in the state space tree that results when Algorithm 4.10 is run on an instance of the Traveling Salesman problem having n vertices.

4.3 Determine the complexity of Algorithm 4.8, with and without the assumption that the objects are sorted according to their profit / weight ratios.

4.4 Use Algorithm 4.9 to solve the following instances of the Knapsack (optimization) problem.

(a) *Profits*   122   2   144 133 52 172 169 50 11   87 127 31   10 132 59
    Weights   63   1   71 73 24   79 82 23   6   43 66 17   5   65 29
    Capacity 323

(b) *Profits*   143 440 120 146 266 574 386 512 106 418 376 124 48 535 55
    Weights   72 202 56 73 144 277 182 240 54 192 183 67 23 244 29
    Capacity 1019

(c) *Profits*   818 460 267 75 621 280 555 214 721 427 78 754 704 44 371
    Weights   380 213 138 35 321 138 280 118 361 223 37 389 387 23 191
    Capacity 1617

4.5 In Algorithm 4.4 and Algorithm 4.14, the vertices are processed according to the prespecified ordering "$<$", always extending a clique with vertices that appear later in the ordering than the vertices already chosen. This trick allows each clique to be generated only once. However, the ordering defined on the vertices can greatly affect the point in the backtrack algorithm when the maximum clique(s) are discovered. Similarly, the speedup provided by pruning using a bounding function may depend strongly on the ordering of the vertices.

There are several natural ways to define the ordering on the vertices. In general, the best choice depends on the graph under consideration. Some of the possible orderings are as follows:

(a) *random*, in which the vertices are arbitrarily ordered;

(b) *increasing*, in which the vertices are sorted from lowest to highest degree;

(c) *decreasing*, in which the vertices are sorted from highest to lowest degree; and

(d) *induced*, in which a minimum degree vertex is placed last in the list. This vertex is then deleted from the graph and the vertex of minimal degree in the new graph is placed next-to-last, and so on.

Investigate the effect of the vertex ordering on Algorithm 4.19, for each of the bounding functions described, using the graphs considered in Tables 4.4 and 4.5 as sample graphs.

4.6 Show that Algorithm 4.8 always produces an optimal solution for the Rational Knapsack problem. **Hint:** Suppose that Algorithm 4.8 generates the solution $X = [x_0, \ldots, x_{n-1}]$, with profit $P = \sum_{i=0}^{n} p_i x_i$. Let $Y = [y_0, \ldots, y_{n-1}]$ be any optimal solution with profit $Q = \sum_{i=0}^{n} p_i y_i$. Since Y is optimal we must have $P \leq Q$. Show that $P = Q$.

4.7 Use Algorithm 4.13 to solve the instance of the Traveling Salesman problem on the vertex set $\mathcal{V} = \{0, 1, 2, \ldots, 9\}$ in which the cost of the edge $\{x, y\}$ is given by the $[x, y]$ entry of the matrix given below.

	0	1	2	3	4	5	6	7	8	9
0	0	22	0	72	56	17	57	13	38	63
1	22	0	95	29	84	75	39	19	26	12
2	0	95	0	87	78	70	39	99	21	12
3	72	29	87	0	95	90	82	33	60	76
4	56	84	78	95	0	39	93	35	2	59
5	17	75	70	90	39	0	36	81	98	25
6	57	39	39	82	93	36	0	28	47	88
7	13	19	99	33	35	81	28	0	78	13
8	38	26	21	60	2	98	47	78	0	72
9	63	12	12	76	59	25	88	13	72	0

4.8 For the Traveling Salesman problem discussed in Section 4.6.2 a Hamiltonian circuit was represented as a permutation

$$X = [0, x_1, x_2, \ldots, x_{n-1}]$$

of the vertices $\mathcal{V}$ starting at 0. This circuit is also represented by

$$X' = [0, x_{n-1}, x_{n-2}, \ldots, x_1].$$

Thus Algorithms 4.10 and 4.13 will consider every Hamiltonian circuit twice. Develop a pruning method so that every Hamiltonian circuit is examined only once and incorporate it into an algorithm for solving the Traveling Salesman problem. Compare your new algorithm to Algorithms 4.10 and 4.13 by running it on the data given in Exercise 4.7 and computing the number of nodes in the corresponding state space trees.

4.9 Use Algorithm 4.19 to find a maximum clique in each of the graphs $\mathcal{G} = (\mathcal{V}, \mathcal{E})$ where $\mathcal{V} = \{0, 1, 2, \ldots, 9, a, b, c, d\}$ and

(a)

$$\mathcal{E} = \left\{ \begin{array}{l} \{0,2\}, \{0,7\}, \{0,a\}, \{0,b\}, \{1,3\}, \{1,8\}, \{1,9\}, \{1,a\}, \{1,e\}, \{2,3\} \\ \{2,a\}, \{2,d\}, \{2,e\}, \{3,4\}, \{3,6\}, \{4,8\}, \{5,7\}, \{5,a\}, \{5,e\}, \{6,c\} \\ \{6,e\}, \{7,8\}, \{7,a\}, \{7,e\}, \{8,b\}, \{9,a\}, \{a,c\}, \{c,d\}, \{c,e\}, \{e,f\} \end{array} \right\}$$

(b)

$$\mathcal{E} = \left\{ \begin{array}{l} \{0,1\}, \{0,2\}, \{0,7\}, \{0,8\}, \{0,9\}, \{0,a\}, \{0,e\}, \{1,4\}, \{1,7\}, \{1,8\} \\ \{1,9\}, \{1,a\}, \{1,b\}, \{1,c\}, \{1,e\}, \{2,3\}, \{2,5\}, \{2,6\}, \{2,7\}, \{2,8\} \\ \{2,a\}, \{2,b\}, \{3,4\}, \{3,5\}, \{3,6\}, \{3,a\}, \{3,f\}, \{4,5\}, \{4,6\}, \{4,8\} \\ \{4,9\}, \{4,a\}, \{4,e\}, \{5,6\}, \{5,8\}, \{5,9\}, \{5,b\}, \{5,d\}, \{5,e\}, \{5,f\} \\ \{6,f\}, \{7,8\}, \{7,a\}, \{7,c\}, \{7,e\}, \{7,f\}, \{8,b\}, \{8,c\}, \{8,e\}, \{8,f\} \\ \{9,d\}, \{9,e\}, \{a,c\}, \{a,d\}, \{a,e\}, \{b,c\}, \{b,e\}, \{b,f\}, \{c,d\}, \{e,f\} \end{array} \right\}$$

(c)

$$\mathcal{E} = \left\{ \begin{array}{l} \{0,2\}, \{0,3\}, \{0,4\}, \{0,5\}, \{0,6\}, \{0,7\}, \{0,8\}, \{0,9\}, \{0,a\}, \{0,c\} \\ \{0,d\}, \{0,e\}, \{1,5\}, \{1,6\}, \{1,7\}, \{1,8\}, \{1,9\}, \{1,a\}, \{1,b\}, \{1,c\} \\ \{1,d\}, \{1,e\}, \{1,f\}, \{2,5\}, \{2,6\}, \{2,7\}, \{2,8\}, \{2,a\}, \{2,b\}, \{2,e\} \\ \{2,f\}, \{3,4\}, \{3,5\}, \{3,6\}, \{3,7\}, \{3,8\}, \{3,a\}, \{3,c\}, \{3,d\}, \{3,e\} \\ \{3,f\}, \{4,6\}, \{4,8\}, \{4,9\}, \{4,b\}, \{4,c\}, \{4,e\}, \{4,f\}, \{5,6\}, \{5,7\} \\ \{5,9\}, \{5,a\}, \{5,b\}, \{5,c\}, \{5,e\}, \{5,f\}, \{6,7\}, \{6,8\}, \{6,9\}, \{6,a\} \\ \{6,b\}, \{6,d\}, \{6,e\}, \{6,f\}, \{7,8\}, \{7,9\}, \{7,b\}, \{7,c\}, \{7,d\}, \{7,e\} \\ \{7,f\}, \{8,9\}, \{8,a\}, \{8,b\}, \{8,f\}, \{9,b\}, \{9,c\}, \{9,d\}, \{9,f\}, \{a,c\} \\ \{a,d\}, \{a,e\}, \{a,f\}, \{b,c\}, \{b,d\}, \{b,e\}, \{b,f\}, \{c,d\}, \{c,e\}, \{d,f\} \end{array} \right\}$$

4.10 Given a graph G, define the *chromatic number* of G to be

$$\chi(G) = \min\{k : G \text{ has a vertex } k\text{-coloring}\},$$

and define the *clique number* of G to be

$$\omega(G) = \max\{k : G \text{ has a clique of size } k\}.$$

Theorem 4.4 shows that $\omega(G) \leq \chi(G)$.

(a) Show that strict inequality can hold in Theorem 4.4. That is, find a graph G such that $\omega(G) < \chi(G)$.

(b) Show that for any integer $d \geq 0$ there is a graph G with $\chi(G) - \omega(G) \geq d$.

4.11 An edge-decomposition of the complete graph K_n into triangles is called a *Steiner triple system* of order n (or, STS(n)). More formally, an STS(n) is a pair (P, B) in which P is an n-element set of *points*; B is a collection of $n(n-1)/6$ 3-element subsets of P called *triples* (or *blocks*); and every pair of points is contained in exactly one triple.

(a) Write a backtracking algorithm to find all STS(n) (on a given set of n vertices), and use your algorithm to determine the number of different STS(7).

(b) Define a graph $G = (V, \mathcal{E})$, where V consists of the $\binom{n}{3}$ 3-subsets of an n-set, and two vertices are adjacent if and only if the intersection of the corresponding subsets has cardinality at most one. Show that an STS(n) is equivalent to a (maximum) clique in G having size $n(n-1)/6$.

(c) Using any of the clique-finding algorithms described in this chapter, determine the number of different STS(7).

4.12 If $x, y \in \{0, 1\}^n$, then recall that dist(x, y) denotes the Hamming distance between x and y. A *non-linear code* of length n and minimum distance d is a subset $C \subseteq \{0, 1\}^n$ such that dist(x, y) $\geq d$ for all $x, y \in C$. Denote by $A(n, d)$ the maximum number of n-tuples in length n non-linear code of minimum distance d.

(a) Use a backtracking algorithm to compute $A(n, 4)$ for $n \leq 8$.

(b) Project: Determine the values of $A(9, 4)$ and $A(10, 4)$ (these values are more difficult to obtain; they are 20 and 40, respectively).

(c) Research problem: Determine the value of $A(11, 4)$ (the exact value of $A(11, 4)$ is unknown, but it is known that $72 \leq A(11, 4) \leq 79$).

4.13 A Latin square on the n-element set $\mathcal{Y} = 1, 2, \ldots, n$ is said to be a *reduced Latin square* if the elements in the first row and in the first column occur in the natural order $1, 2, \ldots, n$. Write a backtracking program to determine the number of reduced Latin squares of order n. Run your algorithm for $n = 2, 3, 4$ and 5.

4.14 The *girth* of a graph is the size of the smallest circuit it contains. An (r, g)-cage is an r-regular graph of minimum order having girth g. Let $f(r, g)$ denote the number of vertices in an (r, g)-cage.

(a) Prove that

$$f(r, g) \geq \begin{cases} \frac{r(r-1)^h - 2}{r - 2} & \text{if } g = 2h + 1 \\ \frac{2(r-1)^h - 2}{r - 2} & \text{if } g = 2h. \end{cases}$$

Show that $f(3, 5) = 10$.

Hint: the Petersen graph, presented in Exercise 1.13, is a $(3, 5)$-cage.

(b) Develop a backtracking algorithm to search for (r, g)-cages. Construct a $(3, 5)$-cage using your algorithm.

(c) Find some other examples of cages using your algorithm. (For example, it is known that $f(3, 6) = 14$ and $f(3, 7) = 24$.)

4.15 The **Minimum Spanning Tree** problem consists of a complete graph K_n with a cost function defined on its edges. The problem is to find a set of $n - 1$ edges that form a tree (i.e., which do not contain a circuit) such that the sum of their costs is minimized. It is well known that this problem can be solved by a greedy algorithm, which considers the edges in increasing order of cost, adding each edge to the tree being constructed if and only if it does not create a circuit.

Suppose that $[x_0, \ldots, x_{\ell-1}]$ is a partial solution for the **Traveling Salesman** problem. Describe a bounding function based on the idea of computing the minimum spanning tree in the subgraph induced by the vertices in the set

$$\{0, \ldots, n - 1\} \setminus \{x_1, \ldots, x_{\ell-2}\}.$$

5

Heuristic Search

5.1   Introduction to heuristic algorithms

Suppose we are trying to solve an optimization problem. We have already discussed backtracking algorithms in Chapter 4, which generate all possible solutions in a certain lexicographic order. Backtracking algorithms are useful both for finding one (optimal) solution and for counting or enumerating all (optimal) solutions to a given problem. If all we want is one optimal solution, then backtracking may not be an efficient approach. For example, much time may be spent before even one optimal solution is found. Further, in order to verify that a given solution is indeed optimal, it may be necessary to examine a large part of the state space tree, even if pruning is used. There are also many situations where it is sufficient to find a feasible solution that is "close to" optimal, and thus exploring the entire state space tree may not be necessary.

If we do not require the generation of solutions in a precise lexicographic order, then it may be faster to proceed through the state space tree in some different way. This may lead to a more efficient algorithm to find an optimal or near-optimal solution. One often inconvenient feature of backtracking is that it is often necessary to "back up" several levels when a partial solution cannot be further extended. This can waste a lot of time, and hence a different method of exploring the state space tree might be preferable.

The term *heuristic algorithm* is used to describe an algorithm (usually a randomized algorithm) that tries to find a certain combinatorial structure or solve an optimization problem by the use of heuristics. The *Oxford Reference Dictionary* defines the adjective *heuristic* as "serving or helping to find out or discover; proceeding by trial and error". In the context of a heuristic algorithm, a heuristic will be a method of performing a minor modification, or a sequence of modifications, of a given solution or partial solution in order to obtain a different solution or partial solution. The actual modifications that are done will involve a *neighborhood search*. A heuristic algorithm will consist of iteratively applying one or more heuristics, in accordance with a certain design strategy.

In order to make these concepts more precise, let's consider a generic combinatorial optimization problem (recall that basic terminology for optimization problems was defined in Section 1.3). Solutions are chosen from a specified finite set, $\mathcal{X}$, which we call a *universe*. An element $X \in \mathcal{X}$ is said to be a *feasible solution* if certain constraints are satisfied. The constraints might be written in the form $g_j(X) \geq 0$ for all j, $1 \leq j \leq m$, where $g_1, \ldots, g_m$ are integer-valued functions. An *optimal solution* is a feasible solution X for which the *profit*, denoted $P(X)$, is as large as possible. Problem 5.1 will be our "generic" optimization problem.

Problem 5.1:   Generic Optimization

Instance: *a finite set* $\mathcal{X}$;

an objective function $P : \mathcal{X} \to \mathbb{Z}$; and

feasibility functions $g_j : \mathcal{X} \to \mathbb{Z}, 1 \leq j \leq m$

Find: the maximum value of $P(X)$

subject to $X \in \mathcal{X}$ and $g_j(X) \geq 0$ for $1 \leq j \leq m$.

The first step in constructing a heuristic is defining a neighborhood function. Formally, a *neighborhood function* is a function

$$N : \mathcal{X} \to 2^{\mathcal{X}}.$$

In plain language, the neighborhood of any element in $\mathcal{X}$ consists of a certain subset of elements of $\mathcal{X}$. We will generally define a *neighborhood* of an element X to consist of certain elements that are "similar" or "close to" X in some sense. Note that a neighborhood $N(X)$ may contain elements Y that are not feasible. Also, we usually do not care if $X \in N(X)$ or not.

Let's present some typical examples of neighborhood functions. First, suppose that $\mathcal{X} = \{0, 1\}^n$ (i.e., $\mathcal{X}$ consists of all binary n-tuples). Then we might define

$$N_{d_0}(X) = \{Y \in \mathcal{X} : \text{dist}(X, Y) \leq d_0\}$$

where d_0 is some positive integer, and $\text{dist}(\cdot, \cdot)$ represents the Hamming distance between two n-tuples. Observe that, in this case, we can easily compute the size of any neighborhood to be

$$|N_{d_0}(X)| = \sum_{i=0}^{d_0} \binom{n}{i}.$$

As a second example, suppose that $\mathcal{X}$ consists of all permutations of the set $\{1, \ldots, n\}$. Given two permutations $\alpha = [\alpha_1, \ldots, \alpha_n]$ and $\beta = [\beta_1, \ldots, \beta_n]$, suppose we define the *distance* between α and β to be

$$\text{dist}(\alpha, \beta) = |\{i : \alpha_i \neq \beta_i\}|.$$

(Essentially, this is the same as the Hamming distance between two vectors.) As we did in the previous example, we could define

$$N_{d_0}(X) = \{Y \in \mathcal{X} : \text{dist}(X, Y) \le d_0\},$$

where d_0 is some positive integer.

We note a couple of facts about this neighborhood function. First, $N_1(X) = X$, since it is impossible for two permutations to differ in exactly one position. However, two permutations can differ in exactly two positions. Given α, we can pick any two positions j and k, and define β as follows:

$$\beta_i = \begin{cases} \alpha_i & \text{if } i \ne j, k \\ \alpha_k & \text{if } i = j \\ \alpha_j & \text{if } i = k. \end{cases}$$

Thus, when $d_0 = 2$, we see that

$$|N_2(X)| = 1 + \binom{n}{2}.$$

It is an interesting exercise to compute $|N_{d_0}(X)|$ for $d_0 \ge 3$.

Once a neighborhood function is defined, we can imagine different ways in which we can try to find a feasible solution in the neighborhood of a given feasible solution X. One obvious approach is to perform an exhaustive search of the neighborhood, trying to find the "best" feasible solution in that neighborhood. However, many heuristic algorithms are instead based on a randomized neighborhood search, which is usually faster than an exhaustive search.

More formally, suppose that N is a neighborhood function. A *neighborhood search* based on N will be an algorithm (possibly a randomized algorithm) in which the input is a feasible solution $X \in \mathcal{X}$, and the output is either a feasible solution $Y \in N(X) \backslash \{X\}$, or *Fail*. Since $N(X)$ may contain elements Y that are not feasible, the neighborhood search must ensure that an element $Y \in \mathcal{X}$ that is produced as output is indeed feasible. This may be done by checking the feasibility functions g_i, or by some other (faster) method, if applicable. Note that there are different reasons why a neighborhood search might return the value *Fail*. One reason might be that there are no feasible solutions in $N(X) \backslash \{X\}$. But it could also be that the neighborhood search looks at only one random element of $N(X)$. In this case, if the search is performed again with the same input X, it might succeed.

The following list enumerates some possible neighborhood search strategies:

1. Find a feasible solution $Y \in N(X)$ such that $P(Y)$ is maximized (return *Fail* if there are no feasible solutions in $N(X) \backslash \{X\}$).

2. Find a feasible solution $Y \in N(X)$ such that $P(Y)$ is maximized. If $P(Y) > P(X)$, then return Y, otherwise return *Fail*. This search method is called *steepest ascent*.

3. Find any feasible solution in $Y \in N(X)$.

4. Find any feasible solution in $Y \in N(X)$. If $P(Y) > P(X)$, then return Y, otherwise return *Fail*.

Strategies 1 and 2 would probably be implemented as exhaustive searches, while strategies 3 and 4 are more likely to be implemented as random searches.

Now, having specified a neighborhood search strategy, based on a given neighborhood function, N, we can proceed to define a heuristic, h_N. The most common way to define h_N is simply to perform the given neighborhood search. However, it may be more convenient in some situations to define the heuristic as a sequence of j neighborhood searches (for some positive integer j), say

$$[X_0 = X, X_1, \ldots, Y = X_j],$$

where each X_i is obtained from X_{i-1} by applying the neighborhood search.

Finally, the heuristic will be incorporated into a heuristic algorithm. We will discuss several popular design strategies for heuristic algorithms in Section 5.2. However, most of these algorithms have a similar basic structure, as presented in Algorithm 5.1, a "generic" heuristic search algorithm.

Algorithm 5.1: GENERICHEURISTICSEARCH (c_{max})

external $N(), h_N(), P()$

$c \leftarrow 0$

Select a feasible solution $X \in \mathcal{X}$

$X_{best} \leftarrow X$

while $c \leq c_{max}$

$\quad$ **do** $\begin{cases} Y \leftarrow h_N(X) \\ \textbf{if } Y \neq \textit{Fail} \\ \quad \textbf{then } \begin{cases} X \leftarrow Y \\ \textbf{if } P(X) > P(X_{best}) \\ \quad \textbf{then } X_{best} \leftarrow X \end{cases} \\ c \leftarrow c + 1 \end{cases}$

return (X_{best})

Algorithm 5.1 uses a given neighborhood function N and a heuristic h_N based on N. The parameter c_{max} is used to specify the number of iterations of h_N that are performed in the algorithm. The variable c keeps track of the number of iterations of h_N, and X_{best} records the best feasible solution found "so far" as the algorithm progresses.

The algorithm begins with an initial feasible solution, X, which must be found by some specified method. Often, X will be taken to be a trivial solution which is far from optimal. In any given iteration of the algorithm, the heuristic h_N is applied and c is increased by one. If h_N does not fail, then X is updated, and

X_{best} is updated if X achieves a higher profit than X_{best}. After c_{max} iterations, the algorithm terminates, returning the feasible solution X_{best}.

5.1.1 Uniform graph partition

In this section, we illustrate the concepts described above with a problem known as the Uniform Graph Partition problem, which we present as Problem 5.2.

Problem 5.2: Uniform Graph Partition

Instance: a complete graph on $2n$ vertices, $\mathcal{G} = (\mathcal{V}, E)$;

a cost function, $\text{cost} : E \rightarrow \mathbb{Z}^+ \cup \{0\}$

Find: the minimum value of

$$C([\mathcal{X}_0, \mathcal{X}_1]) = \sum_{\substack{\{u,v\} \in \mathcal{E} \\ u \in \mathcal{X}_0, v \in \mathcal{X}_1}} \text{cost}(u, v)$$

subject to $\mathcal{V} = \mathcal{X}_0 \,\dot{\cup}\, \mathcal{X}_1$ and $|\mathcal{X}_0| = |\mathcal{X}_1| = n$.

We define some useful notation for this problem. Recall that a weighted graph is a graph $\mathcal{G} = (\mathcal{V}, \mathcal{E})$, together with a cost function

$$\text{cost} : \mathcal{E} \rightarrow \mathbb{Z}^+ \cup \{0\}.$$

For convenience, we denote $\text{cost}(\{u, v\})$ by $\text{cost}(u, v)$ for an edge $\{u, v\}$. If $\{u, v\} \notin \mathcal{E}$, then we define $\text{cost}(u, v) = 0$. We will usually store the cost function as a matrix whose $[u, v]$-entry is $\text{cost}(u, v)$. This matrix is often referred to as a *cost matrix*; see Figure 5.1 for an example.

The cost of the partition $[\{0, 2, 5, 7\}, \{1, 3, 4, 6\}]$ in this weighted graph is

$$\text{cost}(2, 1) + \text{cost}(2, 4) + \text{cost}(2, 6) + \text{cost}(5, 3) = 8 + 7 + 2 + 4 = 21.$$

We can choose the universe $\mathcal{X}$ to be the set of all partitions $[\mathcal{X}_0, \mathcal{X}_1]$ of $\mathcal{V}$ with $|\mathcal{X}_0| = |\mathcal{X}_1|$. Define the neighborhood of $N([\mathcal{X}_0, \mathcal{X}_1])$ of the partition $[\mathcal{X}_0, \mathcal{X}_1]$ to be the set of all partitions that can be obtained from $[\mathcal{X}_0, \mathcal{X}_1]$ by exchanging a vertex of $\mathcal{X}_0$ with a vertex of $\mathcal{X}_1$. For example, the neighborhood of the partition $[\mathcal{X}_0, \mathcal{X}_1] = [\{0, 2, 5, 7\}, \{1, 3, 4, 6\}]$ in the weighted graph given in Figure 5.1 is

$$N([\mathcal{X}_0, \mathcal{X}_1]) = \left\{ \begin{array}{l} [\{1, 2, 5, 7\}, \{0, 3, 4, 6\}], [\{3, 2, 5, 7\}, \{1, 0, 4, 6\}], \\ [\{4, 2, 5, 7\}, \{1, 3, 0, 6\}], [\{6, 2, 5, 7\}, \{1, 3, 4, 0\}], \\ [\{0, 1, 5, 7\}, \{2, 3, 4, 6\}], [\{0, 3, 5, 7\}, \{1, 2, 4, 6\}], \\ [\{0, 4, 5, 7\}, \{1, 3, 2, 6\}], [\{0, 6, 5, 7\}, \{1, 3, 4, 2\}], \\ [\{0, 2, 1, 7\}, \{5, 3, 4, 6\}], [\{0, 2, 3, 7\}, \{1, 5, 4, 6\}], \\ [\{0, 2, 4, 7\}, \{1, 3, 5, 6\}], [\{0, 2, 6, 7\}, \{1, 3, 4, 5\}], \\ [\{0, 2, 5, 1\}, \{7, 3, 4, 6\}], [\{0, 2, 5, 3\}, \{1, 7, 4, 6\}], \\ [\{0, 2, 5, 4\}, \{1, 3, 7, 6\}], [\{0, 2, 5, 6\}, \{1, 3, 4, 7\}] \end{array} \right\}$$

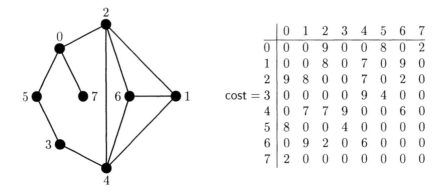

	0	1	2	3	4	5	6	7
0	0	0	9	0	0	8	0	2
1	0	0	8	0	7	0	9	0
2	9	8	0	0	7	0	2	0
3	0	0	0	0	9	4	0	0
4	0	7	7	9	0	0	6	0
5	8	0	0	4	0	0	0	0
6	0	9	2	0	6	0	0	0
7	2	0	0	0	0	0	0	0

$\text{cost} = $

FIGURE 5.1
A weighted graph.

The *gain* (i.e., change in cost) obtained from exchanging $u \in \mathcal{X}_0$ with $v \in \mathcal{X}_1$ is

$$G_{[\mathcal{X}_0, \mathcal{X}_1]}(u, v)$$
$$= C(\mathcal{X}_0, \mathcal{X}_1) - C(\mathcal{X}_0 \setminus \{u\} \cup \{v\}, \mathcal{X}_1 \setminus \{v\} \cup \{u\})$$
$$= \sum_{y \in \mathcal{X}_1} \text{cost}(u, y) + \sum_{x \in \mathcal{X}_0} \text{cost}(x, y) - \sum_{y \in \mathcal{X}_1} \text{cost}(v, y) - \sum_{x \in \mathcal{X}_0} \text{cost}(x, u).$$

Note that the gain can be negative, zero or positive.

Given a partition $[\mathcal{X}_0, \mathcal{X}_1]$, one simple neighborhood search is to find the partition $[\mathcal{Y}_0, \mathcal{Y}_1]$, where $\mathcal{Y}_0 = (\mathcal{X}_0 \setminus \{u\}) \cup \{v\}, \mathcal{Y}_1 = (\mathcal{X}_1 \setminus \{v\}) \cup \{u\}$ and $u \in \mathcal{X}_0$ and $v \in \mathcal{X}_1$ are chosen so that the gain of exchanging $u \in \mathcal{X}_0$ with $v \in \mathcal{X}_1$ is positive and maximum. If this is not possible, the output is *Fail*. Note that this neighborhood search is an example of steepest ascent.

For example, suppose we are given the weighted graph in Figure 5.1, and the partition

$$[\{0, 2, 5, 7\}, \{1, 3, 4, 6\}].$$

Then the partition $[\{0, 3, 5, 7\}, \{1, 2, 4, 6\}]$ would be returned, since exchanging 2 and 3 gives the largest gain. See Table 5.1.

5.2   Design strategies for heuristic algorithms

In this section, we introduce the main *design strategies* for heuristic algorithms, i.e., the means by which we design a neighborhood search and incorporate it into a heuristic search algorithm.

TABLE 5.1
The gain of exchanging u with v in the partition $[\mathcal{X}_0, \mathcal{X}_1]$.

$$[\mathcal{X}_0, \mathcal{X}_1] = [\{0, 2, 5, 7\}, \{1, 3, 4, 6\}] \qquad C([\mathcal{X}_0, \mathcal{X}_1]) = 21$$

u	v	$[\mathcal{Y}_0, \mathcal{Y}_1]$	$G_{[\mathcal{X}_0, \mathcal{X}_1]}(u, v)$	$C([\mathcal{Y}_0, \mathcal{Y}_1])$
0	1	$[\{1, 2, 5, 7\}, \{0, 3, 4, 6\}]$	-27	48
0	3	$[\{2, 3, 5, 7\}, \{0, 1, 4, 6\}]$	-24	45
0	4	$[\{2, 4, 5, 7\}, \{0, 1, 3, 6\}]$	-34	55
0	6	$[\{2, 5, 6, 7\}, \{0, 1, 3, 4\}]$	-32	53
2	1	$[\{0, 1, 5, 7\}, \{2, 3, 4, 6\}]$	-16	37
2	3	$[\{0, 3, 5, 7\}, \{1, 2, 4, 6\}]$	$+3$	18
2	4	$[\{0, 4, 5, 7\}, \{1, 2, 3, 6\}]$	-21	42
2	6	$[\{0, 5, 6, 7\}, \{1, 2, 3, 4\}]$	-9	30
5	1	$[\{0, 1, 2, 7\}, \{3, 4, 5, 6\}]$	-12	33
5	3	$[\{0, 2, 3, 7\}, \{1, 4, 5, 6\}]$	-17	38
5	4	$[\{0, 2, 4, 7\}, \{1, 3, 5, 6\}]$	-19	40
5	6	$[\{0, 2, 6, 7\}, \{1, 3, 4, 5\}]$	-17	38
7	1	$[\{0, 1, 2, 5\}, \{3, 4, 6, 7\}]$	-10	31
7	3	$[\{0, 2, 3, 5\}, \{1, 4, 6, 7\}]$	-7	28
7	4	$[\{0, 2, 4, 5\}, \{1, 3, 6, 7\}]$	-17	38
7	6	$[\{0, 2, 5, 6\}, \{1, 3, 4, 7\}]$	-15	36

There is a tradeoff that must be considered in any heuristic algorithm. If we use "large" neighborhoods, then we would expect that any given neighborhood is more likely to contain a good solution than if we use "small" neighborhoods. But, particularly in the case of an exhaustive neighborhood search, we pay a penalty in computation time if the neighborhoods are too large. We will later see other tradeoffs that arise in the design of particular types of heuristic algorithms.

5.2.1 Hill-climbing

The conceptually simplest design strategy is *hill-climbing*. When hill-climbing, we require that $P(Y) > P(X)$ for any $Y \in N(X)$ returned as the output of the neighborhood search. If the neighborhood search does not find any such Y, it must return *Fail*. Thus, we are attempting to proceed toward an optimal solution by finding a sequence of feasible solutions, each of which is better than the previous one. The analogy of climbing a hill is used here to suggest that the fastest way to the top of a hill is to climb continually upward (of course, this may or may not be true in practice!). If the neighborhood search strategy is exhaustive, then we are using the steepest ascent idea that we described earlier.

We present a "generic" hill-climbing algorithm as Algorithm 5.2.

Algorithm 5.2: GENERICHILLCLIMBING (c_{max})

external $N(), h_N(), P()$
comment: if h_N returns a feasible solution Y, then $P(Y) > P(X)$
Select a feasible solution $X \in \mathcal{X}$
$X_{best} \leftarrow X$
$searching \leftarrow$ **true**
while $searching$

$$
\mathbf{do} \begin{cases} Y \leftarrow h_N(X) \\ \mathbf{if}\ Y \neq Fail \\ \qquad \mathbf{then} \begin{cases} X \leftarrow Y \\ \mathbf{if}\ P(X) > P(X_{best}) \\ \qquad \mathbf{then}\ X_{best} \leftarrow X \end{cases} \\ \qquad \mathbf{else}\ searching \leftarrow \mathbf{false} \end{cases}
$$

return (X_{best})

We will give some nice examples of hill-climbing algorithms in Sections 5.3 and 5.4. However, it is probably not surprising that the hill-climbing strategy is often too restrictive to be successful. The reason is that if $P(Y) < P(X)$ for all feasible solutions $Y \in N(X)$, then a hill-climbing algorithm will get stuck whenever it reaches X. Such an element X is called a *locally optimal solution*, and in a typical combinatorial optimization problem, there may be many locally optimal solutions that are not optimal solutions. Hence, it is desirable to develop design strategies for heuristic search algorithms that will not get stuck every time a locally optimal solution is encountered. Strategies of this type are discussed in the rest of this section.

5.2.2 Simulated annealing

One popular method of escaping from locally optimal solutions is based on an analogy with a method of cooling metal which is known as "annealing". The corresponding algorithmic paradigm is therefore called *simulated annealing*. In simulated annealing, we use a randomized neighborhood search strategy. If $h_N(X) = Y$ is feasible and $P(Y) \geq P(X)$, then X is replaced by Y, as in hill-climbing. However, if $h_N(X) = Y$ is feasible and $P(Y) < P(X)$, then we are sometimes allowed to replace X with Y. A *downward move* of this type will be permitted with a certain probability. This allows the algorithm to escape from locally optimal solutions.

Associated with a simulated annealing algorithm is a variable T called the *temperature*. T is initialized to be a value $T_0 > 0$. During the course of the algorithm the value of T is decreased according to a specified *cooling schedule*. At any point in the algorithm, the probability of replacing X with $Y = h_N(X)$, given that $P(X) > P(Y)$, is

$$e^{(P(Y)-P(X))/T}.$$

This is accomplished by generating a random number $r \in [0, 1]$ and replacing the feasible solution X with Y if

$$r < e^{(P(Y)-P(X))/T}.$$

It remains to specify a cooling schedule. Usually, T is decreased after each iteration, according to a formula $T \leftarrow \alpha T$, where $0 < \alpha < 1$ is some constant (usually α is close to 1, e.g., $\alpha = .999$).

Initially, the probability of allowing a downward move, namely $e^{(P(Y)-P(X))/T}$, will be relatively large. However, as T decreases, this probability also decreases. Thus, as we get closer to the optimal solution, downward moves are permitted with smaller probability.

We present a straightforward generic simulated annealing algorithm in Algorithm 5.3. The variable c keeps track of the total number of iterations, and the algorithm terminates when c_{max} iterations have been performed. The variable X_{best} records the best solution as the algorithm progresses. Finally, Random$(0, 1)$ denotes a random real number chosen in the interval $(0, 1)$.

Algorithm 5.3: GENERICSIMULATEDANNEALING (c_{max}, T_0, α)

external N$()$, $h_N()$, Random$()$, P$()$
$c \leftarrow 0$
$T \leftarrow T_0$
Select a feasible solution $X \in \mathcal{X}$
$X_{best} \leftarrow X$
while $c \leq c_{max}$
$\quad$ **do** $\begin{cases} Y \leftarrow h_N(X) \\ \textbf{if } Y \neq Fail \\ \quad \textbf{then} \begin{cases} \textbf{if } P(Y) > P(X) \\ \quad \textbf{then} \begin{cases} X \leftarrow Y \\ \textbf{if } P(X) > P(X_{best}) \\ \quad \textbf{then } X_{best} \leftarrow X \end{cases} \\ \quad \textbf{else} \begin{cases} r \leftarrow \text{Random}(0, 1) \\ \textbf{if } r < e^{(P(Y)-P(X))/T} \\ \quad \textbf{then } X \leftarrow Y \end{cases} \end{cases} \\ c \leftarrow c + 1 \\ T \leftarrow \alpha T \end{cases}$
return (X_{best})

5.2.3   Tabu search

Tabu search can be thought of as a variation on the theme of steepest ascent. The basic idea is to replace an element X with the element $Y \in N(X) \setminus \{Y\}$ such that Y is feasible and $P(Y)$ is maximum among all such feasible elements. This usually involves an exhaustive search of $N(X)$.

It might happen that $P(Y) < P(X)$; so, we can escape from a locally optimal solution X by this method. However, having replaced X by Y in this case, it is then highly likely that the next step would be to replace Y by X. This is clearly not desirable, since the algorithm would enter an infinite loop from which it cannot escape. Hence, we need to find a way to avoid this situation and other similar problems such as cycling (i.e., a sequence of moves such as $X \to Y \to Z \to X$). This is accomplished by means of a *tabu list* which we name *TabuList*.

Suppose we define a function $\text{change}(X, Y)$ which specifies the changes that are made to a feasible solution X in order to obtain a feasible solution Y. Having made a given change at a certain point in the algorithm, we do not want to perform any operations that will "undo" this change (at least until some time has passed). Thus, after any move $X \to Y$, $\text{change}(Y, X)$ is designated as a *forbidden change* and is added to the list *TabuList*. Changes that are on the *TabuList* remain forbidden for some specified *lifetime*, L. The parameter L is a fixed positive integer (e.g., $L = 10$ is a typical choice).

As an example, suppose that $\mathcal{X} = \{0, 1\}^n$, and

$$N(X) = \{Y \subset \mathcal{X} : \text{dist}(X, Y) = 1\}.$$

The neighborhood of X consists of all binary n-tuples in which exactly one entry of X is changed. Hence,

$$|N(X) \setminus X| = n.$$

Suppose we define

$$\text{change}(X, Y) = i \Leftrightarrow x_i \neq y_i$$

whenever $\text{dist}(X, Y) = 1$. Hence, if an entry i of a feasible solution is changed (from 0 to 1 or from 1 to 0), it cannot be changed back again for at least L iterations of the algorithm. From this, it is easily seen that any cycle of moves $X \to Y \to \cdots \to X$ has length at least $2L$.

We can also observe that $N(X)$ will contain $n - L$ vectors that arise from a non-tabu change. Thus we have another example of a tradeoff that needs to be considered when the algorithm is designed. We want L to be "big" in order to eliminate cycling, but if L becomes too big, then there will not be very many allowable moves at any given point in the algorithm. This may make it difficult to find good (optimal or near-optimal) solutions.

TabuList will be implemented as a list where $TabuList[c] = \Delta$ if Δ is the change that is designated as forbidden at iteration c of the algorithm. Now, the heuristic h_N is defined as

$$h_N(X) = Y, \text{ where } \begin{cases} Y \in \mathsf{N}(X), \\ Y \text{ is feasible,} \\ \text{change}(X, Y) \notin \{ TabuList[d] : c - L \le d \le c - 1 \}, \\ \text{and } \mathsf{P}(Y) \text{ is maximum among all such feasible elements} \end{cases}$$

Algorithm 5.4 is a generic tabu search that is obtained by implementing the heuristic h_N as we have described it above. If, at any time in the algorithm, there are no feasible points in $\mathsf{N}(X)$ that are not forbidden, then the algorithm terminates at that point.

Algorithm 5.4: GENERICTABUSEARCH (c_{max}, L)

external $\mathsf{N}(), \text{change}(), \mathsf{P}()$
$c \leftarrow 1$
Select a feasible solution $X \in \mathcal{X}$
$X_{best} \leftarrow X$
while $c \le c_{max}$

$$\mathbf{do} \begin{cases} N \leftarrow \mathsf{N}(X) \backslash \{ TabuList[d] : c - L \le d \le c - 1 \} \\ \textbf{for each } Y \in N \\ \quad \mathbf{do} \begin{cases} \textbf{if } Y \text{ is infeasible} \\ \quad \textbf{then } N \leftarrow N \backslash \{Y\} \end{cases} \\ \textbf{if } N = \emptyset \\ \quad \textbf{then exit} \\ \text{find } Y \in N \text{ such that } \mathsf{P}(Y) \text{ is maximum} \\ TabuList[c] \leftarrow \text{change}(Y, X) \\ X \leftarrow Y \\ \textbf{if } \mathsf{P}(X) > \mathsf{P}(X_{best}) \\ \quad \textbf{then } X_{best} \leftarrow X \\ c \leftarrow c + 1 \end{cases}$$

return (X_{best})

5.2.4 Genetic algorithms

In a hill-climbing, simulated annealing or tabu search algorithm, we begin with an initial feasible solution and proceed to construct from it a sequence of feasible solutions by applying a heuristic, which is in turn based on a neighborhood search technique. In a *genetic algorithm*, we begin with an *initial population* of feasible solutions. Then feasible solutions from this initial population are *mated* (i.e., recombined in pairs) to produce children. After the children are obtained, some types of *mutation* are allowed to occur (usually mutation is a heuristic based on a neighborhood search). This produces the next *generation* of the population. The process can be iterated for as many generations as desired.

A genetic algorithm must specify how children are produced. A common approach is to take two feasible solutions W and X from the population ("parents"), and use a *recombination* operation to generate two children, Y and Z, which "inherit" properties of the two parents.

One simple recombination operation is called *crossover*. Suppose that we have an optimization problem in which the universe $\mathcal{X} = \{0,1\}^n$, and the two parents are $W = [w_1, \ldots, w_n]$ and $X = [x_1, \ldots, x_n]$. Choose a *crossover point* $j \in \{1, \ldots, n\}$ at random. Then define $Y = [y_1, \ldots, y_n]$ and $Z = [z_1, \ldots, z_n]$ as follows:

$$y_i = \begin{cases} w_i & \text{if } 1 \leq i \leq j \\ x_i & \text{if } j + 1 \leq i \leq n \end{cases}$$

and

$$z_i = \begin{cases} x_i & \text{if } 1 \leq i \leq j \\ y_i & \text{if } j + 1 \leq i \leq n. \end{cases}$$

In other words, Y is formed from the first j entries of W and the last $n - j$ entries of X, and Z is formed from the first j entries of X and the last $n - j$ entries of W.

As an example, suppose that

$$W = [1, 1, 0, 1, 1, 0, 1, 0, 0, 1]$$

and

$$X = [1, 0, 0, 1, 0, 0, 0, 1, 0, 1],$$

and the crossover point $j = 3$ is chosen. Then the two children of W and X are

$$Y = [1, 1, 0, 1, 0, 0, 0, 1, 0, 1]$$

and

$$Z = [1, 0, 0, 1, 1, 0, 1, 0, 0, 1].$$

It is more complicated to think of a crossover operation in the case where $\mathcal{X}$ consists of a set of permutations, say all permutations of $\{1, \ldots, n\}$. The trick we used above will not work in general, since the children of two permutations need not be a permutation. A method called *partially matched crossover* is instead often used. The method is described in terms of two crossover points, j and k, where $1 \leq j < k \leq n$. The algorithm is presented in Algorithm 5.5. It takes as input two parents, α and β, and produces as output two children, γ and δ. The elements α, β, γ and δ are all permutations.

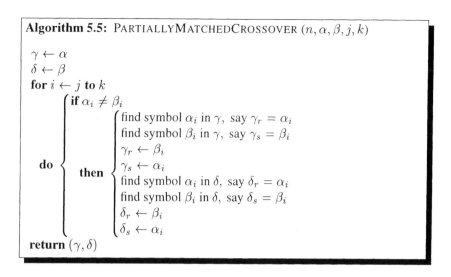

Algorithm 5.5: PARTIALLYMATCHEDCROSSOVER (n, α, β, j, k)

$\gamma \leftarrow \alpha$
$\delta \leftarrow \beta$
for $i \leftarrow j$ **to** k
do $\begin{cases} \textbf{if } \alpha_i \neq \beta_i \\ \textbf{then} \begin{cases} \text{find symbol } \alpha_i \text{ in } \gamma, \text{ say } \gamma_r = \alpha_i \\ \text{find symbol } \beta_i \text{ in } \gamma, \text{ say } \gamma_s = \beta_i \\ \gamma_r \leftarrow \beta_i \\ \gamma_s \leftarrow \alpha_i \\ \text{find symbol } \alpha_i \text{ in } \delta, \text{ say } \delta_r = \alpha_i \\ \text{find symbol } \beta_i \text{ in } \delta, \text{ say } \delta_s = \beta_i \\ \delta_r \leftarrow \beta_i \\ \delta_s \leftarrow \alpha_i \end{cases} \end{cases}$
return (γ, δ)

In Algorithm 5.5, γ and δ are formed from α and β by performing a sequence of transposition of symbols within α and β. First, the two symbols α_j and β_j are transposed within α and β; then the same thing is done with α_{j+1} and β_{j+1}, etc.

We illustrate with an example. Suppose that

$$\alpha = [3, 1, 4, 7, 6, 5, 2, 8] \quad \text{and} \quad \beta = [8, 6, 4, 3, 7, 1, 2, 5].$$

Suppose that the two crossover points are $j = 3$ and $k = 6$. Then we will perform the following sequence of transpositions of symbols: $4 \leftrightarrow 4$, $7 \leftrightarrow 3$, $6 \leftrightarrow 7$ and $5 \leftrightarrow 1$. The first interchange has no effect. After the second interchange, we have

$$\gamma = [7, 1, 4, 3, 6, 5, 2, 8] \quad \text{and} \quad \delta = [8, 6, 4, 7, 3, 1, 2, 5].$$

After the third interchange, we have

$$\gamma = [6, 1, 4, 3, 7, 5, 2, 8] \quad \text{and} \quad \delta = [8, 7, 4, 6, 3, 1, 2, 5].$$

Finally, after the fourth interchange, we have the two children,

$$\gamma = [6, 5, 4, 3, 7, 1, 2, 8] \quad \text{and} \quad \delta = [8, 7, 4, 6, 3, 5, 2, 1].$$

In general, even if we have a recombination operation where the children are always elements of the universe $\mathcal{X}$, there is no guarantee that the children will be feasible solutions for the optimization problem under consideration. Hence, some method needs to be used to ensure that a stable population of feasible solutions is maintained from one generation to the next. This can sometimes be done by tailoring the recombination operation to the problem at hand. Another approach is to redefine the optimization problem in such a way that all elements in the universe are considered to be feasible solutions. This is done by incorporating the constraints into the objective function in such a way that an element that violates one or more of the constraints will have a low profit.

The actual mating can be done in various ways. One method is to randomly partition all the feasible solutions in the population into pairs. A modification is to require that "better" parents produce more children than "poor" parents (where fitness could be measured by the objective function, or by some other method).

We now present a generic genetic algorithm in Algorithm 5.6. The algorithm incorporates a heuristic h_N based on a neighborhood function N, and a recombination operation

$$\mathsf{rec} : \mathcal{X} \times \mathcal{X} \to \mathcal{X} \times \mathcal{X},$$

which produces two children from two parents. We will assume that the children produced from two feasible parents by the function rec are feasible solutions. The parameter *popsize* will denote the population size, and c_{max} will denote the number of generations of the population constructed by the algorithm.

Algorithm 5.6: GENERICGENETICALGORITHM $(popsize, c_{max})$

external $\mathsf{N}(), h_\mathsf{N}(), \mathsf{rec}(), \mathsf{P}()$

$c \gets 1$
Select an initial population $\mathcal{P}$ consisting of *popsize* feasible solutions
Let X_{best} be the element in $\mathcal{P}$ having maximum profit
for each $X \in \mathcal{P}$
  do $X \gets h_\mathsf{N}(X)$
while $c \le c_{max}$

$\mathbf{do} \begin{cases} \text{Construct a pairing of the elements in } \mathcal{P} \\ \mathcal{Q} \gets \mathcal{P} \\ \textbf{for each } \text{ pair } W, X \text{ in the pairing} \\ \quad \mathbf{do} \begin{cases} (Y, Z) \gets \mathsf{rec}(W, X) \\ Y \gets h_\mathsf{N}(Y) \\ Z \gets h_\mathsf{N}(Z) \\ \mathcal{Q} \gets \mathcal{Q} \cup \{Y, Z\} \end{cases} \\ \text{Let the population } \mathcal{P} \text{ consist of the best } popsize \text{ members of } \mathcal{Q} \\ \text{Let } Y \text{ be the element in } \mathcal{P} \text{ having maximum profit} \\ \textbf{if } \mathsf{P}(Y) > \mathsf{P}(X_{best}) \\ \quad \textbf{then } X_{best} \gets Y \\ c \gets c + 1 \end{cases}$

return (X_{best})

5.3   A steepest ascent algorithm for uniform graph partition

In this section, we develop a steepest ascent hill-climbing algorithm for Uniform Graph Partition , Problem 5.2. Recall from Section 5.1.1 that an instance is given by a weighted graph on $2n$ vertices, which we store in the form of a $2n$ by $2n$ cost matrix, M. A feasible solution is a partition $[\mathcal{X}_0, \mathcal{X}_1]$ of the vertices such that $|\mathcal{X}_0| = |\mathcal{X}_1|$ and the objective is to minimize the value

$$C([\mathcal{X}_0, \mathcal{X}_1]) = \sum_{\substack{\{u, v\} \in \mathcal{E} \\ u \in \mathcal{X}_0, v \in \mathcal{X}_1}} \text{cost}(u, v).$$

The neighborhood of $[\mathcal{X}_0, \mathcal{X}_1]$ is the set of all partitions of the vertices that can be obtained by exchanging an element $x \in \mathcal{X}_0$ with an element $y \in \mathcal{X}_1$. The gain resulting from exchanging $u \in \mathcal{X}_0$ with $v \in \mathcal{X}_1$ is denoted by $G_{[\mathcal{X}_0, \mathcal{X}_1]}(u, v)$.

Algorithm 5.7 selects a random initial partition $[\mathcal{X}_0, \mathcal{X}_1]$ by randomly choosing $\mathcal{X}_0$ to be a subset of n vertices and defining $\mathcal{X}_1$ to be the complement of $\mathcal{X}_0$.

Algorithm 5.7: SELECTPARTITION ()

external RandomInteger(), SUBSETLEXUNRANK()
$r \leftarrow$ RandomInteger$(0, \binom{2n}{n} - 1)$
$\mathcal{X}_0 \leftarrow$ SUBSETLEXUNRANK$(2n, r)$
$\mathcal{X}_1 \leftarrow V \setminus \mathcal{X}_0$
return $([\mathcal{X}_0, \mathcal{X}_1])$

Another way to select a random partition of this type is to first generate an array A of $2n$ random numbers and an array B of length $2n$ whose ith entry is initially i. We then sort the array A. Whenever two entries $A[i]$ and $A[j]$ are interchanged, then we also interchange $B[i]$ and $B[j]$. Finally,

$$\mathcal{X}_0 = \{B[0], B[1], B[2], \ldots, B[n-1]\}$$

and

$$\mathcal{X}_1 = \{B[n], B[n+1], B[n+2], \ldots, B[2n-1]\}$$

is a random partition.

The heuristic is presented as Algorithm 5.8. This is simply a steepest ascent neighborhood search.

Algorithm 5.8: ASCEND $([\mathcal{X}_0, \mathcal{X}_1])$

external $G()$
global $Fail$
$g \leftarrow 0$
for each $i \in \mathcal{X}_0$
$$\mathbf{do} \begin{cases} \mathbf{for\ each\ } j \in \mathcal{X}_1 \\ \quad \mathbf{do} \begin{cases} t \leftarrow G_{[\mathcal{X}_0, \mathcal{X}_1]}(i, j) \\ \mathbf{if\ } t > g \\ \quad \mathbf{then} \begin{cases} x \leftarrow i \\ y \leftarrow j \\ g \leftarrow t \end{cases} \end{cases} \end{cases}$$
if $g > 0$
$$\mathbf{then} \begin{cases} \mathcal{Y}_0 \leftarrow (\mathcal{X}_0 \cup \{y\}) \setminus \{x\} \\ \mathcal{Y}_1 \leftarrow (\mathcal{X}_1 \cup \{x\}) \setminus \{y\} \\ Fail \leftarrow \mathbf{false} \\ \mathbf{return\ } ([\mathcal{Y}_0, \mathcal{Y}_1]) \end{cases}$$
$$\mathbf{else} \begin{cases} Fail \leftarrow \mathbf{true} \\ \mathbf{return\ } ([\mathcal{X}_0, \mathcal{X}_1]) \end{cases}$$

Using Algorithms 5.7 and 5.8, it is a simple matter to construct Algorithm 5.9. Note that, once the flag *Fail* is set to have the value **true** by the heuristic (Algorithm 5.8), then the search has reached a local minimum and no further improvement can be achieved. At this point the search is terminated.

Algorithm 5.9: UGP (c_{max})

external SELECTPARTITION(), ASCEND()
global $Fail$
$\mathcal{X} = [\mathcal{X}_0, \mathcal{X}_1] \leftarrow$ SELECTPARTITION()
$c \leftarrow 1$
while $c \leq c_{max}$
$$\mathbf{do} \begin{cases} [\mathcal{Y}_0, \mathcal{Y}_1] \leftarrow \text{ASCEND}(\mathcal{X}) \\ \mathbf{if\ not\ } Fail \\ \quad \mathbf{then} \begin{cases} \mathcal{X}_0 \leftarrow \mathcal{Y}_0 \\ \mathcal{X}_1 \leftarrow \mathcal{Y}_1 \end{cases} \\ \quad \mathbf{else\ return} \\ c \leftarrow c + 1 \end{cases}$$

To test the algorithm, we generated a random cost matrix for the complete graph on 50 vertices. The edge costs were chosen in the range $[0, 99]$. We performed 100 runs of Algorithm 5.9. The number of iterations varied from a minimum of

7 to a maximum of 17, with an average of 11.4. The minimum best cost that was found in the 100 runs was 28103 and the maximum was 28766. The average was 28303.81.

5.4   A hill-climbing algorithm for Steiner triple systems

In this section, we present Stinson's algorithm, which is a hill-climbing algorithm for constructing Steiner triple systems. We will begin with some definitions and discussion of these objects, before proceeding to the algorithm.

A *Steiner triple system* is a set system $(\mathcal{V}, \mathcal{B})$ in which every block has size three, and every pair of points from $\mathcal{V}$ is contained in a unique block. If $|\mathcal{V}| = v$, then we denote such a system as an STS(v).

An example of an STS(7) is

$$\mathcal{V} = \{1, 2, 3, 4, 5, 6, 7\}$$

$$\mathcal{B} = \left\{ \begin{array}{l} \{1,2,4\}, \{2,3,5\}, \{3,4,6\}, \\ \{4,5,7\}, \{1,5,6\}, \{2,6,7\}, \\ \{1,3,7\} \end{array} \right\}$$

An example of an STS(9) is

$$\mathcal{V} = \{1, 2, 3, 4, 5, 6, 7, 8, 9\}$$

$$\mathcal{B} = \left\{ \begin{array}{l} \{1,2,3\}, \{1,4,7\}, \{1,5,9\}, \\ \{1,6,8\}, \{4,5,6\}, \{2,5,8\}, \\ \{2,6,7\}, \{2,4,9\}, \{7,8,9\}, \\ \{3,6,9\}, \{3,4,8\}, \{3,5,7\} \end{array} \right\}$$

Here are two fundamental properties of Steiner triple systems, which can be proved by elementary counting.

LEMMA 5.1   *Let* $(\mathcal{V}, \mathcal{B})$ *be an* STS(v). *Then every point in* $\mathcal{V}$ *occurs in exactly* $r = (v - 1)/2$ *blocks, and* $|\mathcal{B}| = v(v - 1)/6$.

PROOF   To see that $r = (v-1)/2$, it suffices to observe that a point x must occur with each of the other $v - 1$ points in a block, and x occurs with two other points in each of the r blocks in which it occurs.

Let $b = |\mathcal{B}|$. To see that $b = v(v - 1)/6$, observe that $3b = rv$, since each block contains three points and each of the v points occurs in r blocks. The result follows. ∎

Clearly, the numbers r and b defined above must both be integers if an STS(v) is to exist. From this it follows that $v \equiv 1$ or 3 mod 6 is a necessary condition for the existence of an STS(v). It was in fact proved over 100 years ago that an STS(v) exists if and only if $v \equiv 1$ or $3 \bmod 6$. Thus the simple necessary condition for existence turns out to be sufficient. The proof of sufficiency is constructive, and leads to an efficient method of constructing (at least) one STS(v) of every admissible order v.

It is also known that the number of non-isomorphic STS(v) on a specified point set $\mathcal{V}$ grows exponentially quickly. One nice feature about the hill-climbing algorithm we are going to describe is that it is very fast, and it is an effective method of constructing apparently random STS(v).

In order to use a hill-climbing approach, we formulate the problem as an optimization problem. We first need a definition. A *partial Steiner triple system* is a set system $(\mathcal{V}, \mathcal{B})$ in which every block has size three, and every pair of points from $\mathcal{V}$ is contained in at most one block. Such a set system is denoted PSTS(v), where $v = |\mathcal{V}|$. The *size* of a PSTS(v) is the number of blocks it contains. It is easy to see that any PSTS(v) has size at most $v(v-1)/6$; and a PSTS(v) of size $v(v-1)/6$ is in fact an STS(v).

We now present the problem of constructing an STS(v) in the form of a combinatorial optimization problem.

Problem 5.3:   Construct Steiner Triple System

Instance: *a positive integer* $v \equiv 1$ or $3 \bmod 6$;

                    a finite set $V, |\mathcal{V}| = v$;

Find: the maximum value of $|\mathcal{B}|$

                    subject to $(\mathcal{V}, \mathcal{B})$ is a PSTS(v).

Given v and $\mathcal{V}$, we will define our universe, $\mathcal{X}$, to consist of all sets of blocks $\mathcal{B}$ such that $(\mathcal{V}, \mathcal{B})$ is a PSTS(v). Therefore any set $\mathcal{B} \in \mathcal{X}$ is a feasible solution. An optimal solution is any feasible solution having size $v(v-1)/6$.

Instead of explicitly defining a neighborhood function, we instead proceed directly to the description of the heuristic, SWITCH, to be used in the hill-climbing algorithm. The heuristic SWITCH will transform any PSTS(v) into a different PSTS(v), such that the size either remains the same or is increased by one. This is done by a randomized search strategy, which we describe now.

Let $(\mathcal{V}, \mathcal{B})$ be any PSTS(v). A point $x \in \mathcal{V}$ is called a *live point* if $r_x < (v-1)/2$, where r_x is the number of blocks in $\mathcal{B}$ that contain the point x. A pair of distinct points, $\{x, y\}$, is called a *live pair* if there is no block $B \in \mathcal{B}$ such that $\{x, y\} \subseteq B$.

Now, if $(\mathcal{V}, \mathcal{B})$ has size less than $v(v-1)/6$, then there must exist a live point, say x. If x is a live point, then there must exist at least two points $y, z \in \mathcal{V}$ ($y \neq z$), such that the pairs $\{x, y\}$ and $\{x, z\}$ are both live pairs. (This is because $r_x \leq (v-3)/2$, and hence x has occurred in a block with at most $v-3$ other points.)

Here is the description of the heuristic SWITCH .

Algorithm 5.10: SWITCH ()

global *NumBlocks*

let x be any live point
let y, z be points such that $\{x, y\}$ and $\{x, z\}$ are live pairs
if $\{y, z\}$ is a live pair

then $\begin{cases} \mathcal{B} \leftarrow \mathcal{B} \cup \{\{x, y, z\}\} \\ NumBlocks \leftarrow NumBlocks + 1 \end{cases}$

else $\begin{cases} \text{let } \{w, y, z\} \in \mathcal{B} \text{ be the block containing the pair } \{y, z\} \\ \mathcal{B} \leftarrow \mathcal{B} \cup \{\{x, y, z\}\} \setminus \{\{w, y, z\}\} \end{cases}$

The heuristic SWITCH constructs a candidate block, $\{x, y, z\}$, such that either two or three of the pairs contained in it are live pairs. If all three pairs are live, then we can simply add the new block $\{x, y, z\}$ to the system, increasing the size by one. If only two of the three pairs are live, then we add the new block $\{x, y, z\}$, and remove another block $\{w, y, z\}$ (the unique block containing the pair $\{y, z\}$), so the size stays the same.

Here now is the hill-climbing algorithm, which keeps applying the heuristic SWITCH until a Steiner triple system is finally constructed. The variable *NumBlocks* records the size of the PSTS(v) during the course of the algorithm.

Algorithm 5.11: STINSON'S ALGORITHM (v)

global *NumBlocks*

$NumBlocks \leftarrow 0$
$\mathcal{V} \leftarrow \{1, \ldots, v\}$
$\mathcal{B} \leftarrow \emptyset$
while $NumBlocks < v(v - 1)/6$
  do SWITCH
output $(\mathcal{V}, \mathcal{B})$

There is of course no guarantee that the algorithm will ever terminate. But if the choices made by the heuristic SWITCH are random, it seems in practice that the algorithm always terminates successfully by constructing an STS(v), and it usually runs very quickly.

We illustrate the execution of the algorithm by constructing an STS(9), which has 12 blocks. This is an actual run performed by a computer program with randomly generated choices made in the heuristic SWITCH .

x	y	z	w	$NumBlocks$
2	1	9		1
9	4	7		2
6	7	8		3
1	7	3		4
3	9	6		5
7	5	2		6
6	1	2	9	
4	8	5		7
2	3	4		8
8	1	9		9
6	5	4	8	
9	5	2	7	
7	2	5	9	
9	5	2	7	
5	7	3	1	
3	8	1	9	
9	8	1	3	
2	8	7	6	
6	8	7	2	
1	7	3	5	
7	5	2	9	
3	8	5		10
5	9	1	8	
2	9	8		11
8	1	4		12

The blocks of the STS(9) are as follows:

$$
\mathcal{B} = \left\{
\begin{array}{l}
\{1,2,6\},\ \{1,3,7\},\ \{1,4,8\}, \\
\{1,5,9\},\ \{2,3,4\},\ \{2,5,7\}, \\
\{2,8,9\},\ \{3,5,8\},\ \{3,6,9\}, \\
\{4,5,6\},\ \{4,7,9\},\ \{6,7,8\}
\end{array}
\right\}
$$

5.4.1   Implementation details

When we implement the algorithm, it is advantageous to maintain a table (or array) of all the live points. This table does not need to be ordered. When a point ceases to live, the last point in the table can be moved to occupy its place. If a *dead point* becomes live, it is added to the end of the table. In order to make these updating operations efficient (and to eliminate the need for linear searches of this table), we also maintain an indexing array, which keeps track of the position of every element in the first table.

In a similar fashion, we will maintain for each live point x a table of all the points y such that $\{x, y\}$ is a live pair. Also, for each such table, we have an indexing array.

Thus we will have two arrays of length v, which we call *LivePoints* and *IndexLivePoints*, and a variable *NumLivePoints*. We have three further arrays which we call *LivePairs*, *IndexLivePairs* and *NumLivePairs*. Each element of *LivePairs* and *IndexLivePairs* is an array of length v.

We need one more array, which we name *Other*. For each pair of distinct points $\{x, y\}$, this array keeps track of the "other point" in a block containing x and y. More formally, for any PSTS(v), say $(\mathcal{V}, \mathcal{B})$, and for any $x, y \in \mathcal{V}$ ($x \neq y$), we define $Other[x, y] = z$ if and only if $\{x, y, z\} \in \mathcal{B}$ (and $Other[x, y]$ is undefined if x, y is a live pair).

With the use of the array *Other*, it is unnecessary to explicitly keep track of the block set $\mathcal{B}$ during the course of the algorithm. At the end of the algorithm, it is straightforward to computer $\mathcal{B}$ from *Other*. This is done by the procedure CONSTRUCTBLOCKS, as shown in Algorithm 5.12.

Algorithm 5.12: CONSTRUCTBLOCKS $(v, Other)$

$\mathcal{B} \leftarrow \emptyset$
for $x \leftarrow 1$ **to** v
$\quad$**do** $\begin{cases} \textbf{for } y \leftarrow x + 1 \textbf{ to } v \\ \quad \textbf{do } \begin{cases} z \leftarrow Other[x, y] \\ \textbf{if } z > y \\ \quad \textbf{then } \mathcal{B} \leftarrow \mathcal{B} \cup \{\{x, y, z\}\} \end{cases} \end{cases}$
return $\mathcal{B}$

An initialization of the arrays is performed at the beginning of the hill-climbing algorithm, as follows.

Algorithm 5.13: INITIALIZE (v)

global $\begin{cases} NumLivePoints, \\ LivePoints[x], IndexLivePoints[x], x = 1, 2, ldots, v \\ NumLivePairs[x], x = 1, 2, \ldots, v \\ LivePairs[x, y], Other[x, y], x = 1, 2, \ldots, v, y = 1, 2, \ldots, v \end{cases}$

$NumLivePoints \leftarrow v$
for $x \leftarrow 1$ **to** v
$\quad$**do** $\begin{cases} LivePoints[x] \leftarrow x \\ IndexLivePoints[x] \leftarrow x \\ NumLivePairs[x] \leftarrow v - 1 \\ \textbf{for } y \leftarrow 1 \textbf{ to } v - 1 \\ \quad \textbf{do } LivePairs[x, y] \leftarrow (y + x - 1) \pmod{v} + 1 \\ \textbf{for } y \leftarrow 1 \textbf{ to } v \\ \quad \textbf{do } \begin{cases} IndexLivePairs[x, y] \leftarrow (y - x) \pmod{v} \\ Other[x, y] \leftarrow 0 \end{cases} \end{cases}$

It will be necessary to perform "insert" and "delete" operations on these arrays. Thus we define procedures INSERTPAIR and DELETEPAIR, as follows.

Algorithm 5.14: INSERTPAIR (x, y)

$$\textbf{global} \begin{cases} NumLivePoints, \\ LivePoints[x], IndexLivePoints[x], x = 1, 2, \dots, v \\ NumLivePairs[x], x = 1, 2, \dots, v \\ LivePairs[x, y], x = 1, 2, \dots, v, y = 1, 2, \dots, v \end{cases}$$

$\textbf{if } NumLivePairs[x] = 0$

$\quad \textbf{then } \begin{cases} NumLivePoints \leftarrow NumLivePoints + 1 \\ LivePoints[NumLivePoints] \leftarrow x \\ IndexLivePoints[x] \leftarrow NumLivePoints \end{cases}$

$NumLivePairs[x] \leftarrow NumLivePairs[x] + 1$

$posn \leftarrow NumLivePairs[x]$

$LivePairs[x, posn] \leftarrow y$

$IndexLivePairs[x, y] \leftarrow posn$

Algorithm 5.15: DELETEPAIR (x, y)

$$\textbf{global} \begin{cases} NumLivePoints, \\ LivePoints[x], IndexLivePoints[x], x = 1, 2, \dots, v \\ NumLivePairs[x], x = 1, 2, \dots, v \\ LivePairs[x, y], x = 1, 2, \dots, v, y = 1, 2, \dots, v \end{cases}$$

$posn \leftarrow IndexLivePairs[x, y]$

$num \leftarrow NumLivePairs[x]$

$z \leftarrow LivePairs[x, num]$

$LivePairs[x, posn] \leftarrow z$

$IndexLivePairs[x, z] \leftarrow posn$

$LivePairs[x, num] \leftarrow 0$

$IndexLivePairs[x, y] \leftarrow 0$

$NumLivePairs[x] \leftarrow NumLivePairs[x] - 1$

$\textbf{if } NumLivePairs[x] = 0$

$\quad \textbf{then } \begin{cases} posn \leftarrow IndexLivePoints[x] \\ z \leftarrow LivePoints[NumLivePoints] \\ LivePoints[posn] \leftarrow z \\ IndexLivePoints[z] \leftarrow posn \\ LivePoints[NumLivePoints] \leftarrow 0 \\ IndexLivePoints[x] \leftarrow 0 \\ NumLivePoints \leftarrow NumLivePoints - 1 \end{cases}$

These two procedures are used in two higher-level procedures called ADDBLOCK and EXCHANGEBLOCK.

Algorithm 5.16: ADDBLOCK (x, y, z)

external DELETEPAIR()
global $Other[x, y], x = 1, 2, \ldots, v, y = 1, 2, \ldots, v$
$Other[x, y] \leftarrow z$
$Other[y, x] \leftarrow z$
$Other[x, z] \leftarrow y$
$Other[z, x] \leftarrow y$
$Other[y, z] \leftarrow x$
$Other[z, y] \leftarrow x$
DELETEPAIR(x, y)
DELETEPAIR(y, x)
DELETEPAIR(x, z)
DELETEPAIR(z, x)
DELETEPAIR(y, z)
DELETEPAIR(z, y)

Algorithm 5.17: EXCHANGEBLOCK (x, y, z, w)

external DELETEPAIR(), INSERTPAIR()
global $Other[x, y], x = 1, 2, \ldots, v, y = 1, 2, \ldots, v$
$Other[x, y] \leftarrow z$
$Other[y, x] \leftarrow z$
$Other[x, z] \leftarrow y$
$Other[z, x] \leftarrow y$
$Other[y, z] \leftarrow x$
$Other[z, y] \leftarrow x$
$Other[w, y] \leftarrow 0$
$Other[y, w] \leftarrow 0$
$Other[w, z] \leftarrow 0$
$Other[z, w] \leftarrow 0$
INSERTPAIR(w, y)
INSERTPAIR(y, w)
INSERTPAIR(w, z)
INSERTPAIR(z, w)
DELETEPAIR(x, y)
DELETEPAIR(y, x)
DELETEPAIR(x, z)
DELETEPAIR(z, x)

Now, we present a more detailed version of the heuristic SWITCH , in which we include the necessary updating for these arrays.

Algorithm 5.18: REVISEDSWITCH ( )

external ADDBLOCK(), EXCHANGEBLOCK()

global $\begin{cases} NumLivePoints, \\ LivePoints[x], NumLivePairs[x], x = 1, 2, \ldots, v \\ LivePairs[x, y], Other[x, y], x = 1, 2, \ldots, v, y = 1, 2, \ldots, v \end{cases}$

let r be a random integer, $1 \le r \le NumLivePoints$
$x \leftarrow LivePoints[r]$
let s, t be random integers, $1 \le s < t \le NumLivePairs[x]$
$y \leftarrow LivePairs[x, s]$
$z \leftarrow LivePairs[x, t]$
if $Other[y, z] = 0$
$\quad$ **then** $\begin{cases} \text{ADDBLOCK}(x, y, z) \\ NumBlocks \leftarrow NumBlocks + 1 \end{cases}$
$\quad$ **else** $\begin{cases} w \leftarrow Other[y, z] \\ \text{EXCHANGEBLOCK}(x, y, z, w) \end{cases}$

Here now is the final version of Stinson's algorithm.

Algorithm 5.19: REVISEDSTINSON'SALGORITHM (v)

external CONSTRUCTBLOCKS(), REVISEDSWITCH()
global $NumBlocks, Other[x, y], x = 1, 2, \ldots, v, y = 1, 2, \ldots, v$
$NumBlocks \leftarrow 0$
INITIALIZE(v)
while $NumBlocks < v(v - 1)/6$
$\quad$ **do** REVISEDSWITCH()
$\mathcal{B} \leftarrow$ CONSTRUCTBLOCKS$(v, Other)$
output $(\mathcal{V}, \mathcal{B})$

5.4.2   Computational results

For each $v \in \{31, 61, \ldots, 301\}$, we performed ten runs of Algorithm 5.19. In each run of the algorithm, we computed the number of iterations required. This information is summarized below.

v	b	avg. # iterations	avg. # iterations $\frac{\text{}}{b}$	avg. # iterations $\frac{\text{}}{b \ln b}$
31	155	551.8	3.56	.706
61	610	2276.9	3.73	.582
91	1365	5396.6	3.95	.547
121	2420	9882.0	4.08	.523
151	3775	15594.8	4.13	.501
181	5430	23755.5	4.37	.508
211	7385	32286.6	4.37	.491
241	9640	42702.5	4.43	.483
271	12195	55071.5	4.52	.480
301	15050	68398.9	4.54	.472

From this data, it appears that the average number of iterations per block constructed grows very slowly as v increases. In fact, this data is consistent with the hypothesis that the average number of iterations is $O(b \ln b)$.

5.5 Two heuristic algorithms for the knapsack problem

In this section, we develop simulated annealing and tabu search algorithms for the Knapsack (optimization) problem. Recall that an instance is defined by profits $p_0, \ldots, p_{n-1}$, weights $w_0, \ldots, w_{n-1}$, and a capacity, M. The universe is $\mathcal{X} = \{0, 1\}^n$. An n-tuple $X = [x_0, \ldots, x_{n-1}] \in \mathcal{X}$ is feasible if

$$w(X) = \sum_{i=0}^{n-1} x_i w_i \leq M,$$

and the objective is to maximize

$$\mathsf{P}(X) = \sum_{i=0}^{n-1} x_i p_i.$$

5.5.1 A simulated annealing algorithm

Suppose we define our neighborhood function to be

$$\mathsf{N}(X) = \mathsf{N}_1(X) = \{Y \in \{0, 1\}^n : \text{dist}(X, Y) = 1\}.$$

That is, the neighborhood of X consists of all binary n-tuples in which exactly one entry of X has been changed.

We can generate a random $Y = [y_0, \ldots, y_{n-1}] \in \mathsf{N}(X)$ by choosing a random integer j such that $0 \leq j \leq n - 1$, and then defining

$$y_i = \begin{cases} x_i & \text{if } i \neq j \\ 1 - x_i & \text{if } i = j. \end{cases}$$

Clearly we have that

$$w(Y) = \begin{cases} w(X) + w_j & \text{if } x_j = 0 \\ w(X) - w_j & \text{if } x_j = 1. \end{cases}$$

A similar formula relates $\mathsf{P}(Y)$ to $\mathsf{P}(X)$.

Suppose that X is feasible. Then we see that Y is feasible whenever $x_j = 1$, or whenever $x_j = 0$ and $w(X) + w_j \leq M$. The heuristic h_N fails if $x_j = 0$ and $w(X) + w_j > M$.

Now suppose that $Y = h_\mathsf{N}(X)$ is feasible. If $x_j = 0$ (so $y_j = 1$), then $\mathsf{P}(Y) > \mathsf{P}(X)$ and X will always be replaced by Y in a simulated annealing algorithm. If $x_j = 1$ (so $y_j = 0$), then $\mathsf{P}(Y) < \mathsf{P}(X)$. In this case, $\mathsf{P}(Y) - \mathsf{P}(X) = -p_j$; so, X will be replaced by Y with probability $c^{-p_j/T}$.

Finally, it is sufficient to begin with the trivial feasible solution $[0, \ldots, 0]$. With these observations, the simulated annealing algorithm for **Knapsack** problem is easily adapted from our generic simulated annealing algorithm, Algorithm 5.3.

Given a problem instance, it remains to determine suitable values for T_0, c_{max} and α. This is basically a matter of experimentation. We look at a specific (small) problem instance to illustrate.

Suppose we have $n = 15$, and profits, weights and capacity as specified in Table 5.2.

An exhaustive search, such as a backtracking algorithm, can be used to find the optimal solution, which is

$$X = [1, 0, 1, 0, 1, 0, 1, 1, 1, 0, 0, 0, 0, 1, 1],$$

yielding an optimal profit of 1458. We will investigate how close to this optimal solution the simulated annealing algorithm will get with various choices of the parameters.

TABLE 5.2
An instance of the Knapsack problem with 15 items.

profits	135	139	149	150	156	163	173	184
	192	201	210	214	221	229	240	
weights	70	73	77	80	82	87	90	94
	98	106	110	113	115	118	120	
capacity	750							

Algorithm 5.20: KNAPSACKSIMULATEDANNEALING (c_{max}, T_0, α)

external $P(), Random()$

$c \leftarrow 0$
$T \leftarrow T_0$
$X \leftarrow [x_0, \ldots, x_{n-1}] = [0, \ldots, 0]$
$CurW \leftarrow 0$
$X_{best} \leftarrow X$
while $c \leq c_{max}$

$\textbf{do} \begin{cases} \text{let } j \text{ be a random integer between 0 and } n-1 \\ Y \leftarrow X \\ y_j \leftarrow 1 - x_j \\ \textbf{if } (y_j = 1) \textbf{ and } (CurW + w_j > M) \\ \quad \textbf{then } Y \leftarrow Fail \\ \textbf{if } Y \neq Fail \\ \quad \textbf{then} \begin{cases} \textbf{if } y_j = 1 \\ \quad \textbf{then} \begin{cases} X \leftarrow Y \\ CurW \leftarrow CurW + w_j \\ \textbf{if } P(X) > P(X_{best}) \\ \quad \textbf{then } X_{best} \leftarrow X \end{cases} \\ \quad \textbf{else} \begin{cases} r \leftarrow Random(0,1) \\ \textbf{if } r < e^{-p_j/T} \\ \quad \textbf{then} \begin{cases} X \leftarrow Y \\ CurW \leftarrow CurW - w_j \end{cases} \end{cases} \end{cases} \\ c \leftarrow c+1 \\ T \leftarrow \alpha T \end{cases}$

return (X_{best})

It is generally a good idea to begin with an initial temperature T_0 which allows downward moves with a fairly high probability. We chose $T_0 = 1000$, so the probability of accepting a downward move at the beginning of the algorithm is at least $.787 = e^{-240/T_0}$, and at most $.874 = e^{-135/T_0}$. A Slow cooling schedule seemed to be most effective in this algorithm, and we chose $\alpha = 0.999, 0.9995$ and 0.9999 to test the algorithm. We also varied the number of iterations, letting

TABLE 5.3
Summary data for the knapsack simulated annealing algorithm.

α	c_{max}	profits found		
		minimum	maximum	average
0.999	1000	1441	1454	1446.8
0.999	5000	1448	1456	1452.1
0.999	20000	1448	1456	1450.9
0.9995	1000	1445	1455	1448.4
0.9995	5000	1450	1458	1454.6
0.9995	20000	1452	1458	1453.9
0.9999	1000	1445	1455	1449.6
0.9999	5000	1450	1458	1454.3
0.9999	20000	1453	1458	1456.1

$c_{max} = 1000, 5000$ and 20000. For each choice of parameters, we performed 10 runs of the algorithm, and the best profits found are summarized in Table 5.3.

There were no optimal solutions found when $\alpha = 0.999$, three optimal solutions were found (out of 30 runs) when $\alpha = 0.9995$, and five optimal solutions were found (out of 30 runs) when $\alpha = 0.9999$.

5.5.2   A tabu search algorithm

In developing a tabu search algorithm for the Knapsack problem, the main issue is the design of the heuristic. We will use the same neighborhood as we did in the simulated annealing algorithm. Thus, a "change" consists of choosing an index i, and replacing x_i by $1 - x_i$. These values of i will be stored in the *TabuList*.

The tabu search algorithm will use an exhaustive search of the neighborhood to find the "best" way to update a feasible solution X, as opposed to the randomized search strategy employed in the simulated annealing algorithm. This could be done by attempting to change a coordinate x_i from 0 to 1 in such a way that the profit is improved as much as possible, as suggested in the generic tabu search algorithm (Algorithm 5.4). However, it turns out that it is better to look at the profit/weight ratios of the items (as was done in Section 4.6.1), rather than just their profits. Thus, the neighborhood search strategy can be described informally as follows:

1. Suppose there exists at least one index i where $x_i = 0$, such that i is not on the current *TabuList* and such that x_i can be changed to 1 without exceeding the capacity M. Among all such i, choose the one such that p_i/w_i is maximum, and change x_i from 0 to 1.

2. Suppose there exists no i satisfying the conditions above. Then consider all i such that $x_i = 1$ and i is not on the current *TabuList*. Among these values of i, choose the one such that p_i/w_i is minimum, and change x_i from 1 to 0.

The idea is that we want to add items having high profit/weight ratios to the knapsack, and delete items having low profit/weight ratios. Note that we can express the above search procedure more succinctly by saying that we wish to maximize the quantity $(-1)^{x_i} p_i / w_i$ over all indices i not on the *TabuList*.

The heuristic described above is deterministic. A convenient way to introduce randomness into the tabu search algorithm is to begin with a random initial feasible solution. For example, we could select a random sequence of values of i, setting x_i to be 1 as long as the total weight of the knapsack does not exceed the capacity. This randomization allows us to run the algorithm several times on a given problem instance, which should increase the probability of finding an optimal or near-optimal solution.

There are two other decisions to be made. One is how long the lifetime L should be. Of course, it is a simple matter to run the algorithm with various values of L and determine the best choice by experimentation. The other issue is how many iterations we should allow the algorithm to run. Many tabu search algorithms turn out to be rather insensitive to the number of iterations of the heuristic that are performed, i.e., the best feasible solutions tend to be found very early in the search. Thus we can take c_{max} to be quite small (much smaller than in a simulated annealing algorithm), and hence perform a larger number of trials of the algorithm in a given amount of time. We found $c_{max} = 200$ to be sufficient for the problem instances we considered. The resulting algorithm is presented as Algorithm 5.21.

We first tried our algorithm on the instance with 15 items that we presented in Table 5.2. We tested values of L ranging from 1 to 8. The data obtained from 25 runs of the tabu search algorithm for each value of L are presented in Table 5.4.

It can be seen that, for this problem instance, a short lifetime is sufficient to allow the optimal solution to be found very easily. In fact, as L increases, the performance of the algorithm gradually degrades.

We also ran the algorithm on the larger instance with $n = 24$ items that is presented in Table 5.5.

This problem instance has optimal solution

$$X = [1, 1, 0, 1, 1, 1, 0, 0, 0, 1, 1, 0, 1, 0, 0, 1, 0, 0, 0, 0, 0, 1, 1, 1],$$

yielding a profit of 13549094. The results of the tabu search algorithm, when run on this problem instance, are presented in Table 5.6.

In this larger problem instance, the algorithm performs better as L is increased. Among the values of L that were tested, $L = 8$ appears to be the best choice.

TABLE 5.4
Summary data for the knapsack tabu search algorithm (15 items).

L	profits found (in 25 runs)			# optimal solutions found
	minimum	maximum	average	
1	1458	1458	1458.0	25
2	1458	1458	1458.0	25
3	1452	1458	1456.8	16
4	1448	1458	1455.6	14
5	1452	1458	1456.6	16
6	1446	1458	1455.1	11
7	1446	1458	1455.1	12
8	1444	1458	1452.1	5

Algorithm 5.21: KNAPSACKTABUSEARCH (c_{max}, L)

external $P()$

$c \leftarrow 1$
Select a random feasible solution $X = [x_0, \ldots, x_{n-1}] \in \{0, 1\}^n$
$CurW \leftarrow \sum x_i w_i$
$X_{best} \leftarrow X$
while $c \leq c_{max}$

$\mathbf{do} \begin{cases} N \leftarrow \{0, \ldots, n-1\} \\ start \leftarrow \max\{0, c - L\} \\ \mathbf{for}\ j \leftarrow start\ \mathbf{to}\ c - 1 \\ \quad \mathbf{do}\ N \leftarrow N \setminus \{TabuList[j]\} \\ \mathbf{for\ each}\ i \in N \\ \quad \mathbf{do} \begin{cases} \mathbf{if}\ (x_i = 0)\ \mathbf{and}\ (CurW + w_i > M) \\ \quad \mathbf{then}\ N \leftarrow N \setminus \{i\} \end{cases} \\ \mathbf{if}\ N = \emptyset \\ \quad \mathbf{then\ exit} \\ \text{find}\ i \in N\ \text{such that}\ (-1)^{x_i} p_i / w_i\ \text{is maximum} \\ TabuList[c] \leftarrow i \\ x_i \leftarrow 1 - x_i \\ \mathbf{if}\ x_i = 1 \\ \quad \mathbf{then}\ CurW \leftarrow CurW + w_i \\ \quad \mathbf{else}\ CurW \leftarrow CurW - w_i \\ \mathbf{if}\ P(X) > P(X_{best}) \\ \quad \mathbf{then}\ X_{best} \leftarrow X \\ c \leftarrow c + 1 \end{cases}$

return (X_{best})

TABLE 5.5
An instance of the Knapsack problem with 24 items.

profits	825594	1677009	1676628	1523970	943972	97426
	69666	1296457	1678693	1902996	1844992	1049289
	1252836	1319836	953277	2067538	675367	853655
	1826027	65731	901489	577243	466257	369261
weights	382745	799601	909247	729069	467902	44328
	34610	698150	823460	903959	853665	551830
	610856	670702	488960	951111	323046	446298
	931161	31385	496951	264724	224916	169684
capacity	6404180					

TABLE 5.6
Summary data for the knapsack tabu search algorithm (24 items).

L	profits found (in 25 runs)			# optimal solutions found
	minimum	maximum	average	
1	13079298	13466838	13388643.5	0
2	13084476	13500943	13415747.5	0
3	13245597	13500943	13456205.2	0
4	13264009	13500943	13446933.8	0
5	13358351	13500943	13458145.8	0
6	13148978	13549094	13427333.6	1
7	13116665	13549094	13462902.4	4
8	13346220	13549094	13497932.2	7

5.6 A genetic algorithm for the traveling salesman problem

In this section, we develop a genetic algorithm for the Traveling Salesman
problem, Problem 4.4. Recall that an instance is given by an n by n cost matrix M. A feasible solution is a Hamiltonian circuit

$$x_0 x_1 x_2 \cdots x_{n-1} x_0$$

in the underlying graph K_n, and the objective is to minimize

$$C(X) = \sum_{i=0}^{n-2} M[x_i, x_{i+1}] + M[x_{n-1}, x_0].$$

We represent the circuit $x_0 x_1 x_2 \cdots x_{n-1} x_0$ as a permutation

$$X = [x_0, x_1, \ldots, x_{n-1}]$$

of $\{0, \ldots, n-1\}$. There are $2n$ permutations that represent a given Hamiltonian
circuit. Since we are assuming that the underlying graph is complete, every per-

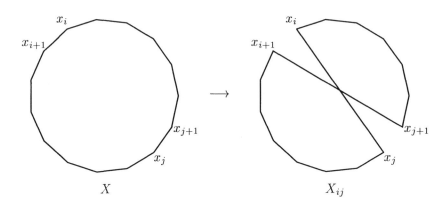

FIGURE 5.2
Illustration of a 2-opt move.

mutation of the n vertices represents a feasible solution, and so the universe $\mathcal{X}$ consists of all $n!$ permutations of $\{0, \ldots, n-1\}$.

In order to design a genetic algorithm, we need recombination and mutation operations, and a method to select an initial population. We first describe a mutation operation, which will be a heuristic consisting of a sequence of steepest ascent neighborhood searches.

Given a Hamiltonian circuit X, we can "cut" two edges, say $\{x_i, x_{i+1}\}$ and $\{x_j, x_{j+1}\}$, and then re-attach the ends of the two resulting paths to create a new Hamiltonian circuit, which we denote X_{ij}. More precisely, given $X = [x_0, x_1, \ldots, x_{n-1}] \in \mathcal{X}$ and indices $0 \le i < j \le n - 1$ such that the distance from x_i to x_j around the circuit X is at least two, then we construct the Hamiltonian circuit

$$X_{ij} = [x_0, x_1, \ldots, x_i, x_j, x_{j-1}, \ldots, x_{i+1}, x_{j+1}, x_{j+2}, \ldots, x_{n-1}].$$

This operation is called a *2-opt move* and is illustrated in Figure 5.2.

The *gain* in applying a 2-opt move is the decrease in cost. This is denoted by

$$\mathsf{G}(X, i, j) = \mathsf{C}(X) - \mathsf{C}(X_{ij})$$
$$= M[x_i, x_{i+1}] + M[x_j, x_{j+1}] - M[x_{i+1}, x_{j+1}] - M[x_i, x_j].$$

The gain can be negative, zero, or positive.

We define our neighborhood function $\mathsf{N}(X)$ to be the set of all $Y \in \mathcal{X}$ that can be obtained from X by a 2-opt move. Our mutation heuristic consists of a sequence of 2-opt moves. Using a steepest ascent strategy, we iteratively apply 2-opt moves until no pair of edges can be found that yield a positive gain. The mutation heuristic is presented as Algorithm 5.22.

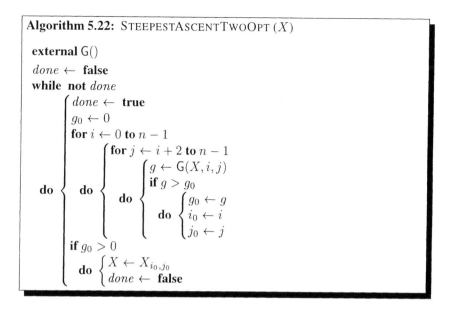

Algorithm 5.22: STEEPESTASCENTTWOOPT (X)

external $G()$
$done \leftarrow$ **false**
while **not** $done$
$\mathbf{do} \begin{cases} done \leftarrow \textbf{true} \\ g_0 \leftarrow 0 \\ \textbf{for } i \leftarrow 0 \textbf{ to } n-1 \\ \quad \mathbf{do} \begin{cases} \textbf{for } j \leftarrow i+2 \textbf{ to } n-1 \\ \quad \mathbf{do} \begin{cases} g \leftarrow G(X,i,j) \\ \textbf{if } g > g_0 \\ \quad \mathbf{do} \begin{cases} g_0 \leftarrow g \\ i_0 \leftarrow i \\ j_0 \leftarrow j \end{cases} \end{cases} \end{cases} \\ \textbf{if } g_0 > 0 \\ \quad \mathbf{do} \begin{cases} X \leftarrow X_{i_0,j_0} \\ done \leftarrow \textbf{false} \end{cases} \end{cases}$

Algorithm 5.23 selects the initial population

$$\mathcal{P} = [P_0, P_1, \ldots, P_{popsize-1}]$$

by first randomly generating *popsize* permutations, each of which is improved to a local minimum by applying Algorithm 5.22.

Algorithm 5.23: SELECT $(popsize)$

external $\begin{cases} \text{RandomInteger}(), \\ \text{PERMLEXUNRANK}(), \\ \text{STEEPESTASCENTTWOOPT}() \end{cases}$
for $i \leftarrow 0$ **to** $popsize - 1$
$\mathbf{do} \begin{cases} r = \text{RandomInteger}(0, n! - 1) \\ P_i \leftarrow \text{PERMLEXUNRANK}(n, r) \\ \text{STEEPESTASCENTTWOOPT}(P_i) \end{cases}$
return $([P_0, P_1, \ldots, P_{popsize-1}])$

Next, we present two recombination operations, Algorithm 5.24 and Algorithm 5.25. For both of them, a random length h is chosen, and then a random substring

$$S = [S_0, S_1, \ldots, S_{h-1}]$$

of length h is chosen from one of the parents by selecting a starting location j. The string length h should not be too "short" nor too "long". If it is too short then the recombination has little effect, and if it is too long, then the variation gets lost. We typically choose h so that $10 \leq h \leq n/2$.

In Algorithm 5.24, we apply Algorithm 5.5 with parameters j, and $k = h + j \bmod n$. Then each of the two resulting children are improved to local minima by applying Algorithm 5.22.

In Algorithm 5.25, the string S is first copied over to the beginning of a new child. The child is completed to a feasible solution by appending the nodes of the other parent that are not in S in the order in which they appear. The resulting child is then improved to a local minimum by applying Algorithm 5.22. This is then repeated, with the roles of the two parents reversed, to generate a second child.

In Tables 5.7 and 5.8, we compare these two recombination operations when used in a genetic algorithm. The genetic algorithm is given as Algorithm 5.26, where REC is either of the two recombination operations MKSREC or PMREC. The tables summarize the output of Algorithm 5.26 when it is run on three random instances on 50 vertices. These instances were generated by randomly choosing integer values in the interval $[0, 99]$ for the entries of a symmetric cost matrix M. The optimal cost of a Hamiltonian circuit was computed by an exhaustive branch and bound algorithm, as described in Section 4.7. We chose *popsize* $= 8, 16$ and 32, and $c_{max} = 50, 100$ and 200. For each choice of parameters, we performed 10 runs of the algorithm, and the best costs obtained are summarized in Tables 5.7 and 5.8. The last column of these tables gives the number of times a Hamiltonian circuit achieving the optimal cost was found. Tables 5.7 and 5.8 suggest that the recombination operation MGKREC performs better than PMREC .

Algorithm 5.24: PMREC (A, B)

$$\textbf{external} \begin{cases} \text{RandomInteger}(), \\ \text{PARTIALLYMATCHEDCROSSOVER}(), \\ \text{STEEPESTASCENTTWOOPT}() \end{cases}$$

$h \leftarrow \text{RandomInteger}(10, \frac{n}{2})$
$j \leftarrow \text{RandomInteger}(0, n - 1)$
$(C, D) \leftarrow \text{PARTIALLYMATCHEDCROSSOVER}(A, B, j, (h + j) \bmod n)$
$\text{STEEPESTASCENTTWOOPT}(C)$
$\text{STEEPESTASCENTTWOOPT}(D)$
return (C, D)

Algorithm 5.25: MGKREC (A, B)

external RandomInteger(), STEEPESTASCENTTWOOPT()
$h \leftarrow$ RandomInteger$(10, \frac{n}{2})$
$j \leftarrow$ RandomInteger$(0, n - 1)$
$T \leftarrow \emptyset$
for $i \leftarrow 0$ **to** $h - 1$ **do** $\begin{cases} D[i] \leftarrow B[(i + j) \bmod n] \\ T \leftarrow T \cup \{D[i]\} \end{cases}$
for $j \leftarrow 0$ **to** $n - 1$
  do if $A[j] \notin T$
  then $\begin{cases} D[i] \leftarrow A[j] \\ i \leftarrow i + 1 \end{cases}$
STEEPESTASCENTTWOOPT(D)
$j \leftarrow$ Random$(0, n - 1)$
$T \leftarrow \emptyset$
for $i \leftarrow 0$ **to** $h - 1$ **do** $\begin{cases} C[i] \leftarrow A[(i + j) \bmod n] \\ T \leftarrow T \cup \{C[i]\} \end{cases}$
for $j \leftarrow 0$ **to** $n - 1$
  do if $B[j] \notin T$
  then $\begin{cases} C[i] \leftarrow B[j] \\ i \leftarrow i + 1 \end{cases}$
STEEPESTASCENTTWOOPT(C)
return (C, D)

Algorithm 5.26: GENETICTSP $(popsize, c_{max})$

external SELECT(), REC()
$c \leftarrow 1$
$[P_0, \ldots, P_{popsize-1}] \leftarrow$ SELECT$(popsize)$
Sort $P_0, P_1, \ldots, P_{popsize}$ in increasing order of cost
$X_{best} \leftarrow P_0$
$BestCost \leftarrow$ C(P_0)
while $c \leq c_{max}$
do $\begin{cases} \textbf{for } i \leftarrow 0 \textbf{ to } popsize/2 - 1 \\ \quad \textbf{do } (P_{popsize+2i}, P_{popsize+2i+1}) \leftarrow \text{Rec}(P_{2i}, P_{2i+1}) \\ \text{Sort } P_0, P_1, \ldots, P_{2 \cdot popsize-1} \text{ in increasing order of cost} \\ CurCost \leftarrow \text{C}(P_0) \\ \textbf{if } CurCost < BestCost \\ \quad \textbf{then } \begin{cases} X_{best} \leftarrow P_0 \\ BestCost \leftarrow CurCost \end{cases} \\ c \leftarrow c + 1 \end{cases}$
return (X_{best})

TABLE 5.7
GENETICTSP data with recombination operation PMREC.

M	n	Opt. Cost	*popsize*	c_{max}	cost found			No. Opt. found
					min	max	avg	
M50a	50	185	8	50	192	214	200.50	0
				100	191	219	200.00	0
				200	190	203	196.60	0
			16	50	187	207	193.20	0
				100	187	206	193.20	0
				200	187	200	193.70	0
			32	50	189	205	194.70	0
				100	186	199	190.70	0
				200	188	200	192.40	0
M50b	50	158	8	50	163	184	175.40	0
				100	163	195	173.70	0
				200	160	191	177.30	0
			16	50	159	176	167.40	0
				100	163	184	171.50	0
				200	161	189	172.10	0
			32	50	161	173	167.60	0
				100	163	178	169.40	0
				200	159	178	166.70	0
M50c	50	155	8	50	162	181	169.40	0
				100	159	186	169.50	0
				200	159	187	169.30	0
			16	50	155	171	161.30	1
				100	155	182	166.10	1
				200	157	182	167.70	0
			32	50	155	170	161.60	1
				100	158	167	161.40	0
				200	157	180	162.50	0

5.7   Notes

Section 5.1

A few books that survey heuristic search techniques are Aarts and Lenstra [1], Rayward-Smith *et al* [87] and Reeves [89].

Section 5.1.1

The Uniform Graph Partition problem is discussed in many books, for example, [1], [87], [89], and [83].

Section 5.2.1

Hill-climbing techniques were first used in the 1950's and 1960's when edge exchange algorithms such as 2-opt were introduced for the Traveling Salesman

TABLE 5.8
GENETICTSP with recombination operation **MGKREC**.

M	n	Opt. Cost	*popsize*	c_{max}	min	max	avg	No. Opt. found
						cost found		
M50a	50	185	8	50	186	196	191.70	0
				100	186	199	190.30	0
				200	186	194	189.20	0
			16	50	186	192	189.20	0
				100	185	192	187.00	3
				200	185	192	187.60	1
			32	50	186	192	188.10	0
				100	185	190	187.30	1
				200	185	190	187.30	1
M50b	50	158	8	50	160	171	165.30	0
				100	159	166	161.60	0
				200	159	170	162.00	0
			16	50	158	164	161.20	1
				100	158	162	159.80	1
				200	159	163	160.70	0
			32	50	158	165	160.70	1
				100	159	163	160.30	0
				200	158	160	158.90	2
M50c	50	155	8	50	156	168	160.50	0
				100	155	167	160.70	2
				200	155	162	157.30	5
			16	50	155	162	157.50	2
				100	155	159	156.30	5
				200	155	159	155.70	8
			32	50	155	159	156.10	6
				100	155	158	155.40	8
				200	155	156	155.10	9

problem; see Croes [24], Lin [65] and Reiter and Sherman [91]. There have been successes in various other applications, for example graph partitioning [52] and scheduling [81]. A good introductory discussion of hill-climbing can be found in [83].

Section 5.2.2

In 1953, Metropolis *et al* [75] developed an algorithm to simulate the cooling of material in a heat bath, a process known as annealing. Three decades later, Kirkpatrick *et al* [53] and Černý [17] observed that this type of simulation could be applied to optimization problems. Interesting surveys on simulated annealing can be found in the books [89], [87] and [1]. A theoretical study can be found in the book by Van Laarhoven and Aarts [108].

Section 5.2.3

Tabu search was originally formulated by Glover [34], and a detailed account is given by Glover and Laguna in [35]. Similar ideas were also developed by Hansen [39]. See Hertz and de Werra [40] for a discussion of successful applications.

Section 5.2.4

Genetic algorithms were developed in the late 60's and early 70's by Holland and his colleagues at the University of Michigan for game playing and pattern recognition in artificial intelligence systems. The first systematic treatment is contained in Holland's book [41]. The method we discuss is called genetic local search and was first described by Mühlenbein, Gorges-Shcleuter and Krämer in [77]; see also Mühlenbein's paper in [1].

Section 5.3

Steepest ascent algorithms are also referred to as local optimization. A steepest ascent algorithm for the Uniform Graph Partition problem can be found in [83], and a variation that uses multiple exchanges can be found in [1]. Johnson *et al* [48] compare steepest ascent with simulated annealing for this problem.

The instance of the Uniform Graph Partition problem that is referred to at the end of this section can be found in the web pages at the following URL:

```
http://www.math.mtu.edu/~kreher/cages/Data.html
```

Section 5.4

The first successful application of a hill-climbing algorithm to the construction of combinatorial structures appears in Dinitz and Stinson [25]. The algorithm in this section was first presented in [102]. Gibbons presents an overview of heuristic algorithms for the construction of combinatorial designs in [33].

Heuristic algorithms for the construction of good error-correcting codes and covering codes have also been studied extensively. El Gamal *et al* [27] were the first to successfully use simulated annealing for this purpose. Honkala and Östergård [42] is a good survey on this topic.

Section 5.6

Early attempts to use genetic algorithms to solve the Traveling Salesman problem can be found in Brady [8], Mühlenbein, Gorges-Shcleuter and Krämer [76] and Jog, Suh and Gucht [47].

The PARTIALLYMATCHEDCROSSOVER algorithm, which is used in the recombination operation Algorithm 5.24, is from [87]. The crossover method used in Algorithm 5.24 is from [77]. A discussion that suggests that Algorithm 5.24 is not suitable for the Traveling Salesman problem is contained in [85] and [86].

The cost matrices M50a, M50b, M50c that are referred to in Tables 5.7 and 5.8 can be found in the web page at the following URL:

```
http://www.math.mtu.edu/~kreher/cages/Data.html
```

Exercises

5.1 Determine the sizes of the following neighborhoods in the associated optimization problems:

(a) an exchange neighborhood for the Uniform Graph Partition problem.

(b) a 2-opt neighborhood for the Traveling Salesman problem.

(c) a neighborhood $N_{d_0}(X)$ for a problem in which the universe consists of permutations of an n-set.

(d) a neighborhood $N_{d_0}(X)$ for a problem in which the universe consists of all $(0, 1)$-vectors of length n having weight w.

5.2 Find an instance of the Uniform Graph Partition problem, together with a feasible solution $[\mathcal{X}_1, \mathcal{X}_2]$ such that $[\mathcal{X}_1, \mathcal{X}_2]$ is a local optimum (with respect to the operation of exchanging one vertex of $\mathcal{X}_1$ with one vertex of $\mathcal{X}_2$), but a better feasible solution can be found by exchanging two vertices of $\mathcal{X}_1$ with two vertices of $\mathcal{X}_2$.

5.3 For sparse graphs, it has been shown that Algorithm 5.9 does not perform very well when solving the Uniform Graph Partition problem. Develop and implement

(a) a simulated annealing algorithm,

(b) a tabu search algorithm, and

(c) a genetic algorithm

for solving the Uniform Graph Partition problem, assuming that the underlying graph is a sparse graph. Compare the effectiveness of your algorithms by running them on random sparse graphs. Do they perform better than Algorithm 5.9? (Note: one convenient way to generate a random sparse graph is to use Algorithm 4.21 with a small value of δ.)

5.4 Develop a hill-climbing algorithm to construct a transversal design $TD(n)$, as defined in Section 1.2.3. (Hint: design a heuristic similar to that used in the algorithm for constructing Steiner triple systems.) Test your algorithm for various values of n.

5.5 Develop a hill-climbing algorithm to embed an $STS(w)$ in an $STS(v)$. In other words, an $STS(w)$ is given, say $(\mathcal{U}, \mathcal{A})$, and we wish to construct an $STS(v)$, say $(\mathcal{V}, \mathcal{B})$, in which $\mathcal{U} \subseteq \mathcal{V}$ and $\mathcal{A} \subseteq \mathcal{B}$.

5.6 Develop

(a) a hill-climbing algorithm,

(b) a simulated annealing algorithm,

(c) a tabu search algorithm, and

(d) a genetic algorithm

for the Maximum Clique problem. Use your algorithms to try to find maximum cliques in each of the graphs $\mathcal{G} = (\mathcal{V}, \mathcal{E})$ given in Exercise 4.9.

5.7 Run Algorithm 5.26 on the instance of the Traveling Salesman problem given in Exercise 4.7.

5.8 Run Algorithms 5.20 and 5.21 on the instances of the Knapsack (optimization) problem given in Exercise 4.4.

5.9 Develop

    (a) a hill-climbing algorithm, and

    (b) a genetic algorithm

for the **Knapsack (optimization)** problem. Test your algorithms with the instances given in Tables 5.2 and 5.5. How do your algorithms compare with the results of Algorithms 5.20 and 5.21 given in tables 5.3, 5.4 and 5.6?

5.10 The recombination operations, Algorithms 5.24 and Algorithms 5.25, chose a random substring

$$S = [S_0, S_1, \dots, S_{h-1}]$$

to crossover from a parent

$$X = [x_0, x_1, \dots, x_{n-1}]$$

to a child. Define the "average cost" of S to be the quantity

$$\text{avgcost}(S) = \frac{\sum_{i=1}^{h-1} M[s_{i-1}, s_i]}{h - 1}.$$

The lower the average cost, the more "fit" the substring is. A substring has "good fitness" if

$$\text{avgcost}(S) < \frac{C(X)}{n}.$$

Modify the recombination operation to randomly choose the substring S among those that have good fitness. Perform experiments to see if this technique improves the performance of Algorithm 5.26.

6

Groups and Symmetry

6.1 Groups

The theory of finite groups can often be helpful in counting the number of certain configurations and in determining when two representations are equivalent or not.

A *binary operation* $*$ on a set G is a function from $G \times G \to G$. If H is a subset of G, then H is *closed* under $*$ if and only if $h_1 * h_2 \in H$ for all $h_1, h_2 \in H$. That is, $*$ is also a binary operation on H. The binary operation $*$ on G is *associative* if

$$(g_1 * g_2) * g_3 = g_1 * (g_2 * g_3)$$

for all $g_1, g_2, g_3 \in G$.

Definition 6.1:   A *group* is a set G of elements together with an associative binary operation $*$ defined on G such that:

1. there is an element $\mathbf{I} \in G$ satisfying $g * \mathbf{I} = g$, for all $g \in G$; and
2. for each $g \in G$ there is an element $g^{-1} \in G$ such that $g^{-1} * g = \mathbf{I}$.

The element $\mathbf{I}$ is called the *identity* and given $g \in G$, the element g^{-1} is called g *inverse* of g.

Example 6.1   *Some examples of groups*

1. $\mathbb{Z}_n$, the integers under addition modulo n.
2. The m by m matrices with non-zero determinant under matrix multiplication.

3.   The matrices

$$G = \left\{ \begin{matrix} \mathbf{I} = \begin{bmatrix} 1 & 0 \\ 0 & 1 \end{bmatrix}, a = \begin{bmatrix} 0 & i \\ i & 0 \end{bmatrix}, b = \begin{bmatrix} -1 & 0 \\ 0 & -1 \end{bmatrix}, c = \begin{bmatrix} 0 & -i \\ -i & 0 \end{bmatrix}, \\ d = \begin{bmatrix} 0 & i \\ -i & 0 \end{bmatrix}, e = \begin{bmatrix} 1 & 0 \\ 0 & -1 \end{bmatrix}, f = \begin{bmatrix} 0 & -i \\ i & 0 \end{bmatrix}, g = \begin{bmatrix} -1 & 0 \\ 0 & 1 \end{bmatrix} \end{matrix} \right\}$$

where $i = \sqrt{-1}$, under matrix multiplication.

                                                                                                        ▯

Examples 6.1.2 and 6.1.3 show that the binary operation need not be *commutative*. (In example 6.1.3 $cd = g$ but $dc = e$.) The identity element $\mathbf{I}$ is unique, for if $g * \mathbf{I}' = g$, then multiplying on the left by g^{-1} shows that $\mathbf{I}' = \mathbf{I}$. Also, we observe that

$$g * g^{-1} = g * \mathbf{I} * g^{-1}$$
$$= g * (g^{-1} * g) * g^{-1}$$
$$= (g * g^{-1}) * (g * g^{-1}).$$

Thus $g * g^{-1} = \mathbf{I}$.

The binary operation $*$ is usually called *multiplication* and is denoted by juxtaposition. That is, if g, h are elements of the group, then gh denotes $g * h$. One method of presenting a group is to give the *multiplication table*. If $G = \{g_1, g_2, \ldots, g_n\}$ is a group, then the multiplication table for G is the n by n array whose $[i, j]$ entry is the product $g_i g_j$. For example, the multiplication table for the group in Example 6.1.3 is

	I	a	b	c	d	e	f	g
I	I	a	b	c	d	e	f	g
a	a	b	c	I	e	f	g	d
b	b	c	I	a	f	g	d	e
c	c	I	a	b	g	d	e	f
d	d	g	f	e	I	c	b	a
e	e	d	g	f	a	I	c	b
f	f	e	d	g	b	a	I	c
g	g	f	e	c	c	b	a	I

If the identity $\mathbf{I}$ in a group G is chosen to be g_1, the first group element, then it is easy to see that the multiplication table for a group is a reduced Latin square. (See Exercise 4.13.) The converse is not true, since a reduced Latin square need not be associative. If M is the multiplication table for the set of elements $G = \{g_1, g_2, \ldots, g_n\}$, then M is the multiplication table of a group with $g_1 = \mathbf{I}$ if and only if

1. M is a reduced Latin square; and

2. $M[M[g_i, g_j], g_k] = M[g_i, M[g_j, g_k]]$, for all choices of g_i, g_j and g_k.

A group can be represented in the computer by simply storing the multiplication table. This is often a good method when the group is small and there is no direct way to compute the product of elements. A method for storing a group that is, in general, more efficient is given in Section 6.2.2.

Definition 6.2:   A *subgroup* H of a group G is a subset of G that is itself a group (with the same multiplication that was provided for G).

For example, the subset of elements $\{\mathbf{I}, a, b, c\}$ of the group in Example 6.1.3 form a subgroup. This is easy to see by checking the upper left hand corner of the above multiplication table.

If G is finite, then we denote the number of elements of G by $|G|$. This is called the *order* of G. The groups of interest to combinatorialists are almost always finite.

THEOREM 6.1   *If H is a non-empty subset of the finite group G, then H is a subgroup of G if and only if H is closed (under the multiplication of G).*

PROOF   If $H = \{\mathbf{I}\}$, there is nothing to prove. Suppose $h_1 h_2 \in H$ for all $h_1, h_2 \in H$, and let $h \in H$, $h \neq \mathbf{I}$. Then

$$h^n = \underbrace{hhh \cdots h}_{n \text{ times}} \in H.$$

The sequence of elements $h^1, h^2, h^3, \ldots$ cannot all be distinct since the number of elements in H is finite. Thus some $h^m = h^n$ where $k = n - m > 0$. Then $h^n h^k = h^m h^k = h^m h^{n-m} = h^n$. Thus $h^k = \mathbf{I}$ and $h^{k-1} = h^{-1}$. Thus H is a subgroup. The converse is obvious.   ∎

Let H be a subgroup of the finite group G. The *left coset* containing the element g of G is

$$gH = \{gh : h \in H\}.$$

THEOREM 6.2   (LAGRANGE)

1. *If H is a subgroup of the finite group G, then G can be written as the disjoint union of left cosets. That is*

$$G = g_1 H \,\dot\cup\, g_2 H \,\dot\cup\, \cdots \,\dot\cup\, g_r H$$

for some $g_1, g_2, \ldots, g_r \in G$.

2.   *If H is a subgroup of the finite group G, then $|H|$ divides $|G|$.*

PROOF   Let H be a subgroup of the finite group G. If g is a fixed element of G, then the mapping $x \mapsto gx$ is a one to one mapping from H to gH, since g has an inverse. Hence $|gH| = |H|$ for all $g \in G$. Thus part (2) follows from the proof of part (1).

Suppose $g_1 H \cap g_2 H \neq \emptyset$, for some $g_1, g_2 \in G$. Then there exists $h_1, h_2 \in H$, so that $g_1 h_1 = g_2 h_2$. Thus,

$$g_1 = g_2 h_2 h_1^{-1}.$$

Hence, for any $h \in H$ we have

$$g_1 h = g_2 h_2 h_1^{-1} h$$
$$= g_2(h_2 h_1^{-1} h) \in g_2 H.$$

Consequently $g_1 H \subseteq g_2 H$, and therefore, because they have the same cardinality, $g_1 H = g_2 H$. Thus left cosets are either identical or disjoint.

Also, for any $g \in G$, $g \in gH$ since $\mathbf{I} \in H$. Therefore G is the union of disjoint left cosets.   ∎

One way to choose the g_i in part (1) of Theorem 6.2 is to first select g_1 arbitrarily, and then for each $i = 2, 3, \ldots$ to select any

$$g_{i+1} \in G \setminus (g_1 H \,\dot\cup\, g_2 H \,\dot\cup\, \cdots \,\dot\cup\, g_i H)$$

until

$$G \setminus (g_1 H \,\dot\cup\, g_2 H \,\dot\cup\, \cdots \,\dot\cup\, g_i H)$$

is empty.

The set $T = \{g_1, g_2, \ldots, g_r\}$ constructed in Theorem 6.2 is called a *system of left coset representatives* or a *left transversal* of H in G. For example, $T = \{\mathbf{I}, d\}$ is a transversal of the subgroup $H = \{\mathbf{I}, a, b, c\}$ of group G in Example 6.1.3. Indeed,

$$\mathbf{I} H \,\dot\cup\, d H = \mathbf{I}\, \{\mathbf{I}, a, b, c\} \,\dot\cup\, d;\, \{\mathbf{I}, a, b, c\}$$
$$= \{\mathbf{I}, a, b, c\} \,\dot\cup\, \{d, g, f, e\}$$
$$= \{\mathbf{I}, a, b, c, d, e, f, g\}$$
$$= G.$$

If $A, B \subseteq \mathsf{Sym}(\mathcal{X})$, then define

$$AB = \{ab : a \in A, b \in A\}.$$

In particular if T is a left transversal of the subgroup H in G, then $G = TH$. Continuing with the above example we see that

$$G = \{\mathbf{I}, d\}\{\mathbf{I}, a, b, c\}.$$

6.2   Permutation groups

In Section 3.2.2, we introduced the concept of permutations as functions on a set of points, and also their representation in cycle notation. If the set of points is understood, then the convention when studying permutation groups is to not write the cycles of length one. (Cycles of length one are called *fixed points*.) For example, the fixed points of the permutation

$$\pi = (0,3,4,1)(2)(5,6)(7,8,9)(10)$$

are 2 and 10, and we write

$$\pi = (0,3,4,1)(5,6)(7,8,9),$$

where it is understood that π is a permutation on the set $V = \{0,1,2,\dots,10\}$.

Multiplication of permutations α and β is defined by *function composition*. That is, we define

$$(\alpha\beta)(x) = \alpha(\beta(x)).$$

For example, if on the set $\mathcal{X} = \{0,1,2,3,4\}$, we have

$$\alpha = (1,2,3)(0,4)$$

and

$$\beta = (1,2,4),$$

then

$$\alpha\beta = (0,4,2)(1,3).$$

It is easy to see that the composition of two permutations is again a permutation. Consequently we have:

THEOREM 6.3   *Let $\mathcal{X}$ be a non-empty set and let $\mathsf{Sym}(\mathcal{X})$ be the set of all permutations on $\mathcal{X}$. Under the operation of composition of functions, $\mathsf{Sym}(\mathcal{X})$ is a group. If $|\mathcal{X}| = n$, then $\mathsf{Sym}(\mathcal{X})$ has $n!$ elements.*

The group $\mathsf{Sym}(\mathcal{X})$ of all permutations on the set $\mathcal{X}$ is called the *symmetric group* on $\mathcal{X}$. A *permutation group* on $\mathcal{X}$ is a subgroup of $\mathsf{Sym}(\mathcal{X})$. For example, the set

$$\{\mathbf{I}, (0,1,2)(3,4), (0,2,1), (3,4), (0,1,2), (0,2,1)(3,4)\}$$

is a permutation group on $\mathcal{X} = \{0,1,2,3,4\}$. Note that we use $\mathbf{I}$ to denote $(0)(1)(2)(3)(4)$, the identity permutation.

Permutation groups are the most important groups in the study of set systems. The reason for this is that the set of permutations that preserve a set system $\mathcal{D}$ is a permutation group called the *automorphism group* of the set system. It is denoted by $\mathrm{Aut}(\mathcal{D})$.

Definition 6.3:   If $\{u, v\}$ is an edge of the graph $\mathcal{G} = (\mathcal{V}, \mathcal{E})$ and α is a permutation of $\mathcal{V}$, then define

$$\alpha(\{u, v\}) = \{\alpha(u), \alpha(v)\}.$$

A permutation α of $\mathcal{V}$ is an *automorphism* of the graph $\mathcal{G} = (\mathcal{V}, \mathcal{E})$ if

$$\alpha(\{u, v\}) \in \mathcal{E} \text{ whenever } \{u, v\} \in \mathcal{E}.$$

The *automorphism group* of a graph $\mathcal{G} = (\mathcal{V}, \mathcal{E})$ denoted by $\mathrm{Aut}(\mathcal{G})$ is the set of all permutations on $\mathcal{V}$ that are automorphisms of $\mathcal{G}$. Thus $\mathrm{Aut}(\mathcal{G})$ consists of the permutations of $\mathcal{V}$ that fix, as a set, the edges $\mathcal{E}$ of $\mathcal{G}$.

Example 6.2   *The automorphism group of a graph.*
Consider the graph $\mathcal{G} = (\mathcal{V}, \mathcal{E})$ where

$$\mathcal{V} = \{0, 1, 2, 3, 4, 5\}, \text{ and}$$

$$\mathcal{E} = \{\{0, 4\}, \{0, 5\}, \{1, 2\}, \{1, 3\}, \{2, 3\}, \{3, 4\}, \{4, 5\}\}.$$

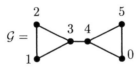

The automorphism group of $\mathcal{G}$ is

$$\mathrm{Aut}(\mathcal{G}) = \left\{ \begin{array}{l} \mathbf{I}, (1, 2), (0, 5), (0, 5)(1, 2), (0, 2)(1, 5)(3, 4), \\ (0, 1)(2, 5)(3, 4), (0, 1, 5, 2)(3, 4), (0, 2, 5, 1)(3, 4) \end{array} \right\}.$$

☐

In Example 6.2 the automorphism group was computed by inspection. In general, to compute $\mathrm{Aut}(\mathcal{G})$, one could use Algorithm 2.14 to run over all permutations of $\mathcal{V}$ and check each one to see if it was an automorphism. This however is an $O(n!)$ algorithm and is consequently impractical when n is large. (Here $n = |\mathcal{V}|$.) Instead a much more efficient algorithm is often used. This algorithm is described in Chapter 7 and uses many of the algorithms described in this chapter.

THEOREM 6.4 *The automorphism group of a graph is a group under the operation of function composition.*

PROOF Let $\mathcal{G} = (\mathcal{V}, \mathcal{E})$ be a graph and let $\mathrm{Aut}(\mathcal{G})$ be the set of all permutations that fix $\mathcal{E}$ as a set. Certainly $\mathbf{I} \in \mathrm{Aut}(\mathcal{G})$ and so $\mathrm{Aut}(\mathcal{G}) \neq \emptyset$. Let $\alpha, \beta \in \mathrm{Aut}(\mathcal{G})$, and suppose $\{u, v\} \in \mathcal{E}$. Then $\beta(\{u, v\}) \in \mathcal{E}$ and thus $(\alpha\beta)(\{u, v\}) = \alpha(\beta(\{u, v\})) \in \mathcal{E}$. Consequently, $\alpha\beta \in \mathrm{Aut}(\mathcal{G})$. Therefore, by Theorem 6.1, $\mathrm{Aut}(\mathcal{G})$ is a group. ∎

FIGURE 6.1
The graph of the pentagon.

In Example 6.3 we give a second example of the automorphism group of a graph. The subgroups of this automorphism group are also displayed.

Example 6.3 *The automorphism group of the pentagon and its subgroups.*
If P denotes the pentagon in Figure 6.1, then

$$\mathrm{Aut}(P) = \left\{ \begin{array}{ll} \mathbf{I} = (0)(1)(2)(3)(4), & (0, 1, 2, 3, 4), \\ (0, 2, 4, 1, 3), & (0, 3, 1, 4, 2), \\ (0, 4, 3, 2, 1), & (0)(1, 4)(2, 3), \\ (0, 2)(1)(3, 4), & (0, 4)(1, 3)(2), \\ (0, 1)(2, 4)(3), & (0, 3)(1, 2)(4) \end{array} \right\}$$

$\mathrm{Aut}(P)$ has 8 subgroups, namely, $\mathrm{Aut}(P)$, C_5, H_0, H_1, H_2, H_3, H_4, and $\{\mathbf{I}\}$, where:

$$\begin{aligned}
C_5 &= \{\mathbf{I}, (0,1,2,3,4), (0,2,4,1,3), (0,3,1,4,2), (0,4,3,2,1)\} \\
H_0 &= \{\mathbf{I}, (0)(1,4)(2,3)\} \\
H_1 &= \{\mathbf{I}, (0,2)(1)(3,4)\} \\
H_2 &= \{\mathbf{I}, (0,4)(1,3)(2)\} \\
H_3 &= \{\mathbf{I}, (0,1)(2,4)(3)\} \\
H_4 &= \{\mathbf{I}, (0,3)(1,2)(4)\}
\end{aligned}$$

□

The concept of the automorphism group of a graph is easily generalized to an arbitrary set system.

Definition 6.4: If B is a block of the set system $\mathcal{S} = (\mathcal{X}, \mathcal{B})$ and α is a permutation of $\mathcal{X}$, then we define

$$\alpha(B) = \{\alpha(x) : x \in B\}.$$

A permutation α of $\mathcal{X}$ is an *automorphism* of the set system $\mathcal{S} = (\mathcal{X}, \mathcal{B})$ if

$$\alpha(B) \in \mathcal{B} \text{ whenever } B \in \mathcal{B}.$$

The *automorphism group* of a set system $\mathcal{S} = (\mathcal{X}, \mathcal{B})$, denoted $\mathsf{Aut}(\mathcal{S})$, is the set of all permutations on $\mathcal{X}$ that are automorphisms of $\mathcal{S}$. Thus $\mathsf{Aut}(\mathcal{S})$ consists of the permutations of $\mathcal{X}$ that fix, as a set, the blocks $\mathcal{B}$ of $\mathcal{S}$.

Example 6.4 *The automorphism group of a set system.*
The set system $\mathcal{S} = (\mathcal{X}, \mathcal{B})$ where

$$\begin{aligned}
\mathcal{X} &= \{0,1,2,3,4,5\} \\
\mathcal{B} &= \{\{0,1,2\}, \{4,5,6\}, \{0,4\}, \{1,5\}, \{2,6\}\}
\end{aligned}$$

has automorphism group

$$\mathsf{Aut}(\mathcal{S}) = \left\{ \begin{array}{l} \mathbf{I}, (1,2)(4,5), (0,1,2)(3,4,5), (0,1)(3,4), \\ (0,2,1)(3,5,4), (0,2)(3,5), (0,3)(1,4)(2,5), \\ (0,3)(1,5)(2,4), (0,4,2,3,1,5), \\ (0,4)(1,3)(2,5), (0,5,1,3,2,4), (0,5)(1,4)(2,3) \end{array} \right\}.$$

□

In Example 6.4 the automorphism group could have been computed by using Algorithm 2.14 to run over all permutations of $\mathcal{X}$. Each permutation would be examined to see if it is an automorphism. A more efficient method is described in Section 7.4.2. It uses several of the algorithms described in this chapter.

In a manner similar to Theorem 6.4 we can establish Theorem 6.5. We leave the proof as an exercise.

THEOREM 6.5 *The automorphism group of a set system is a group under the operation of function composition.*

6.2.1 Basic algorithms

The basic operations that we need to perform on permutations are multiplication, inversion, and conversion between cycle notation and array notation.

An algorithm to multiply permutations is:

Algorithm 6.1: MULT $(n, \alpha, \beta, \gamma)$

for $i \leftarrow 0$ **to** $n - 1$
  do $\pi_0[i] \leftarrow \alpha[\beta[i]]$
for $i \leftarrow 0$ **to** $n - 1$
  do $\gamma[i] \leftarrow \pi_0[i]$

The auxiliary permutation π_0 was used in Algorithm 6.1 to allow calculations of the form MULT(g, h, g) and MULT(g, h, h). Without the temporary permutation in some programming languages such as "C" the inputs α and β would be changed before the calculation is finished, and an incorrect output γ would be obtained.

The identity for this multiplication is the identity function which we denote by **I**. So $\mathbf{I}(x) = x$ for all $x \in \mathcal{X}$. The inverse of a permutation α is the permutation β such that $\beta\alpha = \mathbf{I}$. A procedure to compute α^{-1} is:

Algorithm 6.2: INV (n, α, β)

for $i \leftarrow 0$ **to** $n - 1$
  do $\beta[\alpha[i]] \leftarrow i$

It is easy to see that MULT and INV take $O(n)$ time.

It is useful to have programs that can switch between cycle notation and array notation. Algorithm 6.3 parses the string C, which represents the permutation π in cycle notation, and generates the array A, which is the array representation of π. Algorithm 6.4 does the reverse.

Algorithm 6.3: CYCLETOARRAY (n, C)

for $i \leftarrow 0$ **to** $n - 1$
  do $A[i] \leftarrow i$
$i \leftarrow 0$
Set ℓ to be the length of the string C
while $i < \ell$

$\mathbf{do} \begin{cases} \mathbf{if}\ C[i] = \text{``(''} \\ \quad \mathbf{then} \begin{cases} i \leftarrow i + 1 \\ \mathbf{if}\ C[i] \in \{0, 1, 2, \ldots, 9\} \\ \quad \mathbf{then} \begin{cases} \text{Get the number } x \text{ starting at position } i \\ z \leftarrow x \\ \text{Increment } i \text{ to the position after } x \end{cases} \end{cases} \\ \mathbf{if}\ C[i] = \text{``,''} \\ \quad \mathbf{then} \begin{cases} i \leftarrow i + 1 \\ \mathbf{if}\ C[i] \in \{0, 1, 2, \ldots, 9\} \\ \quad \mathbf{then} \begin{cases} \text{Get the number } y \text{ starting at position } i \\ A[x] \leftarrow y \\ \text{Increment } i \text{ to the position after } y \\ x \leftarrow y \end{cases} \end{cases} \\ \mathbf{if}\ C[i] = \text{``)''} \\ \quad \mathbf{then} \begin{cases} A[x] \leftarrow z \\ i \leftarrow i + 1 \end{cases} \end{cases}$

Algorithm 6.4: ARRAYTOCYCLE (n, A)

for $i \leftarrow 0$ **to** $n - 1$
  do $P[i] \leftarrow$ **true**
$C \leftarrow$ the empty string
for $i \leftarrow 0$ **to** $n - 1$

$\mathbf{do} \begin{cases} \mathbf{if}\ P[i] \\ \quad \mathbf{then} \begin{cases} \text{Append ``(`` to } C \\ \text{Append } i \text{ to } C \\ P[i] \leftarrow \textbf{false} \\ j \leftarrow i \\ \mathbf{while}\ P[A[j]] \\ \quad \mathbf{do} \begin{cases} \text{Append ``,'' to } C \\ j \leftarrow A[j] \\ \text{Append } j \text{ to } C \\ P[j] \leftarrow \textbf{false} \end{cases} \\ \text{Append ``)'' to } C \end{cases} \end{cases}$

return (C)

6.2.2 How to store a group

When searching for a particular set system, it is often advantageous to assume that the set system has certain automorphisms. These automorphisms generate a permutation group. To use a permutation group effectively we must be able to

1. efficiently store the group;
2. test if a given permutation is in the group; and
3. be able to run over all the elements in the group without repetition.

To illustrate the various methods for storing a group in a computer consider the group G of automorphisms of the graph in Figure 6.2. The 48 elements of G are given in Table 6.1. The group G is also generated by taking all possible products of the permutations

$$\alpha = (0, 1, 3, 7, 6, 4)(2, 5)$$

and

$$\beta = (0, 1, 3, 2)(4, 5, 7, 6).$$

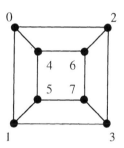

FIGURE 6.2
The graph of the cube.

Definition 6.5: A group is said to be *generated* by the elements $\alpha_1, \alpha_2, \ldots, \alpha_r$ if every element g of the group can be written as a finite product

$$g = \alpha_{i_1} \alpha_{i_2} \cdots \alpha_{i_m}$$

where for each j, $1 \le i_j \le r$. The elements $\alpha_1, \alpha_2, \ldots, \alpha_r$ are called *generators* for the group G.

Suppose we choose to store all of the group elements as a sorted list of permutations, say in the lexicographic order discussed in Section 2.4. Then we observe the following:

TABLE 6.1
Automorphisms of the graph in Figure 6.2.

(0)(1)(2)(3)(4)(5)(6)(7)	(0)(1)(2,4)(3,5)(6)(7)	(0)(1,2)(3)(4)(5,6)(7)
(0)(1,2,4)(3,6,5)(7)	(0)(1,4)(2)(3,6)(5)(7)	(0)(1,4,2)(3,5,6)(7)
(0,1)(2,3)(4,5)(6,7)	(0,1)(2,5)(3,4)(6,7)	(0,1,3,2)(4,5,7,6)
(0,1,3,7,6,4)(2,5)	(0,1,5,4)(2,3,7,6)	(0,1,5,7,6,2)(3,4)
(0,2)(1,3)(4,6)(5,7)	(0,2,6,4)(1,3,7,5)	(0,2,3,1)(4,6,7,5)
(0,2,6,7,5,1)(3,4)	(0,2)(1,6)(3,4)(5,7)	(0,2,3,7,5,4)(1,6)
(0,3)(1)(2)(4,7)(5)(6)	(0,3,5)(1)(2,7,4)(6)	(0,3)(1,2)(4,7)(5,6)
(0,3,6,5)(1,2,7,4)	(0,3,6)(1,7,4)(2)(5)	(0,3,5,6)(1,7,4,2)
(0,4)(1,6)(2,5)(3,7)	(0,4,5,7,3,2)(1,6)	(0,4)(1,5)(2,6)(3,7)
(0,4,6,2)(1,5,7,3)	(0,4,6,7,3,1)(2,5)	(0,4,5,1)(2,6,7,3)
(0,5,3,6)(1,7,2,4)	(0,5,6)(1,7,2)(3)(4)	(0,5)(1,4)(2,7)(3,6)
(0,5,6,3)(1,4,7,2)	(0,5,3)(1)(2,4,7)(6)	(0,5)(1)(2,7)(3)(4)(6)
(0,6,3,5)(1,4,2,7)	(0,6,3)(1,4,7)(2)(5)	(0,6)(1,7)(2,4)(3,5)
(0,6)(1,7)(2)(3)(4)(5)	(0,6,5)(1,2,7)(3)(4)	(0,6,5,3)(1,2,4,7)
(0,7)(1,6)(2,5)(3,4)	(0,7)(1,6)(2,3)(4,5)	(0,7)(1,5)(2,6)(3,4)
(0,7)(1,5,4,6,2,3)	(0,7)(1,3)(2,5)(4,6)	(0,7)(1,3,2,6,4,5)

Generators for this group are:

$$\alpha = (0, 1, 3, 7, 6, 4)(2, 5);$$

and

$$\beta = (0, 1, 3, 2)(4, 5, 7, 6).$$

1. we store $|G|$ elements, in the worst case this is $n!$;
2. we can test if a given permutation is in G by doing a binary search and comparison of permutations, which in the worst case is $O(\log n!)O(n) = O(n^2 \log n)$; and
3. we can easily run through the elements of the group, generating each in $O(1)$ time.

On the other hand if we simply store the generators, then to run through elements of the group or to see if a given permutation is in the group one must generate all products in the generators of length one, length two, length three and so forth crossing out duplicates when they arise until no more elements can be generated. For example, in the group G of automorphisms of the cube, we see that

$$\alpha\alpha\alpha\alpha\beta\beta\beta = \alpha\beta\alpha\alpha.$$

Algorithm 6.5 uses this simple method of generating a group G from a set of generators Γ.

Algorithm 6.5: SIMPLEGEN (n, Γ)

external MULT()
$G \leftarrow \emptyset$
$New \leftarrow \{\mathbf{I}\}$
while $New \neq \emptyset$
$$\mathbf{do} \begin{cases} G \leftarrow G \cup New \\ Last \leftarrow New \\ New \leftarrow \emptyset \\ \mathbf{for\ each}\ g \in \Gamma \\ \mathbf{do} \begin{cases} \mathbf{for\ each}\ h \in Last \\ \mathbf{do} \begin{cases} \text{MULT}(n, g, h, f) \\ \mathbf{if}\ f \notin G \\ \quad \mathbf{then}\ New \leftarrow New \cup \{f\} \end{cases} \end{cases} \end{cases}$$

The fact that the same group element can arise from many different products of generators makes this method of running through the group elements inefficient. Hence, although just storing the generators is economical, all other operations we wish to perform on the group are difficult and costly. We will consider an alternative method in the next section.

6.2.3 Schreier-Sims algorithm

Let G be a permutation group on $\mathcal{X} = \{0, 1, 2, 3, \ldots, n-1\}$ and set

$$G_0 = \{g \in G : g(0) = 0\}$$
$$G_1 = \{g \in G_0 : g(1) = 1\}$$
$$G_2 = \{g \in G_1 : g(2) = 2\}$$
$$\vdots$$
$$G_{n-1} = \{g \in G_{n-2} : g(n-1) = n-1\} = \{\mathbf{I}\}.$$

It is not difficult to show that $G_0, G_1, \ldots, G_{n-1}$ are subgroups and that

$$G \supseteq G_0 \supseteq G_1 \supseteq G_2 \supseteq \cdots \supseteq G_{n-1} = \{\mathbf{I}\}.$$

Let

$$\mathrm{orb}(0) = \{g(0) : g \in G\},$$

the *orbit* of 0 under G. Then $|\mathrm{orb}(0)| = n_0$ for some integer n_0, $0 < n_0 \leq n$. Write

$$\mathrm{orb}(0) = \{x_{0,1}, x_{0,2}, \ldots, x_{0,n_0}\}$$

and for each i, $1 \leq i \leq n_0$, choose some $h_{0,i} \in G$ such that $h_{0,i}(0) = x_{0,i}$. Set

$$\mathcal{U}_0 = \{h_{0,1}, h_{0,2}, \ldots, h_{0,n_0}\}.$$

Example 6.5

Let G be the permutation group on $\mathcal{X} = \{0, 1, \ldots, 7\}$ that is displayed in Table 6.1. Then

$$G_0 = \left\{ \begin{array}{l} \mathbf{I}, (0)(1)(2,4)(3,5)(6)(7), (0)(1,2)(3)(4)(5,6)(7), \\ (0)(1,2,4)(3,6,5)(7), (0)(1,4)(2)(3,6)(5)(7), (0)(1,4,2)(3,5,6)(7) \end{array} \right\}$$

and

$$\text{orb}(0) = \{0, 1, 2, 3, 4, 5, 6, 7\}.$$

Thus, we can choose, for example,

$$\mathcal{U}_0 = \left\{ \begin{array}{l} \mathbf{I}, (0,1,3,7,6,4)(2,5), (0,2,6,4)(1,3,7,5), (0,3,6)(1,7,4)(2)(5), \\ (0,4,6,7,3,1)(2,5), (0,5,3,6)(1,7,2,4), (0,6,3)(1,4,7)(2)(5), \\ (0,7)(1,6)(2,5)(3,4) \end{array} \right\}.$$

$\square$

THEOREM 6.6 *Let $G, \mathcal{U}_0$ and G_0 be as defined above. Then $\mathcal{U}_0$ is a left transversal of G_0 in G.*

PROOF Let $g \in G$. Then $g(0) = x_{0,i}$ for some i, $1 \leq i \leq n_0$. We have $h_{0,i} \in \mathcal{U}_0$ and $h_{0,i}(0) = x_{0,i}$. Hence,

$$(h_{0,i}^{-1}g)(0) = h_{0,i}^{-1}(g(0)) = h_{0,i}^{-1}(x_{0,i}) = 0.$$

Thus $h_{0,i}^{-1}g = g' \in G_0$ for some g'. Therefore $g = h_{0,i}g' \in \mathcal{U}_0 G_0$.

Suppose $h \neq h'$, $h, h' \in \mathcal{U}_0$. Then $h(0) \neq h'(0)$. If $hG_0 \cap h'G_0 \neq \emptyset$, then $hg = h'g'$ for some $g, g' \in G_0$ and

$$h(0) = h(g(0)) = (hg)(0) = (h'g')(0) = (h'(g'(0))) = h'(0),$$

which is a contradiction. So $hG_0 \cap h'G_0 = \emptyset$. $\blacksquare$

Similarly we define for $i = 1, 2, \ldots, n - 1$:

$$\text{orb}(i) = \{g(i) : g \in G_{i-1}\} = \{x_{i,1}, x_{i,2}, \ldots, x_{i,n_i}\}$$

and

$$\mathcal{U}_i = \{h_{i,1}, h_{i,2}, \ldots, h_{i,n_0}\},$$

where $h_{i,j}(i) = x_{i,j}$. A proof, similar to that of Theorem 6.6, gives the following theorem.

THEOREM 6.7 *Let $G, \mathcal{U}_i$ and $\mathcal{G}_i$ be as defined above. Then for all $i = 1, 2, \ldots, n - 1$, $\mathcal{U}_i$ is a left transversal of $\mathcal{G}_i$ in $\mathcal{G}_{i-1}$.*

The data structure

$$\vec{G} = [\mathcal{U}_0, \mathcal{U}_1, \ldots, \mathcal{U}_{n-1}] \tag{6.1}$$

is called the *Schreier-Sims representation* of the group G.

Applying Theorems 6.2, 6.6 and 6.7, it is easy to see that $g \in G$ can be uniquely written as

$$g = h_{0,i_0} h_{1,i_1} h_{2,i_2} \cdots h_{n-1,i_{n-1}}.$$

If we can efficiently construct the sets $\mathcal{U}_i, 0 \leq i \leq n - 1$, then a simple backtracking procedure can be used to run through the elements of the group G. By this method, presented as Algorithm 6.6, we can generate each element of G in $O(n^2)$ time. The variable *DoneEarly* is a globally defined flag, which, when set **true**, terminates running through the group early. For example, it may be the case that only the first group element satisfying a certain property is needed. The procedure USE (g) is a procedure that specifies what is to be done with g, and is passed as an argument to RUN. For example, g could be printed, stored, or applied to some combinatorial object. Typically USE (g) also sets the flag *DoneEarly* when early termination is desired.

Algorithm 6.6: RUN $(n, \vec{G}, \text{USE}())$

external MULT()

global *DoneEarly*

procedure RUNBACKTRACK$(n, \ell, \vec{G}, g, \text{USE}())$
 if *DoneEarly*
  then return
 if $\ell = n$
  then USE(n, g)
  else $\left\{ \begin{array}{l} \textbf{for each } h \in \mathcal{U}_\ell \\ \quad \textbf{do} \left\{ \begin{array}{l} \text{MULT}(n, g, h, f_\ell) \\ \text{RUNBACKTRACK}(n, \ell + 1, \vec{G}, f_\ell, \text{USE}()) \end{array} \right. \end{array} \right.$

main
RUNBACKTRACK$(0, \vec{G}, \mathbf{I}, \text{USE}())$

One application of Algorithm 6.6 is to list the elements of the group without repetition. This is done by specifying that procedure USE simply prints its argument. See Algorithm 6.7.

Algorithm 6.7: LIST $(n, \vec{G})$

external ARRAYTOCYCLE(), RUN()
global *DoneEarly*
procedure LISTUSE(n, g)
  output ARRAYTOCYCLE(n, g)

main
  DoneEarly $\leftarrow$ **false**
  RUN($n, \vec{G}$, LISTUSE())

It is also an easy matter to test membership of a permutation g in the group G, once the sets $\mathcal{U}_i$ have been constructed. The following process will terminate with i and f satisfying the following condition.

$$\left. \begin{array}{l} (f^{-1}g)(x) = x \text{ for all } x < i \text{ and } (f^{-1}g)(i) = j \\ \text{but there is no } h \in \mathcal{U}_i \text{ such that } h(i) = j. \end{array} \right\} \tag{6.2}$$

Step 0. Set $g_0 = g$, the permutation we wish to test, and set $x = g(0)$. If there is an $h_0 \in \mathcal{U}_0$ such that $h_0(0) = x$, then $(h_0^{-1}g_0)(0) = 0$. Consequently if $g_0 \in G$, then $h_0^{-1}g_0 \in G_0$ and so we set

$$f_0 = h_0$$

and

$$g_1 = f_0^{-1}g = h_0^{-1}g$$

and proceed to check if $g_1 \in G_0$; otherwise $i = 0$ and we stop.

Step 1. To check if $g_1 \in G_0$ we set $x = g_1(1)$. If there is an $h_1 \in \mathcal{U}_1$ such that $h_1(1) = x$, then $(h_1^{-1}g_1)(1) = 1$. Consequently if $g_1 \in G_1$, then $h_1^{-1}g_1 \in G_1$ and so we set

$$f_1 = f_0 h_1$$

and

$$g_2 = f_1^{-1}g = h_1^{-1}g_1$$

and proceed to check if $g_2 \in G_1$; otherwise $i = 1$ and we stop.

Step 2. To check if $g_2 \in G_1$ we set $x = g_2(2)$. If there is an $h_2 \in \mathcal{U}_2$ such that $h_2(2) = x$, then $(h_2^{-1}g_2)(2) = 2$. Consequently if $g_2 \in G_2$, then $h_2^{-1}g_2 \in G_2$ and so we set

$$f_2 = f_1 h_2$$

and

$$g_3 = f_2^{-1}g = h_2^{-1}g_2$$

and proceed to check if $g_3 \in G_2$; otherwise $i = 2$ and we stop.

Step 3. ... proceed as above to check if $g_i \in G_{i-1}$.

Since $f_i^{-1}g$ fixes the points $0, 1, \ldots, i$, either this process will terminate with $f_{n-1}^{-1}g = \mathbf{I}$ in which case $g = f_{n-1} = h_0 h_1 \cdots h_{n-1} \in G$, or the process will terminate early. If the latter happens, then $f = f_i$ and i will satisfy Condition 6.2. Example 6.6 illustrates this process with the automorphism group of the cube.

Example 6.6

In Table 6.1 we list the automorphisms of the cube which is displayed in Figure 6.2. The Schreier-Sims representation

$$\vec{G} = [\mathcal{U}_0, \mathcal{U}_1, \ldots, \mathcal{U}_7]$$

of this group is given in Example 6.7. The computations to test if the permutation

$$g = (0,6)(1,7)(2,4)(3,5) \in G,$$

i.e., to see if it is an automorphism of the cube, are given in the table below.

i	g_i	h_i	f_i
0	$(0,6)(1,7)(2,4)(3,5)$	$(0,6,3)(1,4,7)(2)(5)$	$(0,6,3)(1,4,7)(2)(5)$
1	$(0)(1,4,2)(3,5,6)(7)$	$(0)(1,4,2)(3,5,6)(7)$	$(0,6)(1,7)(2,4)(3,5)$
2	$\mathbf{I}$	$\mathbf{I}$	$(0,6)(1,7)(2,4)(3,5)$
3	$\mathbf{I}$	$\mathbf{I}$	$(0,6)(1,7)(2,4)(3,5)$
$\vdots$	$\vdots$	$\vdots$	$\vdots$
$n-1$	$\mathbf{I}$	$\mathbf{I}$	$(0,6)(1,7)(2,4)(3,5)$

Thus $g \in G$ and thus it is an automorphism of the cube. (In fact, g occurs in column three of Table 6.1.)

Now we will check if

$$g = (0,1,2,3,4)(5,6,7)$$

is in the group G.

i	g_i	h_i	f_i
0	$(0,1,2,3,4)(5,6,7)$	$(0,1,3,7,6,4)(2,5)$	$(0,1,3,7,6,4)(2,5)$
1	$(0)(1,5,7,2)(4)(3,6)$	There is no h_i	

Thus, $g \notin G$. $\quad\quad\quad\quad\quad\quad\quad\quad\quad\quad\quad\quad\quad\quad\quad$ ◻

This process is implemented in Algorithm 6.8. It returns the smallest i such that there is an $f \in G$ satisfying the Condition 6.2 and replaces g with this f. Notice that g is replaced by the current f_i at each stage, since it is no longer needed in subsequent tests. Thus in fact there is no need for the actual f_is, and they do not appear in Algorithm 6.8.

Algorithm 6.8: TEST $(n, g, \vec{G} = [\mathcal{U}_0, \ldots, \mathcal{U}_{n-1}])$

external INV$()$, MULT$()$

for $i \leftarrow 0$ **to** $n - 1$

do $\begin{cases} x \leftarrow g[i] \\ \textbf{if } \text{there is an } h \in \mathcal{U}_i \text{ such that } h(i) = x \\ \quad \textbf{then } \begin{cases} \text{INV}(n, h, \pi_2) \\ \text{MULT}(n, \pi_2, g, \pi_3) \\ \textbf{for } j \leftarrow 0 \textbf{ to } n - 1 \\ \quad \textbf{do } g[j] \leftarrow \pi_3[j] \end{cases} \\ \textbf{else return } (i) \end{cases}$

return (n)

If i is returned, then the process performed $i + 1 \leq n$ steps and each step consisted of a permutation product, an inverse and a copy. These each take time $O(n)$. Thus it is clear that TEST runs in $O(n^2)$ time. We now need a procedure to construct the sets $\mathcal{U}_i$, for $0 \leq 0 < n$, given a set $\Gamma = \{\alpha_1, \alpha_2, \ldots, \alpha_r\}$ of generators for G. In order to construct these sets, we consider the subgroups H_ℓ, for $1 \leq \ell \leq r$, where H_ℓ is generated by $\alpha_1, \ldots, \alpha_\ell$. If the sets $\mathcal{U}_i$, $0 \leq i < n$, have been constructed for the subgroup $H_{\ell-1}$, then these sets can be updated to correspond to H_ℓ by entering $g = \alpha_\ell$ in the appropriate place and, then, adjusting the $\mathcal{U}_i$s to achieve closure. Thus we first apply TEST to g. As explained earlier, TEST returns the smallest i such that there is an $f \in G$ satisfying $(f^{-1}g)(x) = x$ for all $x < i$ and $(f^{-1}\alpha_\ell)(i) = j$, but there is no $h \in \mathcal{U}_i$ such that $h(i) = j$. Also, TEST replaces g by $f^{-1}\alpha_\ell$. If $i = n$, then we are done; otherwise, this new g belongs in $\mathcal{U}_i$. Now we need to make sure that each product of the form

$$u_1 u_2 \cdots u_{i-1} g u_{i+1} \ldots u_n,$$

where $u_i \in \mathcal{U}_i$, is in the group represented by the $\mathcal{U}_i$s. That is, this element must cause TEST to return n. If it does not return n, then it needs to be entered in the $\mathcal{U}_i$s. Theorem 6.8 shows that it suffices to only consider products of the form $\alpha_\ell h$ where $h \in \mathcal{U}_j$, for $j \leq i$. This updating is performed recursively by Algorithm 6.9.

Algorithm 6.9: GEN (n, Γ)

external MULT(), TEST()

procedure ENTER$(n, g, \vec{G} = [\mathcal{U}_0, \ldots, \mathcal{U}_{n-1}])$
  $i \leftarrow$ TEST$(n, g, \vec{G} = [\mathcal{U}_0, \ldots, \mathcal{U}_{n-1}])$
  if $i = n$
    then return
    else $\mathcal{U}_i \leftarrow \mathcal{U}_i \cup \{g\}$ $(*)$
  for $j \leftarrow 0$ **to** i
    do $\left\{ \begin{array}{l} \textbf{for each } h \in \mathcal{U}_j \\ \textbf{do } \left\{ \begin{array}{l} \text{MULT}(n, g, h, f) \quad\quad (**) \\ \text{ENTER}(n, f, \vec{G} = [\mathcal{U}_0, \ldots, \mathcal{U}_{n-1}]) \end{array} \right. \end{array} \right.$

main
  for $i \leftarrow 0$ **to** $n - 1$
    do $\mathcal{U}_i \leftarrow \{\mathbf{I}\}$
  for each $\alpha \in \Gamma$
    do ENTER$(n, \alpha, \vec{G} = [\mathcal{U}_0, \ldots, \mathcal{U}_{n-1}])$
  return $(\vec{G} = [\mathcal{U}_0, \ldots, \mathcal{U}_{n-1}])$

THEOREM 6.8 *Let G be a permutation group on $\mathcal{X} = \{0, 1, \ldots, n - 1\}$ with Schreier-Sims representation $\vec{G} = [\mathcal{U}_0, \mathcal{U}_1, \ldots, \mathcal{U}_{n-1}]$, and let $g \in \mathsf{Sym}(\mathcal{X})$ be an arbitrary permutation. Then Procedure ENTER of Algorithm 6.9 correctly updates $\vec{G}$ to one that represents the group generated by G and g.*

PROOF Let $\vec{G}' = [\mathcal{U}_0', \mathcal{U}_1', \ldots, \mathcal{U}_{n-1}']$ be the updated data structure that results when ENTER is run with parameters $\vec{G}$ and g. To show that $\vec{G}'$ represents the group generated by G and g we need only show that

$$G' = \mathcal{U}_0' \mathcal{U}_1' \cdots \mathcal{U}_{n-1}'$$

is a group, since it is clear that $G \subseteq G'$ and every element in G' is a product of elements in $G \cup \{g\}$.

To see that G' is a group, let

$$G_i' = \mathcal{U}_i' \mathcal{U}_{i+1}' \cdots \mathcal{U}_{n-1}'.$$

Then $G_{n-1}' = \mathcal{U}_{n-1}' = \{\mathbf{I}\}$ and so G_{n-1}' is a group. Suppose G_i' is a group for some i, $0 < i \leq n - 1$. We will show that

$$G_{i-1}' = \mathcal{U}_{i-1}' G_i'$$

is a group. Let $x_1, x_2 \in G_{i-1}'$, and write

$$x_1 = h_1 g_1$$

and

$$x_2 = h_2 g_2,$$

where $h_1, h_2 \in \mathcal{U}'_{i-1}$ and $g_1, g_2 \in G'_i$. First, observe that

$$g_1 h_2(j) = j$$

for $j = 0, 1, 2, \ldots, i-1$. Hence, because of line ($**$) of Procedure ENTER , there is an $h_3 \in \mathcal{U}'_{i-1}$ such that $h_3(i) = g_1 h_2(i)$. So, $h_3^{-1} g_1 h_2 = g_3$ for some $g_3 \in G'_i$ and we have

$$\begin{aligned}
x_1 x_2 &= h_1 g_1 h_2 g_2 \\
&= h_1 h_3 h_3^{-1} g_1 h_2 g_2 \\
&= h_1 h_3 g_3 g_2 \\
&= h_1 h_3 g_4,
\end{aligned}$$

where $g_4 = g_3 g_2 \in G'_i$. Again, we see from line ($**$) that there must be an $h_4 \in \mathcal{U}'_{i-1}$ such that $h_4(i) = h_1 h_3(i)$ and so, $h_4^{-1} h_1 h_3 = g_5$ for some $g_5 \in G'_i$. Consequently

$$\begin{aligned}
x_1 x_2 &= h_1 h_3 g_4 \\
&= h_4 h_4^{-1} h_1 h_3 g_4 \\
&= h_4 g_5 g_4 \\
&= h_4 g_6 \in \mathcal{U}_{i-1} G'_i = G'_{i-1}.
\end{aligned}$$

Thus, $x_1 x_2 \in G'_{i-1}$, and so by Theorem 6.1, G'_{i-1} is a group. Therefore, by induction, G'_i is a group for all $i = 0, 1, 2, \ldots, n-1$. In particular, $G' = G'_0$ is a group. ∎

Each $\mathcal{U}_i$ contains at most $n - i$ permutations; so, there are at most $(n+1)n/2$ permutations in

$$\bigcup_{i=0}^{n-1} \mathcal{U}_i.$$

Consequently, in Algorithm 6.9, line ($*$) of ENTER is executed at most $O(n^2)$ times. Following each execution of line ($*$) there are

$$\sum_{j=0}^{i} |\mathcal{U}_i|$$

recursive calls to ENTER . Hence, there are in total $O(n^4)$ calls to ENTER . Each such call is preceded by a multiplication on line ($**$) and followed by a call to TEST at the start of ENTER . The multiplication is an $O(n)$ operation and TEST is $O(n^2)$. Thus Procedure ENTER requires at most $O(n^6)$ elementary operations.

Example 6.7

Applying Algorithm 6.9 to the generators

$$\alpha = (0, 1, 3, 7, 6, 4)(2, 5)$$

and

$$\beta = (0, 1, 3, 2)(4, 5, 7, 6)$$

of the automorphism group of the cube, we obtain:

$$
\mathcal{U}_0 = \left\{
\begin{array}{l}
(0)(1)(2)(3)(4)(5)(6)(7) \\
(0, 1, 3, 7, 6, 4)(2, 5) \\
(0, 2, 6, 4)(1, 3, 7, 5) \\
(0, 3, 6)(1, 7, 4)(2)(5) \\
(0, 4, 6, 7, 3, 1)(2, 5) \\
(0, 5, 3, 6)(1, 7, 2, 4) \\
(0, 6, 3)(1, 4, 7)(2)(5) \\
(0, 7)(1, 6)(2, 5)(3, 4)
\end{array}
\right\},
$$

$$
\mathcal{U}_1 = \left\{
\begin{array}{l}
(0)(1)(2)(3)(4)(5)(6)(7) \\
(0)(1, 2)(3)(4)(5, 6)(7) \\
(0)(1, 4, 2)(3, 5, 6)(7)
\end{array}
\right\},
$$

$$
\mathcal{U}_2 = \left\{
\begin{array}{l}
(0)(1)(2)(3)(4)(5)(6)(7) \\
(0)(1)(2, 4)(3, 5)(6)(7)
\end{array}
\right\}, \text{ and}
$$

$$\mathcal{U}_3 = \mathcal{U}_4 = \cdots = \mathcal{U}_7 = \{(0)(1)(2)(3)(4)(5)(6)(7)\}$$

Observe that $|\mathcal{U}_0| \cdot |\mathcal{U}_1| \cdot |\mathcal{U}_2| = (8)(3)(2) = 48$, the order of the group. □

6.2.4 Changing the base

In Section 6.3 we describe a backtracking algorithm that computes the subset with least rank in the orbit of a subset under a given group. This algorithm requires a slight modification to the Schreier-Sims representation and two of the procedures in Section 6.2.3. These new procedures will also be used in Section 7.3.3.

Let β be a permutation on $\{0, 1, 2, \ldots, n-1\}$, called the *base*, and define

$$G_0 = \{g \in G : g(\beta(0)) = \beta(0)\}$$

and

$$G_i = \{g \in G_{i-1} : g(\beta(i)) = \beta(i)\}$$

for $i = 1, 2, \ldots, n - 1$. Then arguments similar to those in Section 6.2.3 will show that

$$G \supseteq G_0 \supseteq G_1 \supseteq \cdots \supseteq G_{n-1} = \{\mathbf{I}\}$$

are again subgroups. Hence we can choose sets $\mathcal{U}_0, \mathcal{U}_1, \ldots, \mathcal{U}_{n-1}$ of left coset representatives so that

$$G = \mathcal{U}_0 G_0,$$

and

$$G_{i-1} = \mathcal{U}_i G_i$$

for $i = 1, 2, \ldots, n - 1$. Furthermore, if $f \in G$, and $f(\beta(i)) = \beta(i)$ for $i = 0, 1, \ldots, \ell$, then there is a unique $g \in \mathcal{U}_\ell$ such that $g(\ell) = f(\ell)$. Consequently, we modify that data structure to include the base β. That is

$$\vec{G} = (\beta; [\mathcal{U}_0, \mathcal{U}_1, \ldots, \mathcal{U}_{n-1}]).$$

The Procedures TEST and ENTER need to be modified to take into account the base β. This we do in Algorithms 6.10 and 6.11.

Algorithm 6.10: TEST2 $(n, g, \vec{G} = (\beta; [\mathcal{U}_0, \ldots, \mathcal{U}_{n-1}]))$

external MULT$()$, INV$()$

for $i \leftarrow 0$ **to** $n - 1$

$\qquad$ **do** $\begin{cases} x \leftarrow g[\beta[i]] \\ \textbf{if there is an } h \in \mathcal{U}_i \text{ such that } h[\beta[i]] = x \\ \qquad \textbf{then} \begin{cases} \text{INV}(n, h, \pi_2) \\ \text{MULT}(n, \pi_2, g, \pi_3) \\ \textbf{for } j \leftarrow 0 \textbf{ to } n - 1 \\ \qquad \textbf{do } g[j] \leftarrow \pi_3[j] \end{cases} \\ \textbf{else return } (i) \end{cases}$

return (n)

Algorithm 6.11: ENTER2 $(n, g, \vec{G} = (\beta; [\mathcal{U}_0, \ldots, \mathcal{U}_{n-1}]))$

external TEST2$()$, MULT$()$

$i \leftarrow$ TEST2$(n, g, \vec{G} = (\beta; [\mathcal{U}_0, \ldots, \mathcal{U}_{n-1}]))$

if $i = n$

$\qquad$ **then return**

$\qquad$ **else** $\mathcal{U}_i \leftarrow \mathcal{U}_i \cup \{g\}$ $\qquad\qquad\qquad\qquad\qquad$ (*)

for $j \leftarrow 0$ **to** i

$\qquad$ **do** $\begin{cases} \textbf{for each } h \in \mathcal{U}_j \\ \qquad \textbf{do} \begin{cases} \text{MULT}(n, g, h, f) & (**) \\ \text{ENTER2}(n, f, \vec{G} = (\beta; [\mathcal{U}_0, \ldots, \mathcal{U}_{n-1}])) \end{cases} \end{cases}$

To change the base β of $\vec{G} = (\beta; [\mathcal{U}_0, \ldots, \mathcal{U}_{n-1}])$ to a new base β' we can simply construct an identity group H with respect to new base β' and use ENTER2 to enter the members of $\mathcal{U}_0 \cup \cdots \cup \mathcal{U}_s$ into $H = (\beta'; [\mathcal{U}_0', \ldots, \mathcal{U}_{n-1}'])$. This task is accomplished by Algorithm 6.12.

Algorithm 6.12: CHANGEBASE $(n, \vec{G} = (\beta; [\mathcal{U}_0, \ldots, \mathcal{U}_{n-1}]), \beta')$

external ENTER2()
for $j \leftarrow 0$ **to** $n - 1$
  do $\mathcal{U}_j' \leftarrow \{I\}$
$\vec{H} \leftarrow (\beta'; [\mathcal{U}_0', \ldots, \mathcal{U}_{n-1}'])$
for $j \leftarrow 0$ **to** $n - 1$
  do $\begin{cases} \textbf{for each } g \in \mathcal{U}_j \\ \quad \textbf{do } \text{ENTER2}(n, g, \vec{H} = (\beta'; [\mathcal{U}_0', \ldots, \mathcal{U}_{n-1}'])) \end{cases}$
$\vec{G} \leftarrow \vec{H}$

6.3 Orbits of subsets

If G is a permutation group on $\mathcal{X}$ and $S \subseteq \mathcal{X}$, then we define the *image* of S under g to be

$$g(S) = \{g(x) : x \in S\}.$$

Thus G also permutes the subsets of $\mathcal{X}$. The family of subsets

$$G(S) = \{g(S) : g \in G\}$$

is called the *orbit* of S under G. If a set system has G as an automorphism group, then the set of blocks of the set system are a union of orbits under G. Thus the generation of the orbits of subsets is crucial in determining set systems with a given automorphism group.

If $S \subseteq \mathcal{X}$, then the *stabilizer* of S in G is

$$G_S = \{g \in G : g(S) = S\}.$$

It is easy to show that G_S is indeed a subgroup of G.

LEMMA 6.9 *Let G be a permutation group on $\mathcal{X}$ and let $S \subseteq \mathcal{X}$. Then*

$$|G| = |G(S)| \cdot |G_S|.$$

PROOF   Write $G(S) = \{S_1, S_2, \ldots, S_n\}$ and for each i choose a $g_i \in G$ such that $g_i(S) = S_i$. Suppose $g \in G$. Then $g(S) = S_j$ for a unique j, $0 \leq j \leq n$. Set $h = g_j^{-1}g$. Then $h(S) = g_j^{-1}g(S) = g_j^{-1}(S_j) = S$. So $h \in G_S$. Consequently every element g of G can be uniquely written as $g = g_j h$ where $1 \leq j \leq n$ and $h \in G_S$. Therefore, $|G| = n \cdot |G_S| = |G(S)| \cdot |G_S|$, as claimed.   ∎

Example 6.8

In Table 6.1 we list the members of the automorphism group G of the graph of the cube. If $S = \{1, 2, 7\}$, then

$$G_S = \left\{ \begin{array}{l} \mathbf{I}, (1,2)(5,6), (0,5,6)(1,7,2), (0,5)(2,7), \\ (0,6)(1,7), (0,6,5)(1,2,7) \end{array} \right\}$$

and

$$G(S) = \left\{ \begin{array}{l} \{1,2,7\}, \{3,5,6\}, \{2,4,7\}, \{0,3,6\}, \\ \{1,2,4\}, \{0,5,3\}, \{1,4,7\}, \{0,5,6\} \end{array} \right\}.$$

▯

6.3.1   Burnside's lemma

Although our main purpose for introducing Lemma 6.9 is to prove Theorem 6.10, it is also useful for other tasks. For example, let G be the automorphism group of the complete graph K_n; then $|G| = n!$, since G is the symmetric group on the vertices of K_n. But G can also be thought of as a permutation group on the $\binom{n}{2}$ edges $\mathcal{X}$ of K_n. If S is a (spanning) subgraph of K_n, then the edges of S are a subset of the edges $\mathcal{X}$ and the vertices of S are the vertices of K_n. The stabilizer G_S of S in G is the set of permutations of the vertices of K_n that leave S unchanged. That is G_S is $\mathrm{Aut}(S)$, the automorphism group of S. Hence applying Lemma 6.9 we see that there are $n!/|\mathrm{Aut}(S)|$ isomorphic copies of S in K_n. There are for example $6!/10 = 72$ ways to label the pentagon with 6 labels since the automorphism group of the pentagon in K_6 has 10 elements. In K_8 there would be $8!/(10 \cdot 3!)$; the three additional isolated points can be permuted independently of the pentagon. There is a unique Steiner triple system of order 7 and its automorphism group in S_7 has order 168. There are therefore $9!/(168 \cdot 2!) = 1080$ ways to choose a Steiner triple system of order 7 in a set of 9 points. Such calculations are useful for determining the feasibility of a search or for checking results.

   We now use Lemma 6.9 to establish the beautiful and useful theorem of Frobenius, Cauchy and Burnside that counts the number of orbits. First, if $g \in G$, let $\chi_k(g)$ denote the number of k-element subsets fixed by g,

$$\chi_k(g) = |\{S \subseteq \mathcal{X} : |S| = k \text{ and } g(S) = S\}|.$$

THEOREM 6.10 *Let G be a permutation group on $\mathcal{X}$. Then the number of orbits of k-element subsets of $\mathcal{X}$ under G is*

$$N_k = \frac{1}{|G|} \sum_{g \in G} \chi_k(g).$$

PROOF Define an array whose rows are labeled by the elements of G and whose columns are labeled by the k-element subsets of $\mathcal{X}$. The $[g, S]$-entry of the array is a 1 if $g(S) = S$ and is 0 otherwise. Thus the sum of the entries of row g is precisely $\chi_k(g)$ and the sum of the entries in column S is $|G_S|$. Hence

$$\sum_{g \in G} \chi_k(g) = \sum_{S \subseteq \mathcal{X}, |S|=k} |G_S|. \tag{6.3}$$

Now partition the k-element subsets into the N_k orbits $\mathcal{O}_1, \mathcal{O}_2, \ldots, \mathcal{O}_{N_k}$ under G. Choose a fixed representative $S_i \in \mathcal{O}_i$ for each $i = 1, 2, \ldots, N_k$. Then for all $S \in \mathcal{O}_i$, $|G_S| = |G_{S_i}|$ and the right-hand side of Equation 6.3 may be rewritten and Lemma 6.9 can be applied.

$$\sum_{g \in G} \chi_k(g) = \sum_{i=1}^{N_k} |G_{S_i}| \cdot |G(S_i)|$$

$$= \sum_{i=1}^{N_k} |G|$$

$$= N_k |G|.$$

This establishes the result. ∎

In order to present an algorithm for computing the number of orbits of k-subsets, we define the *type of a permutation g* by

$$\text{type}(g) = [t_1, t_2, \ldots, t_n],$$

where t_ℓ is the number of cycles of length ℓ in the cycle decomposition of g. We can compute $\text{type}(g)$ in time $O(n)$ with Procedure TYPE given in Algorithm 6.13. Notice the similarity of this procedure and Algorithm 6.4.

If S is a k-element subset fixed by g, then S is a union of cycles of g. Suppose $\text{type}(g) = [t_1, t_2, \ldots, t_n]$ and S uses c_ℓ cycles of length ℓ, $\ell = 1, 2, \ldots, n$. Then

$$c_\ell \leq t_\ell,$$

$$k = \sum_{\ell=1}^{n} i\, c_\ell,$$

and the number of such fixed subsets is

$$\prod_{\ell=1}^{n} \binom{t_\ell}{c_\ell}.$$

Using this information, Algorithm 6.13 computes the number N_k of orbits of k-element subsets for each $k = 0, 1, 2, \ldots, |\mathcal{X}|$. The procedure RECPARTITION used in this algorithm is a modification of Algorithm 3.3.

Algorithm 6.13: NORB $(n, \vec{G})$

external RUN()
global *DoneEarly*
procedure TYPE(n, g, T)

$\quad$ **for** $i \leftarrow 0$ **to** $n - 1$ **do** $\begin{cases} P[i] \leftarrow \textbf{true} \\ T[i + 1] = 0 \end{cases}$

$\quad$ **for** $i \leftarrow 0$ **to** $n - 1$

$\quad$ **do** $\begin{cases} \textbf{if } P[i] \\ \\ \textbf{then} \begin{cases} \ell \leftarrow 1 \\ j \leftarrow i \\ P[j] \leftarrow \textbf{false} \\ \textbf{while } P[g[j]] \textbf{ do } \begin{cases} \ell = \ell + 1 \\ j \leftarrow g[j] \\ P[j] \leftarrow \textbf{false} \end{cases} \\ T[\ell] \leftarrow T[\ell] + 1 \end{cases} \end{cases}$

procedure RECPARTITION(n, k, m, B, N)

$\quad$ **if** $m = 0$

$\quad$ **then** $\begin{cases} \textbf{for } i \leftarrow 1 \textbf{ to } n \textbf{ do } c[i] \leftarrow 0 \\ \textbf{for } i \leftarrow 1 \textbf{ to } N \textbf{ do } c[V_2[i]] \leftarrow c[V_2[i]] + 1 \\ prod \leftarrow 1 \\ \textbf{for } i \leftarrow 1 \textbf{ to } n \textbf{ do } prod \leftarrow prod \binom{T[i]}{c[i]} \\ V_3[k] = V_3[k] + prod \end{cases}$

$\quad$ **else** $\begin{cases} \textbf{for } i \leftarrow \min(B, N) \textbf{ to } i \\ \textbf{do} \begin{cases} V_2[N + 1] \leftarrow i \\ \text{RECPARTITION}(n, k, m - i, i, N + 1) \end{cases} \end{cases}$

procedure NORBUSE(n, g)

$\quad$ TYPE(n, g, V_1)
$\quad$ **for** $k \leftarrow 0$ **to** $k \leq n$
$\quad$ **do** $\begin{cases} V_2[1] \leftarrow k \\ \text{RECPARTITION}(n, k, k, k, 0) \end{cases}$

main

$\quad$ *DoneEarly* $\leftarrow$ **false**
$\quad$ **for** $k \leftarrow 0$ **to** n **do** $V_3[k] \leftarrow 0$
$\quad$ RUN$(n, \vec{G}, \text{NORBUSE}())$
$\quad$ **for** $k \leftarrow 0$ **to** n **do** $N[k] \leftarrow V_3[k]/|G|$

Example 6.9 *The number of non-isomorphic graphs on four vertices*

To count the number of graphs on four vertices, Theorem 6.10 (or Algorithm 6.13) can be used as follows. Let $G = \mathsf{Sym}(\{1, 2, 3, 4\})$ and label the edges of K_4 as in Figure 6.3.

Then each permutation of the vertices induces a permutation of the edges. For example,

$$g = (1, 2, 3) \mapsto (a, b, c)(d, e, f).$$

Thus, for instance, $\chi_2(g) = 0$ and $\chi_3(g) = 2$. That is, g fixes no subgraphs with two edges and it fixes two subgraphs with three edges. We tabulate this information in Table 6.2 for all elements of G. The last row of Table 6.2 gives N_k, the number of non-isomorphic subgraphs of K_4 with k edges, for $k = 0, 1, 2, \ldots, 6$.
☐

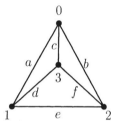

FIGURE 6.3
Edge labeling of K_4.

6.3.2 Computing orbit representatives

Now that we know how to determine the number of orbits of k-element subsets of $\mathcal{X}$ under a permutation group G, we will proceed to compute the orbits themselves. It suffices to have an algorithm that will find one representative of each orbit. The actual orbits can then be constructed by running through all the elements of G, applying them to the orbit representatives, and removing duplicates. First, we compute the number N_k of orbits of k-element subsets under G. Suppose $\mathcal{R}$ is a set of orbit representatives for the orbits of k-element subsets of $\mathcal{X}$ under the permutation group G. Then $|\mathcal{R}| = N_k$. Let

$$\mathcal{S} = \{A \cup \{x\} : A \in \mathcal{R} \text{ and } x \in \mathcal{X} \setminus A\}. \tag{6.4}$$

Then $\mathcal{S}$ contains at least one representative from each orbit of $(k+1)$-element subsets under G and $|\mathcal{S}| \leq (|\mathcal{X}| - k - 1)N_k$. We need only collapse $\mathcal{S}$ down to N_{k+1} elements. The rest of this section will discuss Algorithms 6.14 and 6.16, which are two methods for carrying out this collapsing. Both Algorithm 6.14 and 6.16 maintain $\mathcal{S}$ as a list in lexicographic order, as computed by Algorithm 2.1. (Any ranking algorithm can be used.) Four operations are allowed on this list:

TABLE 6.2
Numbers of non-isomorphic subgraphs in K_4.

g			χ_0	χ_1	χ_2	χ_3	χ_4	χ_5	χ_6
$(0)(1)(2)(3)$	$\mapsto$	$(a)(b)(c)(d)(e)(f)$	1	6	15	20	15	6	1
$(0)(1)(2,3)$	$\mapsto$	$(a)(b,c)(d,e)(f)$	1	2	3	4	3	2	1
$(0)(1,2)(3)$	$\mapsto$	$(a,b)(c)(d)(e,f)$	1	2	3	4	3	2	1
$(0)(1,2,3)$	$\mapsto$	$(a,b,c)(d,f,e)$	1	0	0	2	0	0	1
$(0)(1,3,2)$	$\mapsto$	$(a,c,b)(d,e,f)$	1	0	0	2	0	0	1
$(0)(1,3)(2)$	$\mapsto$	$(a,c)(b)(d,f)(e)$	1	2	3	4	3	2	1
$(0,1)(2)(3)$	$\mapsto$	$(a)(b,d)(c,e)(f)$	1	2	3	4	3	2	1
$(0,1)(2,3)$	$\mapsto$	$(a)(b,e)(c,d)(f)$	1	2	3	4	3	2	1
$(0,1,2)(3)$	$\mapsto$	$(a,d,b)(c,e,f)$	1	0	0	2	0	0	1
$(0,1,2,3)$	$\mapsto$	$(a,d,f,c)(b,e)$	1	0	1	0	1	0	1
$(0,1,3,2)$	$\mapsto$	$(a,e,f,b)(c,d)$	1	0	1	0	1	0	1
$(0,1,3)(2)$	$\mapsto$	$(a,e,c)(b,d,f)$	1	0	0	2	0	0	1
$(0,2,1)(3)$	$\mapsto$	$(a,b,d)(c,f,e)$	1	0	0	2	0	0	1
$(0,2,3,1)$	$\mapsto$	$(a,b,f,e)(c,d)$	1	0	1	0	1	0	1
$(0,2)(1)(3)$	$\mapsto$	$(a,d)(b)(c,f)(e)$	1	2	3	4	3	2	1
$(0,2,3)(1)$	$\mapsto$	$(a,d,e)(b,f,c)$	1	0	0	2	0	0	1
$(0,2)(1,3)$	$\mapsto$	$(a,f)(b)(c,d)(e)$	1	2	3	4	3	2	1
$(0,2,1,3)$	$\mapsto$	$(a,f)(b,d,e,c)$	1	0	1	0	1	0	1
$(0,3,2,1)$	$\mapsto$	$(a,c,f,d)(b,e)$	1	0	1	0	1	0	1
$(0,3,1)(2)$	$\mapsto$	$(a,c,e)(b,f,d)$	1	0	0	2	0	0	1
$(0,3,2)(1)$	$\mapsto$	$(a,e,d)(b,c,f)$	1	0	0	2	0	0	1
$(0,3)(1)(2)$	$\mapsto$	$(a,e)(b,f)(c)(d)$	1	2	3	4	3	2	1
$(0,3,1,2)$	$\mapsto$	$(a,f)(b,c,e,d)$	1	0	1	0	1	0	1
$(0,3)(1,2)$	$\mapsto$	$(a,f)(b,e)(c)(d)$	1	2	3	4	3	2	1
		Sum	24	24	48	72	48	24	24
		Sum/$|G|$	1	1	2	3	2	1	1

1. LISTINSERT(S, A) which inserts A into the list S.

2. LISTDELETE(S, A) which deletes the set A from the list S.

3. PREDECESSOR(S, A) which returns the set $B \in S$ which precedes A in lexicographic order. If there is no such set, then "undefined" is returned.

4. MAXIMUM(S) which returns the last set $A \in S$ in lexicographic order.

We must also be able to test membership of a set in the list. A balanced binary tree can be used to implement this list data structure so that these operations can be processed in $O(\log |S|)$ time.

Algorithm 6.14 repeatedly applies the elements of G to the members of S. If $g \in G$ and A is on the list S, then the rank of $g(A)$ is computed. If it is less than the rank of A, then A is deleted from S and $g(A)$ is inserted. The list either has the same number of elements or has one fewer than before. This process is repeated until $|S| = N_{k+1}$. To run through the elements of G we modify Algorithm 6.6, obtaining the procedure ORBREPBACKTRACK. Each call from

main to this procedure will run through the elements of G at most once.

Algorithm 6.14: ORBREPS1 $(\mathcal{R}, \mathcal{S})$

external $\begin{cases} \text{MAXIMUM}(), \text{PREDECESSOR}(), \\ \text{SUBSETLEXRANK}(), \text{LISTINSERT}(), \text{MULT}() \end{cases}$

global *DoneEarly*

procedure ORBREPBACKTRACK$(n, \ell, \vec{G}, g, \mathcal{S}, N)$

  if *DoneEarly* **then return**

  if $\ell = n$ **then** $A \leftarrow \text{MAXIMUM}(\mathcal{S})$

  while A is defined

  do $\begin{cases} \text{\textbf{then}} \begin{cases} C \leftarrow \text{PREDECESSOR}(\mathcal{S}, A) \\ \textbf{if } \text{SUBSETLEXRANK}(g(A)) < \text{SUBSETLEXRANK}(A) \\ \quad \textbf{then} \begin{cases} \text{LISTDELETE}(\mathcal{S}, A) \\ \textbf{if } A \text{ is not in the list } \mathcal{S} \\ \quad \textbf{then } \text{LISTINSERT}(\mathcal{S}, A) \\ \textbf{if } |\mathcal{S}| = N \\ \quad \textbf{then} \begin{cases} DoneEarly \leftarrow \textbf{true} \\ \textbf{return} \end{cases} \end{cases} \\ A \leftarrow C \end{cases} \\ \textbf{else} \begin{cases} \textbf{for each } h \in \mathcal{U}_\ell \\ \quad \textbf{do} \begin{cases} \text{MULT}(n, g, h, f_\ell) \\ \text{ORBREPBACKTRACK}(n, \ell + 1, \vec{G}, f, \mathcal{S}, N) \end{cases} \end{cases} \end{cases}$

main

  for each A on the list $\mathcal{R}$

  do $\begin{cases} \textbf{for each } x \in \mathcal{X} \setminus A \\ \quad \textbf{do } \text{LISTINSERT}(\mathcal{S}, A \cup \{x\}) \end{cases}$

  DoneEarly $\leftarrow$ **false**

  while $|\mathcal{S}| > N_{k+1}$

  do ORBREPBACKTRACK$(n, 0, \vec{G}, \mathbf{I}, \mathcal{S}, N_{k+1})$

For Algorithm 6.16, we require the concept of a minimum orbit representative. If G is a permutation group on $\mathcal{X}$ and $A \subseteq \mathcal{X}$, then there is a unique $S \in G(A)$ that has smallest rank in lexicographic order. This subset we call the *minimum orbit representative* for the orbit $G(A)$. First we study how to compute the minimum orbit representative in the orbit $G(A)$ when we are only given A and G.

Example 6.10

In Table 6.1, we list the members of the automorphism group G of the graph of the cube. The orbit $G(A)$ of $A = \{1, 2, 7\}$ under G is computed in Example 6.8. The minimum representative for this orbit is $S = \{0, 3, 6\}$. □

Recall, from Section 2.3, that to each k-subset $S \subset \mathcal{X}$, we can associate the k-tuple

$$[s_0, s_1, s_2, ..., s_{k-1}]$$

where $s_0 < s_1 < s_2 < \cdots < s_{k-1}$ and $S = \{s_0, s_1, s_2, \ldots, s_{k-1}\}$. The lexicographic ordering on the k-subsets is induced from the natural lexicographic ordering on the k-tuples. Thus, to compute the minimum orbit representative $G(A)$ we must choose $s_0, s_1, \ldots, s_k$ so that the following properties are satisfied:

1. $s_0 < s_1 < s_2 < \cdots < s_{k-1}$;

2. $h(A) = \{s_0, s_1, s_2, \ldots, s_{k-1}\}$ for some $h \in G$; and

3. $[s_0, s_1, s_2, \ldots, s_{k-1}]$ is smallest in lexicographic order.

We will use the backtracking methods of Chapter 4 to search for the minimum orbit representative and we will prune this search with the automorphisms in the group G.

Suppose that $s_0, s_1, \ldots, s_{\ell-1}$ have been selected and we must choose a value for s_ℓ. Then there is a permutation $h \in G$ and elements $a_0, a_1, \ldots, a_{\ell-1} \in A$ such that $h(a_i) = s_i$, $i = 0, 1, \ldots, \ell - 1$. To ensure that property 3 is satisfied we need to select

$$x \in h(A) \setminus \{s_0, s_1, \ldots, s_{\ell-1}\}$$

and $g \in G$ such that

$$g(s_i) = s_i$$

for $i = 0, 1, \ldots, \ell - 1$ and such that

$$m = g(x)$$

is as small as possible. Once this selection is made, then we set s_ℓ equal to m and continue the search. There are, of course, several such elements x, and to examine each one requires a backtracking algorithm. To reduce the computation in selecting g, we can change the base β of the Schreier-Sims representation

$$\vec{G} = (\beta; [\mathcal{U}_0, \mathcal{U}_1, \ldots, \mathcal{U}_{n-1}])$$

of the group G so that $\beta(i) = s_i$ for $i = 0, 1, \ldots, \ell - 1$ and $\beta(\ell) = x$. The remaining entries of β can be chosen arbitrarily so that β is a permutation. Now there is a $g \in G$ satisfying

$$g(s_i) = s_i$$

for $i = 0, 1, \ldots, \ell - 1$ and

$$m = g(x)$$

if and only if there is a $g' \in \mathcal{U}_\ell$ such that $m = g'(x)$. Hence, we need only consider $g \in \mathcal{U}_\ell$. Furthermore, having chosen $g \in \mathcal{U}_\ell$, if C_ℓ is the set of possible choices for x, then the set of choices for the x that determines $s_{\ell+1}$ is

$$C_{\ell+1} = g(C_\ell) \setminus \{m\}.$$

These observations, together with the techniques described in Chapter 4, lead to the backtracking algorithm, Algorithm 6.15.

Algorithm 6.15: MINREP (n, G, A)

procedure MINREPBT(n, k, G, ℓ)
  external CHANGEBASE()
  global $\mathcal{C}_0, \mathcal{C}_1, \ldots, \mathcal{C}_{n-1}$
  $m \leftarrow n$
  for each $x \in \mathcal{C}_\ell$
  do
$\begin{cases}
r \leftarrow 0 \\
\textbf{while } r < \ell \textbf{ and } s_r \neq OptS[r] \textbf{ do } r \leftarrow r + 1 \\
\textbf{if } r < \ell \textbf{ and } s_r \geq OptS[r] \;\; \textbf{then return} \\
\textbf{for } i \leftarrow 0 \textbf{ to } \ell - 1 \textbf{ do } \beta[i] \leftarrow s_i \\
\beta[\ell] \leftarrow x \\
i \leftarrow \ell \\
\textbf{for each } y \notin \{s_0, s_1, \ldots, s_{\ell-1}, x\} \textbf{ do } \begin{cases} i \leftarrow i + 1 \\ \beta[i] \leftarrow y \end{cases} \\
\text{CHANGEBASE}(n, G, \beta) \\
\text{Choose } g \in \mathcal{U}_\ell \text{ such that } g[x] \text{ is smallest} \\
\textbf{if } g[x] \leq m \\
\textbf{then}
\begin{cases}
m \leftarrow g[x] \\
s_\ell \leftarrow m \\
\textbf{if } k = \ell + 1 \\
\textbf{then}
\begin{cases}
i \leftarrow r \\
\textbf{while } i < k \textbf{ and } s_i = OptS[i] \textbf{ do } i \leftarrow i + 1 \\
\textbf{if } i \neq k \textbf{ and } s_i < OptS[i] \\
\textbf{then } \begin{cases} \textbf{for } i \leftarrow 0 \textbf{ to } k - 1 \\ \quad \textbf{do } OptS[i] \leftarrow s_i \end{cases}
\end{cases} \\
\textbf{else } \begin{cases} \mathcal{C}_{\ell+1} \leftarrow g(\mathcal{C}_\ell) \setminus \{m\} \\ \text{MINREPBT}(n, k, G, \ell + 1) \end{cases}
\end{cases}
\end{cases}$

main
  $k \leftarrow 0$
  for each $i \in A$ **do** $\begin{cases} OptS[k] \leftarrow i \\ k \leftarrow k + 1 \end{cases}$
  $\mathcal{C}_0 \leftarrow A$
  MINREPBT$(n, k, G, 0)$
  return $(\{OptS[i] : i = 0, 1, 2, \ldots, k - 1\})$

Now that we can compute minimum orbit representatives, the set $\mathcal{S}$ given in Equation 6.4 can be collapsed by first replacing each set $B \in \mathcal{S}$ by the minimum orbit representative in $G(B)$, and then deleting duplicates. The computation time required for this method can be reduced if we change the base of the group. Given $A \in \mathcal{R}$, let $c = \mathcal{X} \setminus A$. Then each of the sets $A \cup \{x\}$, where $x \in c$, must be examined for inclusion in A. However, once $A \cup \{x\}$ has been examined, then no set in the orbit $G(A \cup \{x\})$ need be considered. In particular, if the base β of G is defined so that

$$\{\beta[0], \beta[1], \dots, \beta[k-1]\} = A,$$

and

$$\beta[k] = x,$$

where $k = |A|$, then $A \cup \{g[x]\} \in G(A \cup \{x\})$ for all $g \in \mathcal{U}_k$. Hence, the set of points $\{g(x) : g \in \mathcal{U}_k\}$ can be deleted from c. Another observation that helps to reduce the computation time is to note that if $\mathcal{R}$ is a set of minimum representatives of the orbits of k-subsets and $A \in \mathcal{R}$, then the minimum representative of the orbit $G(A \cup \{x\})$, where $x \notin A$, is of the form $A' \cup \{x'\}$ where $A' \in \mathcal{R}$ and $x' > \max\{y : y \in A'\}$. These ideas give us Algorithm 6.16.

Algorithm 6.16: ORBREPS2 (n, R, S)

external MINREP(), LISTINSERT(), CHANGEBASE()

for each A in the list $\mathcal{R}$

$$\mathbf{do} \begin{cases} c \leftarrow \{x \in \mathcal{X} : x > \max\{y : y \in A\}\} \\ \mathbf{for\ each}\ x \in c \\ \quad \mathbf{do} \begin{cases} b \leftarrow A \cup \{x\} \\ \text{MINREP}(n, G, b, a) \\ \mathbf{if}\ a\ \text{is not in the list}\ S \\ \quad \mathbf{then}\ \text{LISTINSERT}(S, a) \\ k \leftarrow 0 \\ \mathbf{for\ each}\ y \in A \\ \quad \mathbf{do} \begin{cases} \beta[k] \leftarrow y \\ k \leftarrow k+1 \end{cases} \\ \beta[k] \leftarrow x \\ j \leftarrow k+1 \\ \mathbf{for\ each}\ y \notin A \cup \{x\} \\ \quad \mathbf{do} \begin{cases} \beta[j] \leftarrow y \\ j \leftarrow j+1 \end{cases} \\ \text{CHANGEBASE}(n, G, \beta) \\ \mathbf{for\ each}\ g \in \mathcal{U}_k \\ \quad \mathbf{do}\ c \leftarrow c \setminus \{g[x]\} \end{cases} \end{cases}$$

6.4 Coset representatives

Suppose that G is a permutation group on the set $\mathcal{X}$ and that $(\mathcal{X}, \mathcal{B})$ is a set system. If $g \in G$ we define the image of $\mathcal{B}$ under g to be the set system on $\mathcal{X}$ whose blocks are

$$g(\mathcal{B}) = \{g(S) : S \in \mathcal{B}\}.$$

Then the stabilizer of $\mathcal{B}$ in G is the subgroup

$$H = \{g \in G : g(\mathcal{B}) = \mathcal{B}\}$$

of G. Thus we can find a left transversal $\{g_1, g_2, \ldots, g_r\}$ of H in G; see Section 6.1.

THEOREM 6.11 *Let G be a permutation group on $\mathcal{X}$ and let $(\mathcal{X}, \mathcal{B})$ be a set system. If H is the subgroup*

$$H = \{g \in G : g(\mathcal{B}) = \mathcal{B}\}$$

and $\{g_1, g_2, \ldots, g_r\}$ is a left transversal of H in G, then the orbit of $\mathcal{B}$ under G has length $r = |G|/|H|$ and is given by

$$G(\mathcal{B}) = \{g_1(\mathcal{B}), g_2(\mathcal{B}), \ldots, g_r(\mathcal{B})\}.$$

PROOF Since $\{g_1, g_2, \ldots, g_r\}$ is a left transversal, then by Theorem 6.2 $|G|/|H| = r$ and we can write G as the disjoint union

$$G = g_1 H \,\dot\cup\, g_2 H \,\dot\cup\, \cdots \,\dot\cup\, g_r H.$$

Thus if $g \in G$, then $g = g_i h$ for some i and $h \in H$. Hence

$$\begin{aligned} g(B) &= (g_i h)(B) \\ &= g_i(h(B)) \\ &= g_i(B). \end{aligned}$$

Furthermore, if $g_i(B) = g_j(B)$ for some $i \neq j$, then $(g_j^{-1} g_i)(B) = B$ and so $g_j^{-1} g_j \in H$. Hence the left cosets $g_j H$ and $g_i H$ are identical, contradicting that $\{g_1, g_2, \ldots, g_r\}$ is a left transversal. ∎

Consequently, computing a left transversal of a subgroup H of G can be very useful, since it can be used to compute the orbit of $\mathcal{B}$ under G.

Algorithm 6.17: TRANSVERSAL $(n, m, \vec{H}, \vec{G})$

external MULT(), TEST()
global *DoneEarly*
procedure USE(n, g)
  for each $f \in R$
    $\mathbf{do} \begin{cases} h \leftarrow \text{MULT}(n, \text{INV}(f), g) \\ \textbf{if } |\mathcal{X}| = \text{TEST}(n, h, \vec{H}) \\ \quad \textbf{then return} \end{cases}$
  $R \leftarrow R \cup \{g\}$
  if $|R| \geq m$
    then *DoneEarly* $\leftarrow$ **true**

main
  DoneEarly $\leftarrow$ **false**
  $m \leftarrow |G|/|H|$
  $R \leftarrow \emptyset$
  RUN$(n, \vec{G}, \text{USE})$
  return (R)

Example 6.11

In Example 6.3, the automorphism group $H = \text{Aut}(P)$ is displayed. It is a subgroup of $G = \text{Sym}(\{0, 1, 2, 3, 4\})$. A transversal of H in G is

$$\left\{ \begin{array}{l} \mathbf{I}, (3,4), (2,3), (2,3,4), (2,4,3), (2,4), \\ (1,2), (1,2)(3,4), (1,3,2), (1,3,4,2), (1,4,3,2), (1,4,2) \end{array} \right\},$$

and the orbit of the pentagon under G is displayed in Figure 6.4. $\square$

6.5 Orbits of k-tuples

A k-*tuple* on the m-element set $\mathcal{X}$ is a sequence $\sigma = [x_1, x_2, \ldots, x_k]$ with entries from $\mathcal{X}$. If all the entries are distinct, then σ is called a k-*permutation* and the set of all k-permutations of $\mathcal{X}$ is denoted by k-$\Pi(\mathcal{X})$. Let G be a permutation group on $\mathcal{X}$. If $\sigma = [x_1, x_2, \ldots, x_k]$ is a k-tuple on $\mathcal{X}$ and $g \in G$, then we define

$$g(\sigma) = [g(x_1), g(x_2), \ldots, g(x_k)].$$

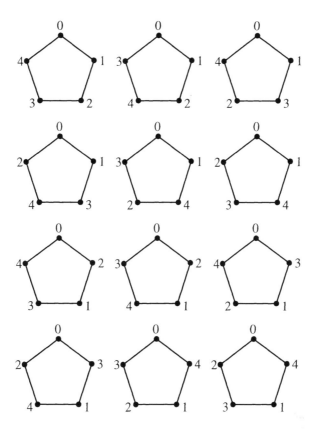

FIGURE 6.4
The orbit of the pentagon.

The orbit of σ is

$$G(\sigma) = \{g(\sigma) : g \in G\}.$$

For $g \in G$, we let $\vec{\chi}_k(g)$ denote the number of k-permutations fixed by g. Then

$$\vec{\chi}_k(g) = |\{\sigma \in k\text{-}\Pi(\mathcal{X}) : g(\sigma) = \sigma\}|.$$

Analogously to Theorem 6.10, we have

THEOREM 6.12 *Let G be a permutation group on $\mathcal{X}$. Then the number of orbits of k-permutations of $\mathcal{X}$ under G is*

$$\vec{N}_k = \frac{1}{|G|} \sum_{g \in G} \vec{\chi}_k(g).$$

It is significantly easier to develop an algorithm to compute the number of orbits of k-permutations (as well as the orbits themselves) than it is for k-subsets. This is because a k-permutation is fixed if and only if all of its points are fixed. We leave these algorithms for exercises.

The number $\vec{N}_k^\star$ of orbits of k-tuples can be related to the number $\vec{N}_k$ of orbits of k-permutations on $\mathcal{X}$ using the Stirling numbers of the second kind, $S(m, n)$. These were introduced in Section 3.2.

Given a k-tuple $\sigma = [x_1, x_2, \ldots, x_k]$, we construct a partition π_σ of $\{1, 2, \ldots, n\}$ by assigning i and j to the same block of π_σ if and only if $x_i = x_j$. If the partition π_σ has n blocks, then σ has n distinct entries. Let $v_\sigma = [x_1', x_2', \ldots x_n']$ be, in the order in which they first appear, the distinct entries in σ. For example, if $\sigma = [3, 1, 4, 5, 3, 2, 4, 1, 4]$, then

$$\pi_\sigma = [\{1, 5\}, \{2, 8\}, \{3, 7, 9\}, \{4\}, \{6\}]$$

and

$$v_\sigma = [3, 1, 4, 5, 2].$$

We now see that two k-tuples σ and τ are in the same orbit under G if and only if

$$\pi_\sigma = \pi_\tau$$

and v_σ and v_τ are in the same orbit of G. If $|\mathcal{X}| = m$, then there are $S(m, n)$ ways to partition $\mathcal{X}$ into n parts, and for each partition there are $\vec{N}_n$ n-permutations. Multiplying and summing over n gives the following result.

THEOREM 6.13

$$\vec{N}_k^\star = \sum_{n=1}^k S(m, n) \vec{N}_n.$$

6.6   Generating objects having automorphisms

When a desired set system is too large or too complicated to search for with backtracking methods, other properties of the set system must be assumed. One approach is to assume that the set system has certain automorphisms. Searching the orbits under the assumed automorphism group is often a much easier task. If the proper group is chosen, the set system may be found. This approach has been very successful and will be discussed in the next few sections. Also, knowing an automorphism group of the set system often provides a much shorter description of the object. For example, a Steiner triple system, STS(25), (see Exercise 4.11 and Section 5.4) is given by the orbits of the three blocks

$$\{0, 8, 13\}, \quad \{0, 2, 3\}, \quad \{0, 4, 11\}, \quad \{0, 6, 15\}$$

under the cyclic group G generated by

$$\alpha = (0, 1, 2, 3, \ldots, 23, 24).$$

This is a much smaller presentation than listing all 100 blocks in an STS(25).

6.6.1 Incidence matrices

One popular method for using automorphisms is to construct an incidence matrix. Recall that if G is a permutation group on $\mathcal{X}$, then, for any positive integer $j \leq |\mathcal{X}|$, the number of orbits of j-element subsets of $\mathcal{X}$ under G is denoted by N_j and can be computed using Theorem 6.10.

LEMMA 6.14 *Let G be a permutation group on $\mathcal{X}$, let Δ be any orbit of t-subsets and let Γ be any orbit of k-subsets. Then the quantity*

$$|\{K \in \Gamma_j : K \supset T_0\}|$$

is independent of the choice of representative $T_0 \in \Delta$.

PROOF If $T_0, T_0' \in \Delta$, then there is a $g \in G$ such that $g(T_0) = T_0'$. If $T_0 \subseteq K \in \Gamma$, then $T_0' \subseteq g(K)$. Since g is one-to-one it follows that the number of k-subsets in Γ that contain T_0 is the same as the number that contain T_0'. ∎

Definition 6.6: Let G be a permutation group on $\mathcal{X}$ and suppose $0 \leq j \leq k \leq |\mathcal{X}|$. Then the *orbit incidence matrix* is the N_t by N_k matrix A_{tk} such that

1. the rows of A_{tk} are labeled by the orbits $\Delta_1, \Delta_2, \ldots, \Delta_{N_t}$ of t-element subsets;

2. the columns of A_{tk} are labeled by the orbits $\Gamma_1, \Gamma_2, \ldots, \Gamma_{N_k}$ of k-element subsets; and

3. the $[\Delta_i, \Gamma_j]$-entry is

$$A_{tk}[\Delta_i, \Gamma_j] = |\{K \in \Gamma_j : K \supseteq T_0\}|$$

with $T_0 \in \Delta_i$ any fixed representative.

Example 6.12 *Two orbit incidence matrices.*

1. Let G be the group generated by the two permutations

$$(0)(1, 2, 3)(4, 5, 6)$$

and

$$(0)(1, 2)(5, 6).$$

The orbits of pairs under G are:

$$\Delta_1 = \{\{4, 5\}, \{4, 6\}, \{5, 6\}\}$$
$$\Delta_2 = \{\{1, 5\}, \{2, 6\}, \{3, 4\}\}$$
$$\Delta_3 = \{\{1, 4\}, \{1, 6\}, \{2, 4\}, \{2, 5\}, \{3, 5\}, \{3, 6\}\}$$
$$\Delta_4 = \{\{1, 2\}, \{1, 3\}, \{2, 3\}\}$$
$$\Delta_5 = \{\{0, 4\}, \{0, 5\}, \{0, 6\}\}$$
$$\Delta_6 = \{\{0, 1\}, \{0, 2\}, \{0, 3\}\}$$

The orbits of triples under G are:

$$\Gamma_1 = \{\{4, 5, 6\}\}$$
$$\Gamma_2 = \{\{1, 4, 6\}, \{2, 4, 5\}, \{3, 5, 6\}\}$$
$$\Gamma_3 = \{\{1, 4, 5\}, \{1, 5, 6\}, \{2, 4, 6\}, \{2, 5, 6\}, \{3, 4, 5\}, \{3, 4, 6\}\}$$
$$\Gamma_4 = \{\{1, 2, 5\}, \{1, 3, 5\}, \{2, 3, 6\}, \{1, 2, 6\}, \{1, 3, 4\}, \{2, 3, 4\}\}$$
$$\Gamma_5 = \{\{1, 2, 4\}, \{2, 3, 5\}, \{1, 3, 6\}\}$$
$$\Gamma_6 = \{\{1, 2, 3\}\}$$
$$\Gamma_7 = \{\{0, 4, 5\}, \{0, 4, 6\}, \{0, 5, 6\}\}$$
$$\Gamma_9 = \{\{0, 1, 5\}, \{0, 2, 6\}, \{0, 3, 4\}\}$$
$$\Gamma_8 = \{\{0, 1, 4\}, \{0, 2, 5\}, \{0, 3, 6\}, \{0, 1, 6\}, \{0, 2, 4\}, \{0, 3, 5\}\}$$
$$\Gamma_{10} = \{\{0, 1, 2\}, \{0, 2, 3\}, \{0, 1, 3\}\}$$

The orbit incidence matrix A_{23} for G is:

	Γ_1	Γ_2	Γ_3	Γ_4	Γ_5	Γ_6	Γ_7	Γ_8	Γ_9	Γ_{10}
Δ_1	1	1	2	0	0	0	1	0	0	0
Δ_2	0	0	2	2	0	0	0	1	0	0
Δ_3	0	1	1	1	1	0	0	0	1	0
Δ_4	0	0	0	2	1	1	0	0	0	1
Δ_5	0	0	0	0	0	0	2	1	2	0
Δ_6	0	0	0	0	0	0	0	1	2	2

2. Let G' be the group generated by the two permutations

$$(0, 1, 14, 11, 2, 4, 8, 5, 13, 7, 3, 10, 12, 6, 9)$$

and

$$(0)(1, 2)(3)(4)(5, 6)(7, 8)(9, 10)(11, 12)(13, 14).$$

For this example, we give one orbit representative for each orbit.

Orbits of Pairs under G'

$\Delta_1 = G'(\{0, 2\})$,	$\Delta_2 = G'(\{0, 1\})$,	$\Delta_3 = G'(\{0, 5\})$,
$\Delta_4 = G'(\{0, 3\})$,	$\Delta_5 = G'(\{0, 6\})$	

Orbits of Triples under G'

$\Gamma_1 = G'(\{0, 2, 5\})$,	$\Gamma_2 = G'(\{0, 1, 13\})$,	$\Gamma_3 = G'(\{0, 1, 2\})$,
$\Gamma_4 = G'(\{0, 1, 8\})$,	$\Gamma_5 = G'(\{0, 1, 4\})$,	$\Gamma_6 = G'(\{0, 1, 5\})$,
$\Gamma_7 = G'(\{0, 2, 9\})$,	$\Gamma_8 = G'(\{0, 2, 7\})$,	$\Gamma_9 = G'(\{0, 1, 7\})$,
$\Gamma_{10} = G'(\{0, 1, 6\})$,	$\Gamma_{11} = G'(\{0, 3, 6\})$,	$\Gamma_{12} = G'(\{0, 2, 8\})$,
$\Gamma_{13} = G'(\{0, 1, 9\})$,	$\Gamma_{14} = G'(\{0, 1, 10\})$,	$\Gamma_{15} = G'(\{0, 3, 9\})$,
$\Gamma_{16} = G'(\{0, 5, 10\})$,	$\Gamma_{17} = G'(\{0, 2, 12\})$,	$\Gamma_{18} = G'(\{0, 1, 12\})$,
$\Gamma_{19} = G'(\{0, 1, 3\})$		

The orbit incidence matrix A_{23} for G' is:

	Γ_1	Γ_2	Γ_3	Γ_4	Γ_5	Γ_6	Γ_7	Γ_8	Γ_9	Γ_{10}	Γ_{11}	Γ_{12}	Γ_{13}	Γ_{14}	Γ_{15}	Γ_{16}	Γ_{17}	Γ_{18}	Γ_{19}
Δ_1	1	1	1	2	0	0	1	2	1	0	0	1	1	0	0	0	1	0	1
Δ_2	0	1	2	1	1	2	0	0	1	1	0	0	1	1	0	0	0	1	1
Δ_3	2	0	0	0	0	2	0	2	0	2	0	0	0	2	0	1	2	0	0
Δ_4	2	2	0	0	1	0	0	0	0	0	2	0	0	0	1	0	2	1	2
Δ_5	0	0	0	0	0	0	1	0	2	2	1	1	2	2	2	0	0	0	0

▯

The orbits in Example 6.12 can be constructed with Algorithms 6.14 or 6.16, and the A_{23} matrix can be obtained with Algorithm 6.18. Before running Algorithm 6.18, it is assumed that a complete set of orbit representatives R and S of t- and k-element subsets (respectively) have been constructed.

Algorithm 6.18: INCIDENCEMATRIX $(n, \vec{G}, R, S)$

external RUN()

global $T, K, stab, A$

procedure MATUSE1(n, g)
  for each $K' \in S$
    do $\begin{cases} \textbf{if } T \subseteq g(K') \\ \quad \textbf{then } A[T, K'] \leftarrow A[T, K'] + 1 \end{cases}$

procedure MATUSE2(n, g)
  if $K = g(K)$
    then $stab = stab + 1$

main
  for each $T \in R$
    do $\begin{cases} \textbf{for each } K \in S \\ \quad \textbf{do } A[T, K] \leftarrow 0 \end{cases}$
  for each $T \in R$
    do RUN$(n, \vec{G}, \text{MATUSE1})$
  for each $K \in S$
    do $\begin{cases} stab \leftarrow 0 \\ \text{RUN}(n, \vec{G}, \text{MATUSE2}) \\ \textbf{for each } T \in R \\ \quad \textbf{do } A[T, K] \leftarrow A[T, K]/stab \end{cases}$

Observe that the columns of the first matrix in Example 6.12 labeled by Γ_1, Γ_5 and Γ_9 sum to $J = [1, 1, 1, 1, 1, 1]^T$. That is, if

$$U = [1, 0, 0, 0, 1, 0, 0, 0, 1, 0]^T,$$

then $A_{23}U = J$ and consequently the subsets in the union of the orbits Γ_1, Γ_5 and Γ_9 form a Steiner triple system of order 7 with automorphism group G.

Also, observe that the columns of the second matrix in Example 6.12 labeled by Γ_7, Γ_{16} and Γ_{18} sum to $J = [1, 1, 1, 1, 1]^T$. That is, if

$$U = [0, 0, 0, 0, 0, 0, 1, 0, 0, 0, 0, 0, 0, 0, 0, 1, 0, 1, 0]^T,$$

then $A_{23}U = J$ and consequently the subsets in the union of the orbits Γ_7, Γ_{16} and Γ_{18} form a Steiner triple system of order 15 with automorphism group G'.

Definition 6.7:   Let $\mathcal{T}$ be a collection of subsets of the set $\mathcal{X}$. A set system $(\mathcal{X}, \mathcal{B})$ is called an *exact cover* of $\mathcal{T}$ if every subset in $\mathcal{T}$ is contained in exactly one member of $\mathcal{B}$. If in addition $\mathcal{T}$ is the collection of all t-element subsets of $\mathcal{X}$, then $(\mathcal{X}, \mathcal{B})$ is called a *Steiner system*.

Let $\mathcal{K}$ be the collection of subsets that are admissible as members of $\mathcal{B}$. Suppose further that $\mathcal{T}$ and $\mathcal{K}$ are both the union of orbits under the subgroup G of $\mathrm{Sym}(\mathcal{X})$, say

$$\mathcal{T} = \Delta_1 \cup \Delta_2 \cup \cdots \cup \Delta_{N_t}$$

and

$$\mathcal{K} = \Gamma_1 \cup \Gamma_2 \cup \cdots \cup \Gamma_{N_k},$$

then it is conceivable that G could be an automorphism group of the desired exact covering system $(\mathcal{X}, \mathcal{B})$. If this is the case, then an $A_{\mathcal{T}\mathcal{K}}$ orbit incidence matrix can be defined in a similar fashion as Definition 6.6. An exact covering of $\mathcal{T}$ having G as an automorphism group would correspond to a solution to the equation

$$A_{\mathcal{T}\mathcal{K}} U = J,$$

where $J = [1, 1, \ldots, 1]^T$. There will often be orbits Γ_j that will contain some member of $\mathcal{T}$ more than once. The corresponding column of $A_{\mathcal{T}\mathcal{K}}$ would have a non-$(0, 1)$-entry and the orbit cannot be used in the construction of the desired exact covering system. Consequently, we may remove these columns from $A_{\mathcal{T}\mathcal{K}}$ leaving a $(0, 1)$-matrix M for which we seek a solution $MU = J$. This new problem can be transformed to the **Exact Cover** problem. This is Problem 4.2 and was described in Chapter 4, where Algorithm 4.6 was developed to solve it. Think of the rows of M as a set $\mathcal{R}$ and the columns of M as subsets $\{C_1, C_2, \ldots, C_n\}$ of $\mathcal{R}$, where

$$C_j = \{i \in \mathcal{R} : M[i, j] = 1\}.$$

The transformation is now complete: a solution corresponds to a subcollection $C_{j_1}, C_{j_2}, \ldots, C_{j_\ell}$ such that

$$C_{j_{h_1}} \cap C_{j_{h_2}} = \emptyset$$

for all $1 \leq h_1 < h_2 \leq \ell$, and

$$\bigcup_{h=1}^{\ell} C_{j_h} = \mathcal{R}.$$

When the problem becomes too large for Algorithm 4.6 to solve, another approach is to use the method of basis reduction we describe in Chapter 8.

6.7   Notes

Section 6.1

A recommended general book on group theory is Rotman [93].

Section 6.2

Two recommended books on permutation groups are Dixon and Mortimer [26] and Butler [15]. A book on combinatorics that includes a discussion of algorithms for automorphism groups and permutation groups is Cameron [16].

Algorithm 6.9 is our own design but originates from the work of Schreier and Sims [99]. A faster but more complex method was developed by Knuth [54]. It runs in $O(n^5)$ time. A version that reduces the worst case storage requirement to $O(n^2)$ at the sacrifice of speed was given by Jerrum [46].

Section 6.3

There are many group theory and combinatorics books that discuss applications of Theorem 6.10, for example Brualdi [14], Cameron [16], van Lint and Wilson [67], Roberts [92], Straight [105], Rotman [93], and Tucker [107] to name a few.

Algorithm 6.14 was first described by Kreher and Radziszowski [60]. Algorithms 6.15 and 6.16 are apparently new.

Section 6.5

Theorem 6.13 can also be found in Cameron [16].

Section 6.6

The method of using orbit incidence matrices to find certain systems is described in Kramer and Mesner [57]. Algorithm 6.18 can be found in the article by Kreher and Radziszowski [60].

Exercises

6.1 Write out the multiplication table for the group given in Example 6.2. Find all the subgroups of this group.

6.2 Let p be a prime and define for each $a, b \in \mathbb{Z}_p$, $a \neq 0$, the function

$$f_{(a,b)}(X) = aX + b.$$

Let

$$G = \{f_{(a,b)} : a, b \in \mathbb{Z}_p \, a \neq 0\}.$$

(a) Show that G is a group under function composition.

    (b) What is the order of G?

    (c) Let

$$H = \{f_{(a,0)} : a \in \mathbb{Z}_p, \ a \neq 0\}.$$

    Show that H is a subgroup of G.

    (d) What is the order of H?

    (e) Find a left transversal of H in G.

6.3 Prove Theorem 6.5.

6.4 Use Algorithm 2.14 to compute the automorphism group of the Steiner triple system with blocks

$$\left\{ \begin{array}{llll} \{0,1,2\}, & \{3,4,5\}, & \{6,7,8\}, & \{0,3,6\}, \\ \{1,4,7\}, & \{2,5,8\}, & \{0,4,8\}, & \{1,5,6\}, \\ \{2,3,7\}, & \{0,5,7\}, & \{1,3,8\}, & \{2,4,6\} \end{array} \right\}.$$

6.5 Compute the automorphism group of the graph depicted below by inspection. Check your work with Algorithm 2.14.

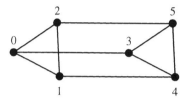

6.6 Use Algorithm 6.5 to compute the group generated by

$$\alpha = (0,1,2)(3,4,5)$$

and

$$\beta = (1,2)(4,5)$$

6.7 Consider the permutation group G on $\{0,1,\ldots,9\}$ with Schreier-Sims representation

$$\vec{G} = (\beta : [\mathcal{U}_0, \mathcal{U}_1, \ldots, \mathcal{U}_9])$$

where $\beta = \mathbf{I}$ and

$$\mathcal{U}_0 = \left\{ \begin{array}{l} \mathbf{I}, \ (0,1,3,6)(2,5,9,7)(4,8), \ (0,2,5,9,6)(1,4,8,3,7), \\ (0,3)(1,6)(2,9)(5,7), \ (0,4,9,6)(1,2,5,8)(3,7), \\ (0,5,6,2,9)(1,8,7,4,3), \ (0,6,9,5,2)(1,7,3,8,4), \\ (0,7,6)(1,2,8)(3,4,9), \ (0,8,4,1,9)(2,5,3,6,7), \\ (0,9,2,6,5)(1,3,4,7,8) \end{array} \right\}$$

$$\mathcal{U}_1 = \left\{ \begin{array}{l} \mathbf{I}, \ (1,2)(3,4)(6,7), \ (1,3,6)(2,4,7)(5,9,8), \\ (1,4)(2,3)(6,7)(8,9), \ (1,6)(2,7)(5,9), \ (1,7,3,2,6,4)(5,8,9) \end{array} \right\}$$

$$\mathcal{U}_2 = \{\mathbf{I}\}$$

$$\mathcal{U}_3 = \{\mathbf{I}, \ (3,6)(4,7)(5,8)\}$$

$$\mathcal{U}_4 = \mathcal{U}_5 = \mathcal{U}_6 = \mathcal{U}_7 = \mathcal{U}_8 = \mathcal{U}_9 = \{\mathbf{I}\}$$

Determine which of the following permutations are members of the group G.

    (a) $\alpha = (0,1,2,3,4,5,6)$

    (b) $\beta = (0,1,2,3,4)(5,6,7,8,9)$

    (c) $\gamma = (0,3,5,8,7)(1,9,2,6,4)$.

6.8 Find a Schreier-Sims representation with base

$$\beta = [3, 1, 4, 2, 0, 6, 7, 5] = (0, 3, 2, 4)(6, 7, 5)$$

of the automorphism group of the cube given in Figure 6.2.

6.9 Let G be a permutation group on $\mathcal{X} = \{0, 1, 2, \ldots, n - 1\}$ and let $S \subseteq \mathcal{X}$. Show that

$$G_S = \{g \in G : g(S) = S\}$$

is a subgroup of G, where

$$g(S) = \{g(x) : x \in S\}.$$

6.10 Let G be a permutation group on $\mathcal{X} = \{0, 1, 2, \ldots, n - 1\}$. Show that

$$G_0 = \{g \in G : g(0) = 0\}$$

$$G_1 = \{g \in G_0 : g(1) = 1\}$$

$$G_2 = \{g \in G_1 : g(2) = 2\}$$

$$\vdots$$

$$G_{n-1} = \{g \in G_{n-2} : g(n - 1) = n - 1\} = \{\mathbf{I}\}$$

are subgroups of G and that

$$G \supseteq G_0 \supseteq G_1 \supseteq G_2 \supseteq \cdots G_{n-1} = \{\mathbf{I}\}.$$

6.11 Use Theorem 6.10 to compute the number of non-isomorphic graphs on 6 vertices with 4 edges.

6.12 Use Algorithm 6.13 to compute the number of non-isomorphic graphs on 10 vertices with 7 edges.

6.13 What is the minimum orbit representative in the orbit of $\{3, 6, 4, 2\}$ under the group G given in Exercise 7?

6.14 Prove Theorem 6.12. (The proof is similar to that used for Theorem 6.10.)

6.15 Given a permutation group G on a set $\mathcal{X}$, develop and implement algorithms to compute
    (a) the number of orbits of k-permutations; and
    (b) the number of orbits of k-tuples.

6.16 Given a permutation group G on a set $\mathcal{X}$, develop and implement algorithms that find orbit representatives for
    (a) each of the orbits of k-permutations; and
    (b) each of the orbits of k-tuples.

6.17 Let G be a permutation group on $\mathcal{X}$ and consider the matrix A_{tk} defined in Definition 6.6.
    (a) Show that A_{tk} has a constant row sum, and that this sum does not depend on the group G.
    (b) Suppose U is a vector with entries either 0 or 1 that solves the matrix equation $A_{tk}U = J$, where $J = [1, 1, 1, \ldots, 1]^{\mathsf{T}}$. Show that U also satisfies the *orbit length equation*

$$\sum_{j=1}^{N_t} |\Gamma_j| \, U[j]. \tag{6.5}$$

    (c) Write Equation 6.5 for the two situations described in Example 6.12.

6.18 Let G be the permutation group on $\mathcal{X} = \{0, 1, \ldots, 8\}$ generated by

$$\alpha = (0, 1, 2)(3, 4, 5)(6, 7, 8)$$

(a) Find all the orbits of pairs under the action of G.
(b) Find all the orbits of triples under the action of G.
(c) Compute the A_{23} matrix of G.
(d) Find a vector U with entries either 0 or 1 that solves the matrix equation

$$A_{23}U = J,$$

where $J = [1, 1, 1, \ldots, 1]^{\mathsf{T}}$. The orbits of triples corresponding to the entries of U that have value 1 will form a Steiner triple system of order 9. Verify this fact!

7

Computing Isomorphism

7.1 Introduction

Among the most important concepts common to all areas of mathematics are isomorphism and symmetry. In particular, these ideas pervade combinatorial algorithms to an exceptional degree. Not only is the enumeration of equivalence classes and the selection of representative configurations important, but also is the elimination of repeated computation. The detection of isomorphic structures is essential in the construction of practical algorithms. To illustrate these ideas, we will discuss in the next two sections the graph isomorphism problem, culminating with an algorithm for determining when two graphs are isomorphic. This will be followed with applications of the algorithms to other structures.

We say that two graphs are isomorphic if there is a one-to-one correspondence between their vertex sets that sends edges to edges. More precisely, we have:

Definition 7.1: Two graphs $\mathcal{G}_1 = (\mathcal{V}_1, \mathcal{E}_1)$ and $\mathcal{G}_2 = (\mathcal{V}_2, \mathcal{E}_2)$ are *isomorphic* if there is a bijection $f : \mathcal{V}_1 \to \mathcal{V}_2$ such that

$$\{f(x), f(y)\} \in \mathcal{E}_2 \text{ if and only if } \{x, y\} \in \mathcal{E}_1.$$

The mapping f is said to be an *isomorphism* between $\mathcal{G}_1$ and $\mathcal{G}_2$.

If f is an isomorphism from a graph $\mathcal{G}$ to itself, then f is called an *automorphism*. The set of all automorphisms of a graph is a permutation group called the *automorphism group* $\text{Aut}(\mathcal{G})$ of the graph. The automorphism group of a graph was discussed in Chapter 6.

The problem of determining if two graphs are isomorphic is, in general, very difficult, although most researchers believe that it is not NP-complete. If additional structural properties are assumed, then often one can find a polynomial algorithm for the restricted set of graphs. For example, it is known that the isomorphism of graphs whose maximum degree is bounded by a given constant can

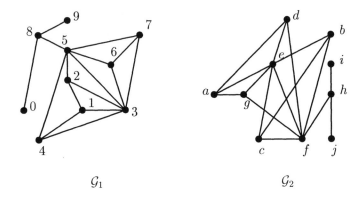

FIGURE 7.1

The function $f = \begin{pmatrix} 0 & 1 & 2 & 3 & 4 & 5 & 6 & 7 & 8 & 9 \\ i & a & d & e & g & f & c & b & h & j \end{pmatrix}$ **is an isomorphism**

between $\mathcal{G}_1$ and $\mathcal{G}_2$.

be determined in polynomial time. We will present in Section 7.3.1 a polynomial time algorithm to determine isomorphism of trees.

In Section 7.2 we will give a general graph isomorphism algorithm that uses the method of invariants or repeated refinement. An algorithm that is, in general, superior to this one is presented in Section 7.3. It uses the method of certificates or canonical labeling. Isomorphism of other combinatorial structures is studied in Section 7.4.

7.2 Invariants

Informally a function Φ is an invariant of the graph $\mathcal{G} = (\mathcal{V}, \mathcal{E})$ if Φ does not depend on the presentation of $\mathcal{G}$. For example, consider the set of all graphs having vertex set $\mathcal{V} = \{v_1, v_2, \ldots, v_n\}$. The list

$$[\deg(v_1), \deg(v_2), \deg(v_3), \ldots, \deg(v_n)],$$

called the *degree sequence* of the graph, is not an invariant. However, if the degree sequence is sorted in non-decreasing order, then it is an invariant.

Other invariants are, for example, the number of triangles contained in $\mathcal{G}$, the determinant of the adjacency matrix of $\mathcal{G}$, and the number of spanning trees of $\mathcal{G}$. Formally we define an invariant for a specified family of graphs as follows:

Definition 7.2: Let $\mathcal{F}$ be a family of graphs. An *invariant* on $\mathcal{F}$ is a function Φ with domain $\mathcal{F}$ such that

$$\Phi(\mathcal{G}_1) = \Phi(\mathcal{G}_2) \text{ if } \mathcal{G}_1 \text{ is isomorphic to } \mathcal{G}_2.$$

Observe that if $\Phi(\mathcal{G}_1) \neq \Phi(\mathcal{G}_2)$, then we may conclude that $\mathcal{G}_1$ and $\mathcal{G}_2$ are non-isomorphic. However, if $\Phi(\mathcal{G}_1) = \Phi(\mathcal{G}_2)$, then no conclusion can be drawn. For example, there are two non-isomorphic graphs on 6 vertices that have the same degree sequence.

There are countless invariants to choose from. One should choose invariants that are easy to compute which distinguish many graphs. The function Φ given by $\Phi(\mathcal{G}) = 1$ for all graphs $\mathcal{G}$ is an invariant, but not a very good one — it distinguishes no graphs. A convenient way to specify an invariant is to use ordered partitions.

Definition 7.3: An *ordered partition* B of $\mathcal{V}$ is an ordered list

$$B = [B[0], B[1], B[2], \ldots, B[k-1]],$$

where $\{B[0], B[1], \ldots, B[k-1]\}$ is a partition of $\mathcal{V}$. The *size* k of the partition is denoted by $|B|$.

For example, if $\mathcal{G}$ is a graph on the n-element vertex set $\mathcal{V}$, and if $\deg_G(v)$ denotes the degree of vertex v in $\mathcal{G}$, then

$$B = [B[0], B[1], B[2], \ldots, B[n-1]],$$

defined by

$$B[i] = \{v \in \mathcal{V} : \deg_G(v) = i\},$$

is an ordered partition of the vertices of $\mathcal{G}$. Furthermore the function

$$\Phi(\mathcal{G}) = [|B[0]|, |B[1]|, \ldots, |B[n-1]|]$$

is an invariant for the family of graphs. This is because $|B[i]|$ is the number of vertices of degree i in $\mathcal{G}$, and if two graphs on n vertices are isomorphic, then they have the same number of vertices of degree i, for each $i = 0, 1, \ldots, n-1$.

Definition 7.4:   Let $\mathcal{F}$ be a family of graphs on the vertex set $\mathcal{V}$ and let D be a function with domain $(\mathcal{F} \times \mathcal{V})$. Then the *partition induced* by D is

$$B = [B[0], B[1], B[2], \ldots, B[n-1]],$$

where

$$B[i] = \{v \in \mathcal{V} : G(\mathcal{G}, v) = i\}.$$

If the function

$$\Phi_D(\mathcal{G}) = [|B[0]|, |B[1]|, \ldots, |B[n-1]|]$$

is an invariant, then we say that D is an *invariant inducing function*.

Let x denote an arbitrary vertex of $\mathcal{G}$. Some invariant inducing functions are: the degree of x; the number of triangles in $\mathcal{G}$ that contain x; the determinant of the adjacency matrix of $\mathcal{G} \setminus x$; and the number of spanning trees of $\mathcal{G}$ rooted at x.

Let $\mathcal{I} = [D_1, D_2, ..., D_n]$ be a list of invariant inducing functions. We will use $\mathcal{I}$ to construct a sequence of partitions of $\mathcal{V}$ that become finer and finer. Set $X_0(\mathcal{G}) = \mathcal{V}$, the partition of $\mathcal{V}$ consisting of one block. For $i > 0$, define the partition $X_i(\mathcal{G})$ by stipulating that x and y belong to the same block of $X_i(\mathcal{G})$ if and only if

1. x and y belong to the same block of $X_{i-1}(\mathcal{G})$; and

2. x and y belong to the same block of the partition induced by D_i, i e , if $D_i(\mathcal{G}, x) = D_i(\mathcal{G}, y)$.

To determine if two graphs $\mathcal{G}_1 = (\mathcal{V}_1, \mathcal{E}_1)$ and $\mathcal{G}_2 = (\mathcal{V}_2, \mathcal{E}_2)$ are isomorphic by this method we can proceed as follows. Let f be a bijection from $\mathcal{V}_1$ to $\mathcal{V}_2$. (There must exist such a bijection for otherwise $\mathcal{G}_1$ and $\mathcal{G}_2$ do not have the same number of vertices and are hence non-isomorphic.) Compute the sequence $X_i(\mathcal{G}_1)$ and the sequence $X_i(\mathcal{G}_2)$, for each $i = 0, 1, 2, \ldots, n$. If, for any i, the structure of the corresponding partitions do not agree, then the graphs cannot be isomorphic. If they agree for all i, then $\mathcal{G}_1$ is isomorphic to $\mathcal{G}_2$ if and only if there is a bijection f from $\mathcal{V}_1$ to $\mathcal{V}_2$ that respects the final partitions $X_n(\mathcal{G}_1)$ and $X_n(\mathcal{G}_2)$ and is also an isomorphism. We give an example to illustrate.

Example 7.1 *Graph isomorphism by invariant inducing functions*
For any graph $\mathcal{G}$, define

$$D_1(\mathcal{G}, x) = \deg_G(x) \text{ and}$$
$$D_2(\mathcal{G}, x) = [d_j(x) : j = 1, 2, ..., d_{n-1}(x)],$$

where

$$d_j(x) = |\{y : y \text{ is adjacent to } x \text{ and } \deg_G(y) = j\}|$$

and n is the number of vertices. Let $\mathcal{G}_1$ and $\mathcal{G}_2$ be the graphs in Figure 7.1. Then we compute the following.

$$X_0(\mathcal{G}_1) = \{0, 1, 2, 3, 4, 5, 6, 7, 8, 9\}.$$

$$X_0(\mathcal{G}_2) = \{a, b, c, d, e, f, g, h, i, j\}.$$

x	0 1 2 3 4 5 6 7 8 9
$D_1(\mathcal{G}_1, x)$	1 3 3 6 3 6 3 3 3 1

$$\Downarrow$$

$$X_1(\mathcal{G}_1) = \{0, 9\}, \{1, 2, 4, 6, 7, 8\}, \{3, 5\}$$

$\bar{x}$	$a\ b\ c\ d\ e\ f\ g\ h\ i\ j$
$D_1(\mathcal{G}_2, \bar{x})$	3 3 3 3 6 6 3 3 1 1

$$\Downarrow$$

$$X_1(\mathcal{G}_2) = \{i, j\}, \{a, b, c, d, g, h\}, \{e, f\}.$$

$$D_2(\mathcal{G}_1, 0) = (0, 0, 1, 0, 0, 0, 0, 0, 0, 0)$$
$$D_2(\mathcal{G}_1, 1) = (0, 0, 2, 0, 0, 1, 0, 0, 0, 0)$$
$$D_2(\mathcal{G}_1, 2) = (0, 0, 1, 0, 0, 2, 0, 0, 0, 0)$$
$$D_2(\mathcal{G}_1, 3) = (0, 0, 5, 0, 0, 1, 0, 0, 0, 0)$$
$$D_2(\mathcal{G}_1, 4) = (0, 0, 1, 0, 0, 2, 0, 0, 0, 0)$$
$$D_2(\mathcal{G}_1, 5) = (0, 0, 5, 0, 0, 1, 0, 0, 0, 0)$$
$$D_2(\mathcal{G}_1, 6) = (0, 0, 1, 0, 0, 2, 0, 0, 0, 0)$$
$$D_2(\mathcal{G}_1, 7) = (0, 0, 1, 0, 0, 2, 0, 0, 0, 0)$$
$$D_2(\mathcal{G}_1, 8) = (2, 0, 0, 0, 0, 1, 0, 0, 0, 0)$$
$$D_2(\mathcal{G}_1, 9) = (0, 0, 1, 0, 0, 0, 0, 0, 0, 0)$$

$$\Downarrow$$

$$X_2(\mathcal{G}_1) = \{0, 9\}, \{8\}, \{2, 4, 6, 7\}, \{1\}, \{3, 5\}.$$

$$D_2(\mathcal{G}_2, a) = (0, 0, 2, 0, 0, 1, 0, 0, 0, 0)$$
$$D_2(\mathcal{G}_2, b) = (0, 0, 1, 0, 0, 2, 0, 0, 0, 0)$$
$$D_2(\mathcal{G}_2, c) = (0, 0, 1, 0, 0, 2, 0, 0, 0, 0)$$
$$D_2(\mathcal{G}_2, d) = (0, 0, 1, 0, 0, 2, 0, 0, 0, 0)$$
$$D_2(\mathcal{G}_2, e) = (0, 0, 5, 0, 0, 1, 0, 0, 0, 0)$$
$$D_2(\mathcal{G}_2, f) = (0, 0, 5, 0, 0, 1, 0, 0, 0, 0)$$
$$D_2(\mathcal{G}_2, g) = (0, 0, 1, 0, 0, 2, 0, 0, 0, 0)$$
$$D_2(\mathcal{G}_2, h) = (2, 0, 0, 0, 0, 1, 0, 0, 0, 0)$$
$$D_2(\mathcal{G}_2, i) = (0, 0, 1, 0, 0, 0, 0, 0, 0, 0)$$
$$D_2(\mathcal{G}_2, j) = (0, 0, 1, 0, 0, 1, 0, 0, 0, 0)$$

$$\Downarrow$$

$$X_2(\mathcal{G}_2) = \{i, j\}, \{h\}, \{b, c, d, g\}, \{a\}, \{e, f\}.$$

This restricts a possible isomorphism to bijections between the following sets:

$$\begin{aligned}
\{0,9\} &\longleftrightarrow \{i,j\} \\
\{8\} &\longleftrightarrow \{h\} \\
\{2,4,6,7\} &\longleftrightarrow \{b,c,d,g\} \\
\{1\} &\longleftrightarrow \{a\} \\
\{3,5\} &\longleftrightarrow \{e,f\}
\end{aligned}$$

There are $96 = (2!)(1!)(4!)(1!)(2!)$ bijections giving the possible isomorphisms. Examination of each of these possible isomorphisms shows that only the following eight bijections are isomorphisms.

$$\begin{pmatrix} 0 & 1 & 2 & 3 & 4 & 5 & 6 & 7 & 8 & 9 \\ i & a & d & e & g & f & c & b & h & j \end{pmatrix} \qquad \begin{pmatrix} 0 & 1 & 2 & 3 & 4 & 5 & 6 & 7 & 8 & 9 \\ j & a & d & e & g & f & c & b & h & i \end{pmatrix}$$

$$\begin{pmatrix} 0 & 1 & 2 & 3 & 4 & 5 & 6 & 7 & 8 & 9 \\ i & a & d & e & g & f & b & c & h & j \end{pmatrix} \qquad \begin{pmatrix} 0 & 1 & 2 & 3 & 4 & 5 & 6 & 7 & 8 & 9 \\ j & a & d & e & g & f & b & c & h & i \end{pmatrix}$$

$$\begin{pmatrix} 0 & 1 & 2 & 3 & 4 & 5 & 6 & 7 & 8 & 9 \\ i & a & e & d & g & f & c & b & h & j \end{pmatrix} \qquad \begin{pmatrix} 0 & 1 & 2 & 3 & 4 & 5 & 6 & 7 & 8 & 9 \\ j & a & e & d & g & f & c & b & h & i \end{pmatrix}$$

$$\begin{pmatrix} 0 & 1 & 2 & 3 & 4 & 5 & 6 & 7 & 8 & 9 \\ i & a & e & d & g & f & b & c & h & j \end{pmatrix} \qquad \begin{pmatrix} 0 & 1 & 2 & 3 & 4 & 5 & 6 & 7 & 8 & 9 \\ j & a & e & d & g & f & b & c & h & i \end{pmatrix}$$

□

In order to successfully implement this method for graph isomorphism an appropriate set $\mathcal{I}$ of invariant inducing functions should be chosen. After choosing these functions, a data structure for the graphs can then be chosen that optimizes the speed of evaluating functions in $\mathcal{I}$. Once all of the induced invariants are evaluated for the two graphs and the final partitions of the vertices are reached, a backtracking algorithm can be used to determine which of the correspondences are actual isomorphisms.

Algorithm 7.1 can also be used to compute the automorphism group of a graph $\mathcal{G}$ by specifying both parameters $\mathcal{G}_1$ and $\mathcal{G}_2$ to be equal to $\mathcal{G}$. However, it is usually more efficient to write a new algorithm that uses the fact that $\mathcal{G}_2$ is the same as $\mathcal{G}_1$. If this is done, then Algorithm 6.9 of Section 6.2.3 can be used to store the automorphism group. This is particularly helpful if the group is large.

Algorithms 7.1 and 7.2, when used with the two invariants suggested in Example 7.1, are particularly bad when the graph is extremely regular.

Algorithm 7.1: ISO $(\mathcal{I}, \mathcal{G}_1, \mathcal{G}_2)$

global n, W, X, Y

procedure GETPARTITIONS()
$\quad X[0] \leftarrow V(\mathcal{G}_1)$
$\quad Y[0] \leftarrow V(\mathcal{G}_2)$
$\quad N \leftarrow 1$
$\quad$ **for each** $D \in \mathcal{I}$

$\quad$ **do** $\begin{cases} \textbf{for } i \leftarrow 0 \textbf{ to } N - 1 \\ \quad \textbf{do} \begin{cases} \text{Partition } X[i] \text{ into sets } X_1[i], X_2[i], \dots, X_{m_i}[i], \\ \quad \text{where } x, x' \in X_j[i] \Leftrightarrow D(x) = D(x') \\ \text{Partition } Y[i] \text{ into sets } Y_1[i], Y_2[i], \dots, Y_{n_i}[i], \\ \quad \text{where } y, y' \in Y_j[i] \Leftrightarrow D(y) = D(y') \\ \textbf{if } m_i \neq n_i \\ \quad \textbf{then exit} \ (\mathcal{G}_1 \text{ and } \mathcal{G}_2 \text{ are not isomorphic.}) \\ \text{Order } Y_1[i], Y_2[i], \dots, Y_{m_i}[i] \text{ so that for all } j \\ \quad D(x) = D(y) \text{ whenever } x \in X_j[i] \text{ and } y \in Y_j[i] \\ \textbf{if } \text{ordering is not possible} \\ \quad \textbf{then exit} \ (\mathcal{G}_1 \text{ and } \mathcal{G}_2 \text{ are not isomorphic.}) \end{cases} \\ \text{Order the partitions so that:} \\ \quad |X[i]| = |Y[i]| \leq |X[i+1]| = |Y[i+1]| \text{ for all } i \\ N \leftarrow N + m - 1 \end{cases}$

$\quad$ **return** (N)

procedure FINDISOMORPHISM(ℓ)
$\quad$ **if** $\ell = n$ **then output** (f)
$\quad j \leftarrow W[\ell]$
$\quad$ **for each** $y \in Y[j]$

$\quad$ **do** $\begin{cases} OK \leftarrow \textbf{true} \\ \textbf{for } u \leftarrow 0 \textbf{ to } \ell - 1 \\ \quad \textbf{do if} \begin{cases} (\{u, \ell\} \in \mathcal{E}(\mathcal{G}_1) \textbf{ and } \{f[u], y\} \notin \mathcal{E}(\mathcal{G}_2)) \\ \textbf{or} \\ (\{u, \ell\} \notin \mathcal{E}(\mathcal{G}_1) \textbf{ and } \{f[u], y\} \in \mathcal{E}(\mathcal{G}_2)) \end{cases} \\ \quad \textbf{then } OK \leftarrow \textbf{false} \\ \textbf{if } OK \ \textbf{ then } \begin{cases} f[\ell] \leftarrow y \\ \text{FINDISOMORPHISM}(\ell + 1) \end{cases} \end{cases}$

main
$\quad N \leftarrow$ GETPARTITIONS()
$\quad$ **for** $i \leftarrow 0$ **to** N **do for each** $x \in X[i]$ **do** $W[x] \leftarrow i$
$\quad$ FINDISOMORPHISM(0)

Algorithm 7.2: AUT $(\mathcal{I}, \mathcal{G})$

external ENTER()

global n, W, X

procedure GETPARTITION()
$X[0] \leftarrow V(\mathcal{G}_1)$
$N \leftarrow 1$
for each $D \in \mathcal{I}$
$\quad$**do** $\begin{cases} \textbf{for } i \leftarrow 0 \textbf{ to } N - 1 \\ \quad \textbf{do } \begin{cases} \text{Partition } X[i] \text{ into sets } X_1[i], X_2[i], \ldots, X_m[i] \\ \text{where } x, x' \in X_j[i] \Leftrightarrow D(x) = D(x') \end{cases} \\ \text{Order the partition so that } |X[i]| \leq |X[i+1]| \text{ for all } i \\ N \leftarrow N + m - 1 \end{cases}$
return (N)

procedure FINDAUTOMORPHISMS(ℓ)
$\textbf{if } \ell = n$
$\quad \textbf{then } \text{ENTER}(f)$
$j \leftarrow W[\ell]$
for each $y \in X[j]$
$\quad$**do** $\begin{cases} OK \leftarrow \textbf{true} \\ \textbf{for } u \leftarrow 0 \textbf{ to } \ell - 1 \\ \quad \textbf{do if } \begin{cases} (\{u, \ell\} \in \mathcal{E}(\mathcal{G}) \textbf{ and } \{f[u], y\} \notin \mathcal{E}(\mathcal{G})) \\ \textbf{or} \\ (\{u, \ell\} \notin \mathcal{E}(\mathcal{G}) \textbf{ and } \{f[u], y\} \in \mathcal{E}(\mathcal{G})) \end{cases} \\ \quad \textbf{then } OK \leftarrow \textbf{false} \\ \textbf{if } OK \\ \quad \textbf{then } \begin{cases} f[\ell] \leftarrow y \\ \text{FINDAUTOMORPHISM}(\ell + 1) \end{cases} \end{cases}$

main
$N \leftarrow \text{GETPARTITION}()$
for $i \leftarrow 0$ **to** $n - 1$
$\quad$**do** $\mathcal{U}_i \leftarrow \{\textbf{I}\}$
for $i \leftarrow 0$ **to** N
$\quad$**do** $\begin{cases} \textbf{for each } x \in X[i] \\ \quad \textbf{do } W[x] \leftarrow i \end{cases}$
FINDAUTOMORPHISM(0)
$s \leftarrow 0$
while $|\mathcal{U}_s| > 1$
$\quad$**do** $s \leftarrow s + 1$
return $([\mathcal{U}_0, \mathcal{U}_1, \ldots, \mathcal{U}_s])$

7.3   Computing certificates

The last section determined when two graphs are isomorphic by actually attempt-ing to construct an isomorphism between them. Another way to determine iso-morphism is to compute from any graph in a given isomorphism class a unique representative. This leads us to the concept of certificates. Currently, the fastest general graph isomorphism algorithms use this method.

Definition 7.5:   A *certificate* Cert() for family $\mathcal{F}$ of graphs is a function such that for any $\mathcal{G}_1, \mathcal{G}_2 \in \mathcal{F}$,

$$\mathsf{Cert}(\mathcal{G}_1) = \mathsf{Cert}(\mathcal{G}_2) \text{ if and only if } \mathcal{G}_1 \text{ and } \mathcal{G}_2 \text{ are isomorphic.}$$

Note that a certificate is an invariant.

7.3.1   Trees

In this section we will develop a certificate for the family of trees. Recall a that tree is a connected graph $\mathcal{G} = (\mathcal{V}, \mathcal{E})$ with no circuits. A vertex $x \in \mathcal{V}$ is a *leaf* if $\deg_{\mathcal{G}}(x) = 1$. The certificate for trees will be a string of 0s and 1s of length $2n$, where $n = |\mathcal{V}|$. To compute the certificate we follow the following steps.

1. Label all the vertices of $\mathcal{G}$ with the string 01.

2. While there are more than two vertices of $\mathcal{G}$:

For each non-leaf x of G,

(a) let Y be the set of labels of the leaves adjacent to x and the label of x, with the initial 0 and trailing 1 deleted from x;

(b) replace the label of x with the concatenation of the labels in Y sorted in increasing lexicographic order, with a 0 prepended and a 1 appended;

(c) remove all leaves adjacent to x.

3. If there is only one vertex x left, report the label of x as the certificate.

4. If there are two vertices x and y left, then report the labels of x and y, concatenated in increasing lexicographic order, as the certificate.

When the leaves of a tree are repeatedly removed, then eventually either one or two vertices will be left. These vertices are called the *center* of the tree. Thus, because the algorithm works by repeated pruning of the leaves, we will be left with one of two possible final cases. Examples of the computation for both types of trees appear in Examples 7.2 and 7.3. A formal description of the procedure is given as Algorithm 7.3.

Example 7.2 *A tree with one center*

Number of Vertices	Non-leaves	Current Tree
9	1 : $Y = \{\}$ 2 : $Y = \{01, 01\}$ 3 : $Y = \{01\}$ 4 : $Y = \{\}$ 5 : $Y = \{\}$ 6 : $Y = \{01\}$	
6	1 : $Y = \{001011\}$ 4 : $Y = \{0011\}$ 5 : $Y = \{0011\}$	
3	4 : $Y = \{00010111, 000111, 0011\}$	
1		• 4 : 00001011100011100111

Certificate = 00001011100011100111.

〔

Example 7.3 *A tree with two centers*

Number of Vertices	Non-leaves	Current Tree
9	$0:\ Y = \{01\}$ $1:\ Y = \{01\}$ $2:\ Y = \{01\}$ $4:\ Y = \{01\}$ $5:\ Y = \{01\}$	
5	$1:\ Y = \{0011, 0011, 01\}$ $4:\ Y = \{0011, 01\}$	
2		

Certificate = 00011001101100011011.

□

Algorithm 7.3: TREETOCERTIFICATE $(\mathcal{G})$

external SORT()

global $N, n, Label, Leaves, Children, LastParent$

procedure FINDLEAVESANDCHILDREN()

  for $j \leftarrow 0$ **to** $n - 1$

    do $Children[j] \leftarrow \emptyset$

  $Num \leftarrow 0$

  for $j \leftarrow 0$ **to** $n - 1$

$$\mathbf{do} \begin{cases} \mathbf{if}\ |\mathsf{N}_{\mathcal{G}}[j]| = 1 \\ \quad \mathbf{then} \begin{cases} Leaves[Num] \leftarrow j \\ Children[k] \leftarrow Children[k] \cup \{j\}\ \text{where}\ \mathsf{N}[j] = \{k\} \\ Num \leftarrow Num + 1 \end{cases} \end{cases}$$

  return (Num)

procedure REDUCE()

  for $i \leftarrow 0$ **to** $n - 1$

$$\mathbf{do} \begin{cases} \mathbf{if}\ Children[i] \neq \emptyset \\ \quad \mathbf{then} \begin{cases} j \leftarrow 0 \\ LastParent \leftarrow i \\ A[i] \leftarrow A[i] \setminus Children[i] \\ \mathbf{for\ each}\ u \in Children[i] \\ \quad \mathbf{do} \begin{cases} A[u] \leftarrow \emptyset \\ Y[j] \leftarrow Label[u] \\ j \leftarrow j + 1 \end{cases} \\ Y[j] \leftarrow Label[i]\ \text{with first and last symbols deleted} \\ j \leftarrow j + 1 \\ \text{SORT}(Y) \\ Label[i] \leftarrow\ \text{the concatenation of}\ 0, Y[u],\ \text{and}\ 1 \end{cases} \end{cases}$$

main

  for $i \leftarrow 0$ **to** $n - 1$

    do $Label[i] \leftarrow 01$

  $N \leftarrow n$

  while $N > 2$

    $\mathbf{do} \begin{cases} N \leftarrow N - \text{FINDLEAVESANDCHILDREN}() \\ \text{REDUCE}() \end{cases}$

  FINDLEAVESANDCHILDREN()

  if $N = 2$

    then return $(\text{SORT}([Label[Leaves[0]], Label[Leaves[1]]]))$

    else return $Label[LastParent]$

Given a certificate $S = s_1 \cdots s_{2n}$ it may be desirable to determine what tree it came from. It will, of course, be impossible to determine the precise tree but it is possible to compute the tree up to isomorphism. Observe that Algorithm 7.3 initially labels the vertices of the input tree by 01. Thus the resulting certificate $S = s_1 \cdots s_{2n}$ will be a totally balanced sequence, as defined in Section 3.4, and we can take advantage of the mountain range description described there. To review, consider the function $f : \{0, 1, \ldots, 2n\} \to \mathbb{Z}$ defined by

$$f(0) = 0$$

$$f(x + 1) = \begin{cases} f(x) + 1 \text{ if } s_x = 0 \\ f(x) - 1 \text{ if } s_x = 1 \end{cases}$$

Then $f(x)$ is the excess of 0s over 1s in the certificate up to position x. The graph of this function will look like a mountain range rising from $(0, 0)$ having several peaks and returning to $(0, 2n)$. An example is given in the first iteration of Example 7.4. Call *sea level* the line $y = 0$. Then the mountain range will hit sea level either 2 or 3 times depending on whether the tree has 1 or 2 centers. That is, the equation $f(x) = 0$ will have 2 or 3 solutions and the graph will either look like one mountain with several peaks or two mountains each with several peaks. Mountains correspond to vertices of the tree. In the case of sea level, $y = 0$, if there are two mountains we have two adjacent vertices; otherwise, there is one mountain and one corresponding vertex.

Letting the water rise to $y = 1$ will divide the mountains into sub-mountains. That is, the graph of $f(x)$ splits into segments according to how it crosses the line $y = 1$. If a mountain M corresponding to vertex v splits into sub-mountains $M_1, M_2, \ldots, M_k$, $k \geq 1$, then we introduce new vertices $v_1, v_2, \ldots, v_k$ that are adjacent to v.

The water continues to rise and at each level $y = 2, 3, 4, \ldots$, we check to see how the mountains divide and introduce new vertices and edges as we go until there are no more mountains left. A formal description of this procedure is given in Algorithm 7.4.

Example 7.4

Initial certificate: 00001011100011100111

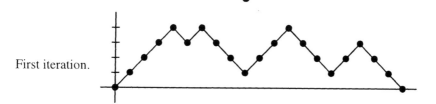

First iteration.

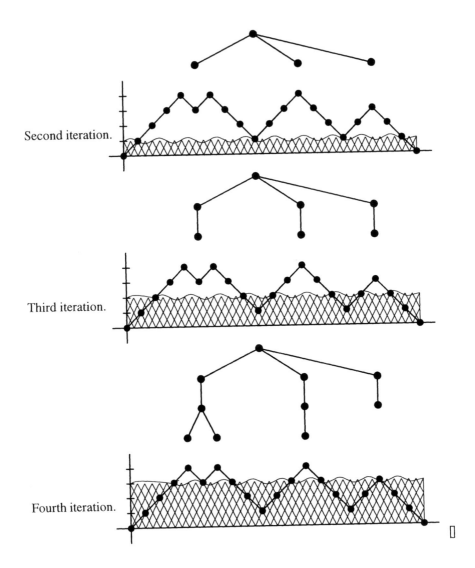

Second iteration.

Third iteration.

Fourth iteration.

Example 7.5

Initial certificate: 00011001101100011011

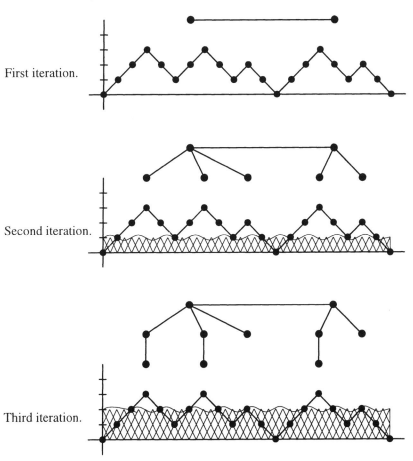

First iteration.

Second iteration.

Third iteration.

□

Algorithm 7.4: CERTIFICATETOTREE (S)

procedure FINDSUBMOUNTAINS(ℓ, S)
  comment: if $\ell = 2$, then the first and last symbols of S are
                dropped; otherwise ℓ should be 1.
  $m \leftarrow |S|$
  $k \leftarrow 0$
  $M[k] \leftarrow$ the empty string
  $M[k] \leftarrow M[k], S_{\ell-1}$
  $f \leftarrow 1$
  for $x \leftarrow \ell$ **to** $m - \ell$

$$
\mathbf{do}
\begin{cases}
\textbf{if } S[x] = 0 \\
\quad \textbf{then } f \leftarrow f + 1 \\
\quad \textbf{else } f \leftarrow f - 1 \\
M[k] \leftarrow M[k], S_x \\
\textbf{if } f = 0 \\
\quad \textbf{then }
\begin{cases}
k \leftarrow k + 1 \\
M[k] \leftarrow \text{ the empty string} \\
f \leftarrow 0
\end{cases}
\end{cases}
$$

  return (k)

main
  $n \leftarrow |S|/2$
  $\mathcal{G} = (\mathcal{V}, \mathcal{E}) \leftarrow$ the empty graph of order n
  $v \leftarrow 0$
  $k \leftarrow$ FINDSUBMOUNTAINS$(1, S)$
  if $k = 1$

$$
\textbf{then }
\begin{cases}
Label[v] \leftarrow M[0] \\
v \leftarrow v + 1
\end{cases}
$$

$$
\textbf{else }
\begin{cases}
Label[v] \leftarrow M[0] \\
v \leftarrow v + 1 \\
Label[v] \leftarrow M[1] \\
v \leftarrow v + 1 \\
\mathcal{E} \leftarrow \mathcal{E} \cup \{\{0, 1\}\}
\end{cases}
$$

  for $i \leftarrow 0$ **to** $n - 1$

$$
\mathbf{do}
\begin{cases}
\textbf{if } |Label[v]| > 2 \\
\quad \textbf{then }
\begin{cases}
k \leftarrow \text{FINDSUBMOUNTAINS}(2, Label[i]) \\
Label[i] \leftarrow 01 \\
\textbf{for } j \leftarrow 0 \textbf{ to } k - 1 \\
\quad \mathbf{do}
\begin{cases}
Label[v] \leftarrow M[j] \\
\mathcal{E} \leftarrow \mathcal{E} \cup \{\{i, v\}\} \\
v \leftarrow v + 1
\end{cases}
\end{cases}
\end{cases}
$$

  return $(\mathcal{G} = (\mathcal{V}, \mathcal{E}))$

7.3.2 Graphs

The most popular method of defining a certificate for the family of all graphs is to consider the adjacency matrix of a given graph $\mathcal{G} = (\mathcal{V}, \mathcal{E})$. Each possible ordering or permutation $\pi : \mathcal{V} \to \mathcal{V}$ of the vertices of a graph determines a particular adjacency matrix $A_\pi(\mathcal{G})$.

$$A_\pi(\mathcal{G})[u, v] = \begin{cases} 1 \text{ if } \{\pi(u), \pi(v)\} \in \mathcal{E} \\ 0 \text{ otherwise.} \end{cases}$$

The $n(n-1)/2$ entries above the main diagonal of $A_\pi(\mathcal{G})$ form an $n(n-1)/2$ bit binary number $\mathsf{Num}_\pi(\mathcal{G})$, when written column by column. Of course, when n is large, an $n(n-1)/2$ bit binary number will exceed the largest integer that can be represented by the computer and thus in practice the $n(n-1)/2$ bits are partitioned into an array of several integers, or into a string of characters. The smallest number that can be obtained via the different possible orderings defines a certificate of the graph.

$$\mathsf{Cert}(\mathcal{G}) = \min\{\mathsf{Num}_\pi(\mathcal{G}) : \pi \in \mathsf{Sym}(V)\} \qquad (7.1)$$

Unfortunately this certificate is difficult to compute. If it were in fact computed, then necessarily there will be as many leading bits that arc 0 as is possible. Consequently, the first k vertices in the ordering that achieves this certificate are pairwise non-adjacent and k is as large as possible. Thus these k vertices form a maximal independent set. They would be a maximum clique in the complement of the graph $\mathcal{G}$. Consequently, any algorithm that actually computes this certificate also solves the **Maximum Clique** problem; see Section 4.6.3. This problem is known to be NP-complete. On the other hand, it is generally believed (but not proven) that problem of determining is two given graphs are isomorphic is not NP-complete. If this is indeed true, then computing this certificate may be more work than is necessary. Indeed many modern graph isomorphism programs, including the one that we will present, define the certificate to be

$$\mathsf{Cert}(\mathcal{G}) = \min\{\mathsf{Num}_\pi(\mathcal{G}) : \pi \in \Pi_\mathcal{G}\} \qquad (7.2)$$

where $\Pi_\mathcal{G}$ is a set of permutations determined by the structure of $\mathcal{G}$ but not by any particular predefined ordering of $\mathcal{V}$.

Definition 7.6:   A partition B is a *discrete partition* if $|B[j]| = 1$ for all j, $0 \le j < k$. It is a *unit partition* if $|B| = 1$.

A partition B is an *equitable partition* with respect to the graph $\mathcal{G} = (\mathcal{V}, \mathcal{E})$ if for all i and j

$$|N_\mathcal{G}(u) \cap B[j]| = |N_\mathcal{G}(v) \cap B[j]|$$

for all $u, v \in B[i]$, where

$$N_\mathcal{G}(u) = \{x \in \mathcal{V} : \{u, v\} \in \mathcal{E}\}$$

is the *neighborhood* of u in $\mathcal{G}$.

Suppose B is an ordered equitable partition of size k with respect to the graph $\mathcal{G}$. Define the k by k matrix M_B as follows:

$$M_B[i, j] = |N_\mathcal{G}(v) \cap B[j]|,$$

where $v \in B[i]$. The value $M_B[i, j]$ does not depend on the choice of $v \in V[i]$, because the partition is equitable. If B is discrete, then M_B is an adjacency matrix of $\mathcal{G}$. The entries of M_B are non-negative integers, and we define $\mathsf{Num}(B)$ to be the sequence of $k(k - 1)/2$ entries above the main diagonal written column by column. Furthermore, if B is discrete, then B determines a permutation π of $\mathcal{V}$ by the relation $B[i] = \{\pi[i]\}$. Thus, in this case, we have

$$\mathsf{Num}(B) = \mathsf{Num}_\pi(\mathcal{G}),$$

when $\mathsf{Num}(B)$ is interpreted as a binary number.

Example 7.6   *An equitable partition*

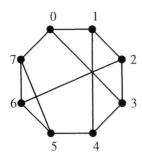

$$B = [\{0\}, \{2, 4\}, \{5, 6\}, \{7\}, \{1, 3\}]$$

is an equitable partition,

$$M_B = \begin{bmatrix} 0 & 0 & 0 & 1 & 2 \\ 0 & 0 & 1 & 0 & 2 \\ 0 & 1 & 1 & 1 & 0 \\ 1 & 0 & 2 & 0 & 0 \\ 1 & 2 & 0 & 0 & 0 \end{bmatrix}, \text{ and}$$

$$\mathsf{Num}(B) = [0, 0, 1, 1, 0, 1, 2, 2, 0, 0].$$

[]

Definition 7.7:   An ordered partition B is a *refinement* of the ordered partition A if

1. every block $B[i]$ of B is contained in some block $A[j]$ of A; and
2. if $u \in A[i_1]$ and $v \in A[j_1]$ with $i_1 \leq j_1$, then $u \in B[i_2]$ and $v \in B[j_2]$ with $i_2 \leq j_2$.

For example

$$B = [\{0\}, \{2, 4\}, \{5, 6\}, \{7\}, \{1, 3\}]$$

is a refinement of

$$A = [\{0\}, \{1, 2, 3, 4, 5, 6, 7\}].$$

However, $[\{2, 4\}, \{5, 6\}, \{0\}, \{7\}, \{1, 3\}]$ is not a refinement of A, because the blocks are out of order, with respect to A.

Let A be an ordered partition and consider any block T of A. Define a function $D_T : V \to \{0, 1, \ldots, n-1\}$ as follows:

$$D_T(v) = |N_G(v) \cap T|.$$

The function D_T induces the invariant Φ_{D_T}, which can be used to refine the ordered partition A to a ordered partition B, as follows:

1. Set B equal to A.
2. Let S be a list containing the blocks of B.
3. While $S \neq \emptyset$ do
4.   remove a block T from the list S;
5.   for each block $B[i]$ of B do
6.   for each h, set $L[h] = \{v \in B[i] : D_T(v) = h\}$;
7.   if there is more than one non-empty block in L, then
8.   replace the block $B[i]$ with the non-empty blocks in L, in order of the index h, $h = 0, 1, \ldots, n-1$;
9.   add the non-empty blocks in L to the end of the list S.

After Step 5 of this procedure is repeated for each block of B, either B will not have changed, or B will be a refinement with more blocks. In the latter case, we add the new blocks of B to the list S and use them for possible further refinement. The process continues until all blocks in S have been considered. Observe that we may ignore a set T in the list S if we have already considered sets that partition T. The partition B that results will, because of Step 6, be equitable and Step 8 will guarantee that $\mathsf{Num}(B)$ will be minimal among all arrangements of the blocks of the partition B that are a refinement of A.

Algorithm 7.5 gives a more detailed presentation of this refinement procedure. An example is given in Example 7.7.

Algorithm 7.5: REFINE $(n, \mathcal{G}, A, B)$

global L, U, S, T, N

procedure SPLITANDUPDATE$(n, \mathcal{G}, B, j)$

$\quad L \leftarrow$ empty list

$\quad$**for each** $u \in B[j]$

$\qquad$**do** $\begin{cases} h \leftarrow |T \cap \mathsf{N}_{\mathcal{G}}(u)| \\ L[h] \leftarrow L[h] \cup \{u\} \end{cases}$

$\quad m \leftarrow 0$

$\quad$**for** $h \leftarrow 0$ **to** $n - 1$

$\qquad$**do if** $L[h] \neq \emptyset$

$\qquad$**then** $m \leftarrow m + 1$

$\quad$**if** $m > 1$

$\qquad$**then** $\begin{cases} \textbf{for } h \leftarrow |B| - 1 \textbf{ downto } j + 1 \\ \quad \textbf{do } B[m - 1 + h] \leftarrow B[h] \\ k \leftarrow 0 \\ \textbf{for } h \leftarrow 0 \textbf{ to } n - 1 \\ \quad \textbf{do } \begin{cases} \textbf{if } L[h] \neq \emptyset \\ \quad \textbf{then } \begin{cases} B[j + k] \leftarrow L[h] \\ S[N + k] \leftarrow L[h] \\ U \leftarrow U \cup L[h] \\ k \leftarrow k + 1 \end{cases} \end{cases} \\ j = j + m - 1 \\ N = N + m \end{cases}$

main

$\quad B \leftarrow A$

$\quad$**for** $N \leftarrow 0$ **to** $|B|$

$\qquad$**do** $S[N] = B[N]$

$\quad U \leftarrow \mathcal{V}$

$\quad$**while** $N \neq 0$

$\qquad$**do** $\begin{cases} N \leftarrow N - 1 \\ T = S[N] \\ \textbf{if } T \subset U \\ \quad \textbf{do } \begin{cases} U \leftarrow U \setminus T \\ j \leftarrow 0 \\ \textbf{while } j < |B| \textbf{ and } |B| < n \\ \quad \textbf{do } \begin{cases} \textbf{if } |B| \neq 1 \\ \quad \textbf{then } \text{SPLITANDUPDATE}(n, \mathcal{G}, B, j) \\ j \leftarrow j + 1 \end{cases} \\ \textbf{if } |B| = n \\ \quad \textbf{then exit} \end{cases} \end{cases}$

If $A = [A[0], A[1], \ldots, A[k]]$ is an ordered partition of the set $\mathcal{V}$ and f is a bijection with domain $\mathcal{V}$, then the image of A under f is the ordered partition

$$f(A) = [B[0], B[1], \ldots, B[k]]$$

where $B[i] = \{f(x) : x \in A[i], i = 0, 1, \ldots, k$. Observe that the above procedure and Algorithm 7.5 do not depend on the ordering of the vertices. A consequence of this is Theorem 7.1 whose proof we leave as an exercise.

THEOREM 7.1 *Let f be an isomorphism from the graph $\mathcal{G}_1 = (\mathcal{V}_1, \mathcal{E}_1)$ to the graph $\mathcal{G}_2 = (\mathcal{V}_2, \mathcal{E}_2)$. If B_1 is the partition that results when Algorithm 7.5 is run with input $\mathcal{G}_1$ and A_1, then $f(B_1)$ is the partition that results when Algorithm 7.5 is run with input $\mathcal{G}_2$ and $f(A_1)$,*

Example 7.7 *Refining to an equitable partition*
We illustrate the refinement procedure using the graph given in 7.6 and the initial partition $A = [\{0\}, \{1, 2, 3, 4, 5, 6, 7\}]$.

$$B = [\{0\}, \{1, 2, 3, 4, 5, 6, 7\}]$$
$$S = [\{1, 2, 3, 4, 5, 6, 7\}, \underbrace{\{0\}}_{T}]$$

$$D_{\{0\}} : B = [\{0\}, \{2, 4, 5, 6\}, \{1, 3, 7\}]$$
$$S = [\{1, 2, 3, 4, 5, 6, 7\}, \{1, 3, 7\}, \underbrace{\{2, 4, 5, 6\}}_{T}]$$

$$D_{\{2,4,5,6\}} : B = [\{0\}, \{2, 4\}, \{5, 6\}, \{1, 3, 7\}]$$
$$S = [\{1, 2, 3, 4, 5, 6, 7\}, \{1, 3, 7\}, \{5, 6\}, \underbrace{\{2, 4\}}_{T}]$$

$$D_{\{2,4\}} : B = [\{0\}, \{2, 4\}, \{5, 6\}, \{7\}, \{1, 3\}]$$
$$S = [\{1, 2, 3, 4, 5, 6, 7\}, \{1, 3, 7\}, \{5, 6\}, \{1, 3\}, \underbrace{\{7\}}_{T}]$$

$$D_{\{7\}} : B = [\{0\}, \{2, 4\}, \{5, 6\}, \{7\}, \{1, 3\}]$$
$$S = [\{1, 2, 3, 4, 5, 6, 7\}, \{1, 3, 7\}, \{5, 6\}, \underbrace{\{1, 3\}}_{T}]$$

$$D_{\{1,3\}} : B = [\{0\}, \{2, 4\}, \{5, 6\}, \{7\}, \{1, 3\}]$$
$$S = [\{1, 2, 3, 4, 5, 6, 7\}, \{1, 3, 7\}, \underbrace{\{5, 6\}}_{T}]$$

$$D_{\{5,6\}} : B = [\{0\}, \{2, 4\}, \{5, 6\}, \{7\}, \{1, 3\}]$$

$$S = [\{1, 2, 3, 4, 5, 6, 7\}, \underbrace{\{1, 3, 7\}}_{T}]$$

$$S = [\underbrace{\{1, 2, 3, 4, 5, 6, 7\}}_{T}]$$

$$S = [\quad]$$

The final refined equitable partition is $B = [\{0\}, \{2, 4\}, \{5, 6\}, \{7\}, \{1, 3\}]$. ▯

Example 7.8 shows that Algorithm 7.5 cannot be expected to lead to a procedure that produces a certificate of the type given in Equation 7.1. This is because, starting with the unit partition, Algorithm 7.5 first partitions the vertices in order of their degrees, and this partitioning may not necessarily lead to an arrangement with smallest Num.

Example 7.8 *Refining to discrete partitions*
We illustrate the refinement procedure using the graph

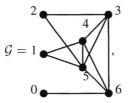

and the initial partition $A = [\{0, 1, 2, 3, 4, 5, 6\}]$. The adjacency matrix of $\mathcal{G}$ is

$$
\begin{array}{ccccccc}
0 & 0 & 0 & 0 & 0 & 0 & 1 \\
0 & 0 & 0 & 0 & 1 & 1 & 0 \\
0 & 0 & 0 & 1 & 0 & 1 & 0 \\
0 & 0 & 1 & 0 & 1 & 1 & 1 \\
0 & 1 & 0 & 1 & 0 & 1 & 1 \\
0 & 1 & 1 & 1 & 1 & 0 & 1 \\
1 & 0 & 0 & 1 & 1 & 1 & 0 \\
\end{array}
$$

and

$$\mathsf{Num_I}(\mathcal{G}) = (000001010101111100111)_{\text{binary}}.$$

$$B = [\{0, 1, 2, 3, 4, 5, 6\}]$$

$$S = [\underbrace{\{0, 1, 2, 3, 4, 5, 6\}}_{T}]$$

$$D_{\{0,1,2,3,4,5,6\}} : B = [\{0\}, \{1,2\}, \{3,4,6\}, \{5\}]$$
$$S = [\{5\}, \{3,4,6\}, \{1,2\}, \underbrace{\{0\}}_{T}]$$

$$D_{\{0\}} : B = [\{0\}, \{1,2\}, \{3,4\}, \{6\}, \{5\}]$$
$$S = [\{5\}, \{3,4,6\}, \{1,2\}, \{6\}, \underbrace{\{3,4\}}_{T}]$$

$$D_{\{3,4\}} : B = [\{0\}, \{1,2\}, \{3,4\}, \{6\}, \{5\}]$$
$$S = [\{5\}, \{3,4,6\}, \{1,2\}, \underbrace{\{6\}}_{T}]$$

$$D_{\{6\}} : B = [\{0\}, \{1,2\}, \{3,4\}, \{6\}, \{5\}]$$
$$S = [\{5\}, \{3,4,6\}, \underbrace{\{1,2\}}_{T}]$$

$$D_{\{1,2\}} : B = [\{0\}, \{1,2\}, \{3,4\}, \{6\}, \{5\}]$$
$$S = [\{5\}, \underbrace{\{3,4,6\}}_{T}]$$

$$D_{\{3,4,6\}} : B = [\{0\}, \{1,2\}, \{3,4\}, \{6\}, \{5\}]$$
$$S = [\underbrace{\{5\}}_{T}]$$

$$D_{\{5\}} : B = [\{0\}, \{1,2\}, \{3,4\}, \{6\}, \{5\}]$$
$$S = [\quad]$$

This results in the equitable partition

$$B = [\{0\}, \{1,2\}, \{3,4\}, \{6\}, \{5\}].$$

This partition is not discrete and thus does not determine an ordering of the vertices. The first block containing more than one vertex is $\{1,2\}$. Thus any discrete partition that refines B must either look like $[\{0\}, \{1\}, \{2\}, \ldots]$ or like $[\{0\}, \{2\}, \{1\}, \ldots]$. We try both possibilities.

First we refine $B = [\{0\}, \{1\}, \{2\}, \{3,4\}, \{6\}, \{5\}]$.

$$B = [\{0\}, \{1\}, \{2\}, \{3,4\}, \{6\}, \{5\}]$$
$$S = [\{5\}, \{6\}, \{3,4\}, \{2\}, \{1\}, \underbrace{\{0\}}_{T}]$$

$$D_{\{0\}} : B = [\{0\}, \{1\}, \{2\}, \{3,4\}, \{6\}, \{5\}]$$

$$S = [\{5\}, \{6\}, \{3, 4\}, \{2\}, \underbrace{\{1\}}_{T}]$$

$$D_{\{1\}} : B = \{0\}, \{1\}, \{2\}, \{3\}, \{4\}, \{6\}, \{5\}$$

This results in the discrete partition

$$B = [\{0\}, \{1\}, \{2\}, \{3\}, \{4\}, \{6\}, \{5\}]$$

which determines the ordering

$$\pi_1 = [0, 1, 2, 3, 4, 6, 5]$$

of the vertices. The corresponding adjacency matrix is

$$\begin{matrix}
0 & 0 & 0 & 0 & 0 & 1 & 0 \\
0 & 0 & 0 & 0 & 1 & 0 & 1 \\
0 & 0 & 0 & 1 & 0 & 0 & 1 \\
0 & 0 & 1 & 0 & 1 & 1 & 1 \\
0 & 1 & 0 & 1 & 0 & 1 & 1 \\
1 & 0 & 0 & 1 & 1 & 0 & 1 \\
0 & 1 & 1 & 1 & 1 & 1 & 0
\end{matrix}$$

and

$$\mathsf{Num}_{\pi_1}(\mathcal{G}) = (000001010110011011111)_{\text{binary}}.$$

The other possibility is to refine $B = [\{0\}, \{2\}, \{1\}, \{3, 4\}, \{6\}, \{5\}]$.

$$B = [\{0\}, \{2\}, \{1\}, \{3, 4\}, \{6\}, \{5\}]$$

$$S = [\{5\}, \{6\}, \{3, 4\}, \{1\}, \{2\}, \underbrace{\{0\}}_{T}]$$

$$D_{\{0\}} : B = [\{0\}, \{2\}, \{1\}, \{3, 4\}, \{6\}, \{5\}]$$

$$S = [\{5\}, \{6\}, \{3, 4\}, \{1\}, \underbrace{\{2\}}_{T}]$$

$$D_{\{2\}} : B = \{0\}, \{2\}, \{1\}, \{4\}, \{3\}, \{6\}, \{5\}$$

This gives the discrete partition

$$B = [\{0\}, \{2\}, \{1\}, \{4\}, \{3\}, \{6\}, \{5\}]$$

which determines the ordering

$$\pi_2 = [0, 2, 1, 4, 3, 6, 5]$$

of the vertices. The corresponding adjacency matrix is

$$
\begin{array}{ccccccc}
0 & 0 & 0 & 0 & 0 & 1 & 0 \\
0 & 0 & 0 & 0 & 1 & 0 & 1 \\
0 & 0 & 0 & 1 & 0 & 0 & 1 \\
0 & 0 & 1 & 0 & 1 & 1 & 1 \\
0 & 1 & 0 & 1 & 0 & 1 & 1 \\
1 & 0 & 0 & 1 & 1 & 0 & 1 \\
0 & 1 & 1 & 1 & 1 & 1 & 0
\end{array}
$$

and

$$
\mathsf{Num}_{\pi_2}(\mathcal{G}) = (00000101011001101111)_{\text{binary}}.
$$

The two computed permutations π_1 and π_2 have

$$
\mathsf{Num}_{\pi_1}(\mathcal{G}) = \mathsf{Num}_{\pi_2}(\mathcal{G}) = (00000101011001101111)_{\text{binary}}.
$$

So we would report the certificate to be:

$$
C = (00000101011001101111)_{\text{binary}}.
$$

In general we would take the smallest $\mathsf{Num}_\pi(\mathcal{G})$ that was obtained as the certificate.

It is interesting to observe that

$$
\mathsf{Num}_I(\mathcal{G}) < C,
$$

which can be seen by examining the eleventh bit. Also, the degree sequence of the original graph is $1, 2, 2, 4, 4, 5, 4$, whereas the degree sequence with a vertex ordering that yields certificate C is $1, 2, 2, 4, 4, 4, 5$. □

Starting with any equitable partition P of the vertices $\mathcal{V}$ of a graph $\mathcal{G}$, a block $P[i]$ of size m greater than 1 can be split into two blocks, the first having size 1 and the second having size $m - 1$. All such splittings must eventually be considered. Given a new partition obtained by splitting, Algorithm 7.5 can be applied to obtain an equitable partition. This process can be repeated until all blocks have size 1. When a discrete partition P' is reached, an ordering π of the vertices of the graph is determined:

$$
\pi[j] = p_j \text{ where the } j^{\text{th}} \text{ block of } P' \text{ is } \{p_j\}. \tag{7.3}
$$

From π we can compute $\mathsf{Num}_\pi(\mathcal{G})$ and compare it to $\mathsf{Num}_\mu(\mathcal{G})$, where μ is the ordering giving the smallest Num that has been found so far. That is, among the orderings discovered during the search, μ is continually updated so that $\mathsf{Num}_\mu(G)$ is smallest.

Using the techniques in Chapter 4, a backtracking algorithm for the ordering that gives the smallest possible Num can be developed as follows. A partial solution or node of the state space tree is an equitable partition. The root node is the partition consisting of a single block. To continue the backtrack search from the current node, a block of it is found that has size greater than one. An entry of this block is chosen; it is split off and the resulting partition is refined. When the search backs up to this node a different entry is chosen. This continues until all entries have been considered. If there are no blocks of size greater than one, then the partition P' is a discrete partition. Using Equation 7.3 an ordering, π, is now determined from which $\text{Num}_\pi(\mathcal{G})$ is calculated and compared to $\text{Num}_\mu(\mathcal{G})$.

One simple method to prune this search is to compare the current partition to μ at each node. That is, we make comparisons as we go and not just when the partition becomes discrete. If P is the current partition and ℓ is the index of the first block $P[\ell]$ such that $|P[\ell]| > 1$, we can define the *partial permutation* π by

$$\pi[j] = p_j \text{ where the } j^{\text{th}} \text{ block of } P' \text{ is } \{p_j\}$$

for $j = 0, 1, \ldots, \ell - 1$. Thus the first ℓ entries of π are the first ℓ entries of the ordering that would result if we were to continue the backtracking until the partition became discrete. We can compare these entries to the first ℓ entries of μ by just checking the number that results from the first ℓ rows and columns of the adjacency matrices A_π and A_μ. It was precisely for this application that we defined $\text{Num}_\pi(\mathcal{G})$ to be the $n(n-1)/2$ bit binary number obtained from the entries above the main diagonal of $A_\pi(\mathcal{G})$ written column by column rather than row by row. The partial ordering π can only complete to an ordering π^* with $\text{Num}_{\pi^*}(\mathcal{G}) < \text{Num}_\mu(\mathcal{G})$ if, the first time that entries differ, we have $A_\pi[i, j] = 0$ and $A_\mu[i, j] = 1$. These observations lead to the procedure given as Algorithm 7.6. Using the techniques of Chapter 4, a backtrack procedure can now be developed. The backtrack procedure with this amount of pruning is Algorithm 7.7. It is invoked by Algorithm 7.8.

Algorithm 7.6: COMPARE $(\mathcal{G}, \pi, \ell)$

for $j \leftarrow 1$ **to** $\ell - 1$

do $\begin{cases} \textbf{for } i \leftarrow 0 \textbf{ to } j - 1 \\ \quad \textbf{do} \begin{cases} x \leftarrow A_\mathcal{G}[\mu[i], \mu[j]] \\ y \leftarrow A_\mathcal{G}[\pi[i], \pi[j]] \\ \textbf{if } x < y \\ \quad \textbf{then return } (Worse) \\ \textbf{if } x > y \\ \quad \textbf{then return } (Better) \end{cases} \\ \textbf{return } (Equal) \end{cases}$

Algorithm 7.7: CANON1 $(\mathcal{G}, P)$

external REFINE(), COMPARE()
REFINE$(\mathcal{G}, P, Q)$
Find the index ℓ of the first block of Q with $|Q[\ell]| > 1$
$Res \leftarrow Better$
if $BestExist$
$\quad$ **then** $\begin{cases} \textbf{for } i \leftarrow 0 \textbf{ to } n - 1 \\ \quad \textbf{do } \pi_1[i] = q_i \text{ where } Q[i] = \{q_i\} \\ Res \leftarrow \text{COMPARE}(\mathcal{G}, \pi_1, \ell) \end{cases}$
if Q has n blocks
$\quad$ **then** $\begin{cases} \textbf{if not } BestExist \\ \quad \textbf{then } \begin{cases} \textbf{for } i \leftarrow 0 \textbf{ to } n - 1 \\ \quad \textbf{do } \mu[i] = q_i \text{ where } Q[i] = \{q_i\} \\ BestExist \leftarrow \textbf{true} \end{cases} \\ \quad \textbf{else } \begin{cases} \textbf{if } Res = Better \\ \quad \textbf{then } \mu \leftarrow \pi_1 \end{cases} \end{cases}$
$\quad$ **else** $\begin{cases} \textbf{if } Res \neq Worse \begin{cases} C \leftarrow Q[\ell] \\ D \leftarrow Q[\ell] \\ \textbf{for } j \leftarrow 0 \textbf{ to } j < \ell \\ \quad \textbf{do } R[j] \leftarrow Q[j] \\ \textbf{for } j \leftarrow \ell + 1 \textbf{ to } size(Q) \\ \quad \textbf{do } R[j + 1] \leftarrow Q[j] \\ \textbf{while } C \neq \emptyset \\ \quad \textbf{do } \begin{cases} u \leftarrow \text{any element of } C \\ R[\ell] \leftarrow \{u\} \\ R[\ell + 1] \leftarrow D \setminus \{u\} \\ \text{CANON1}(\mathcal{G}, R) \\ C = C \setminus \{u\} \end{cases} \end{cases} \end{cases}$

Algorithm 7.8: CERT1 $(\mathcal{G})$

external CANON1()
$P \leftarrow [\{0, 1, \ldots, n\}]$
CANON1$(\mathcal{G}, P)$
$k \leftarrow 0$
$C \leftarrow 0$
for $j \leftarrow n - 1$ **downto** 1
$\quad$ **do** $\begin{cases} \textbf{for } i \leftarrow j - 1 \textbf{ downto } 0 \\ \quad \textbf{do } \begin{cases} \textbf{if } \{\mu[i], \mu[j]\} \in \mathcal{E}(\mathcal{G}) \textbf{ then } C \leftarrow x + 2^k \\ k = k + 1 \end{cases} \end{cases}$
return (C)

The state space tree that results when Algorithm 7.8 is run on the Graph in Example 7.7 is given in Figure 7.2. The nodes labeled by *Worse* give pruned subtrees. The node labeled by *First* gives the ordering π that was obtained the first time the partition became discrete. It also gives the Num_π that results. A smaller Num_π that turns out to be the actual certificate is seen at the node labeled *Better*. The nodes labeled *Equal* occur when the ordering gives a Num_π that equals Num_μ, where μ is the ordering with the smallest Num found so far. In this example we see that the certificate of the graph in Example 7.7 is 5192304.

7.3.3 Pruning with automorphisms

Recall that if $\pi \in \text{Sym}(\mathcal{V})$ is an ordering of the vertices of the graph $\mathcal{G} = (\mathcal{V}, \mathcal{E})$, then A_π denotes the adjacency matrix with respect to this ordering. That is,

$$A_\pi[i,j] = A[\pi[i], \pi[j]],$$

where A is the adjacency matrix of G with respect to the usual ordering of vertices,

$$0, 1, 2, \ldots, n-1.$$

Consequently, $\alpha \in \text{Sym}(\mathcal{V})$ is an automorphism of $\mathcal{G}$ if and only if $A_\alpha = A$.

THEOREM 7.2   *Let $\mathcal{G} = (\mathcal{V}, \mathcal{E})$ be a graph and let $\pi_1, \mu \in \text{Sym}(\mathcal{V})$ be two orderings of the vertices of $\mathcal{G}$. If $\text{Num}_{\pi_1}(\mathcal{G}) = \text{Num}_\mu(\mathcal{G})$, then $\pi_2 = \pi_1 \mu^{-1}$ is an automorphism of $\mathcal{G}$.*

PROOF   Observe that $\text{Num}_{\pi_1}(\mathcal{G}) = \text{Num}_\mu(\mathcal{G})$ if and only if $A_{\pi_1} = A_\mu$. Thus

$$
\begin{aligned}
A_{\pi_2}[i,j] &= A_{\pi_1 \mu^{-1}}[i,j] \\
&= A[\pi_1 \mu^{-1}[i], \pi_1 \mu^{-1}[j]] \\
&= A_{\pi_1}[\mu^{-1}[i], \mu^{-1}[j]] \\
&= A_\mu[\mu^{-1}[i], \mu^{-1}[j]] \\
&= A[\mu\mu^{-1}[i], \mu\mu^{-1}[j]] \\
&= A.
\end{aligned}
$$

Therefore π_2 is an automorphism of $\mathcal{G}$.   ∎

Theorem 7.2 tells us how to obtain the automorphism group of the graph. When a leaf node with a discrete partition π for which Algorithm 7.6 returns *Equal* is reached, an automorphism is discovered, namely $g = \pi\mu^{-1}$. The automorphisms discovered can be managed with Algorithm 6.11 and the methods in Chapter 6.

If g is an automorphism of the graph, then consider the orderings $\pi = g\mu$. Let $P_1, P_2, \ldots, P_k$ be the path from the top node of the state space tree to the

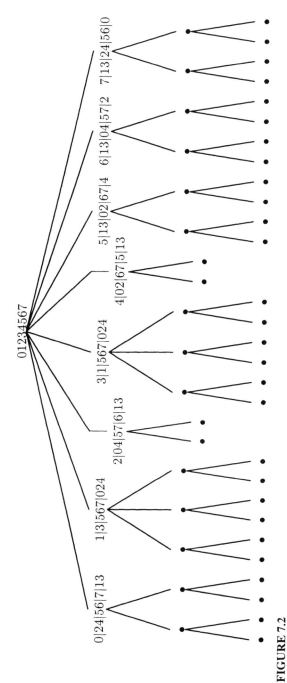

FIGURE 7.2
Overview of the state space tree that results from running Algorithm 7.7 on the graph in Example 7.7.

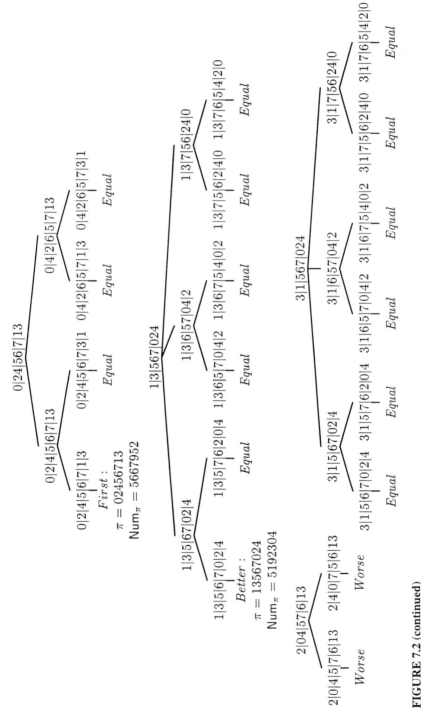

FIGURE 7.2 (continued)
Subtrees with roots 0|24|56|7|13, 1|3|567|024, 2|04|57|13 **and** 3|1|567|024.

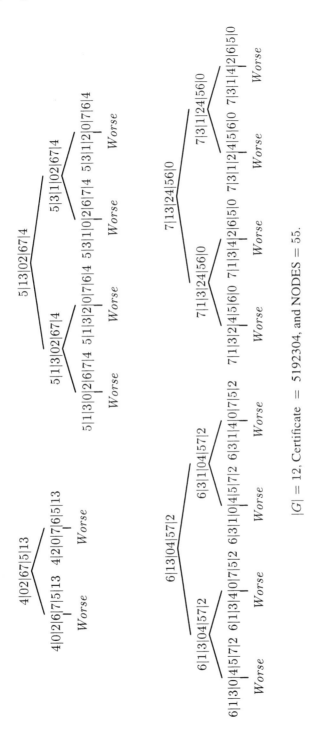

$|G| = 12$, Certificate $= 5192304$, and NODES $= 55$.

FIGURE 7.2 (continued)
Subtrees with roots 4|02|67|5|13, 5|13|02|67|4, 6|13|04|57|2 **and** 7|13|24|56|0.

leaf in which the ordering μ was last updated. Then $g(P_1), g(P_2), \ldots, g(P_k)$ is a path from the top node to a leaf that obtains the ordering $\pi = g\mu$. If $P = [P[0], \ldots, P[\ell]]$ is a partition, then $g(P)$ is the partition $[g(P[0]), \ldots, g(P[\ell])]$. Thus every automorphism of the graph will be obtained using this modification to the backtrack search. Note that some automorphisms may be obtained earlier when μ is not as yet optimal, but they will be discovered again when the optimal ordering is found.

Now that we know how to find automorphisms, we can use them to drastically prune the search. Consider the backtrack tree given in Figure 7.3. When node 1|3|567|024 is reached we know that $(1, 3)$ and $(2, 4)(5, 6)$ are automorphisms and thus the subgroup

$$\{\mathbf{I}, (1, 3), (2, 4)(5, 6), (1, 3)(2, 4)(5, 6)\}$$

which is generated by them is an automorphism group of the graph. Ordinarily we would split the block 567 of this partition in each of the three ways

1. 1|3|5|67|024
2. 1|3|6|57|024
3. 1|3|7|56|024.

When the first of these is searched another automorphism $(0, 2)(6, 7)$ is discovered. This one, together with the first two, generates the subgroup

$$\left\{\begin{array}{l}
\mathbf{I}, \\
(1, 3), \\
(2, 4)(5, 6), \\
(1, 3)(2, 4)(5, 6), \\
(0, 2)(6, 7), \\
(0, 2, 4)(5, 7, 6), \\
(0, 2)(1, 3)(6, 7), \\
(0, 2, 4)(1, 3)(5, 7, 6), \\
(0, 4, 2)(5, 6, 7), \\
(0, 4)(5, 7), \\
(0, 4, 2)(1, 3)(5, 6, 7), \\
(0, 4)(1, 3)(5, 7)
\end{array}\right\}$$

of the automorphism group of the graph. We now see that $(2, 4)(5, 6)$ is a known automorphism and that it carries partition 1 to partition 2. Also the automorphism $(0, 4)(5, 7)$ will carry partition 1 to partition 3. Consequently, the subtrees that would result from searching the nodes labeled by partitions 2 and 3 are isomorphic to the already searched subtree labeled by partition 1. They will produce orderings π for which the Num_π are exactly those found when partition 1 was searched. Thus these subtrees may be pruned.

In order to accomplish the pruning by automorphisms as suggested by the above the algorithm needs to be able to do the following:

1. manage discovered automorphisms; and
2. use known automorphisms to prune the search.

For 1, we use Algorithm 6.11 as described earlier. For 2, suppose

$$Q = q_0|q_1|q_2| \cdots |q_{\ell-1}|Q[\ell]| \cdots$$

is a partition whose first block of size greater than 1 is $Q[\ell]$. Let $u \in Q[\ell]$ and let

$$R = q_0|q_1|q_2| \cdots |q_{\ell-1}|u|Q[\ell] \setminus \{u\}| \cdots .$$

After searching the subtree obtained from the refinement of R, we will have obtained an automorphism group

$$\vec{G} = [\mathcal{U}_0, \mathcal{U}_1, \ldots, \mathcal{U}_\ell, \ldots, \mathcal{U}_n].$$

The partition

$$R' = q_0|q_1|q_2| \cdots |q_{\ell-1}|u'|Q[\ell] \setminus \{u'\}| \cdots ,$$

where $u' \in Q[\ell]$ and $u' \neq u$, can be pruned if there is a $g \in \vec{G}$ such that $g(R) = R'$. Thus g must fix

$$q_0, q_1, \ldots, q_{\ell-1}$$

and map u to u'. If the β of G is

$$q_0, q_1, \ldots, q_\ell, u, \ldots,$$

then g can easily be determined. In this case, there is such a g if and only if there exists $g \in \mathcal{U}_\ell$ such that $g(u) = u'$. Consequently, we need only use Algorithm 6.12 to change the base β of G to

$$\beta' = [q_0, q_1, \ldots, q_\ell, u, \ldots],$$

and then check $\mathcal{U}_\ell$. This pruning by automorphisms is done by Algorithm 7.9, which is invoked by Algorithm 7.10. There is a dramatic improvement, as is shown by Figure 7.3.

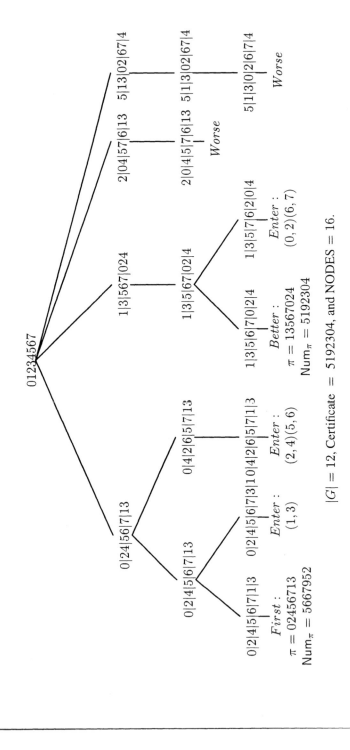

FIGURE 7.3
The state space tree that results from running Algorithm 7.9 on the Graph in Example 7.7.

Algorithm 7.9: CANON2 $(\mathcal{G}, \vec{G}, P)$

external REFINE(), COMPARE(), ENTER2(), CHANGEBASE()

REFINE$(\mathcal{G}, P, Q)$

Find the index ℓ of the first block of Q with $|Q[\ell]| > 1$

$Res \leftarrow Better$

if $BestExist$

then $\begin{cases} \textbf{for } i \leftarrow 0 \textbf{ to } n - 1 \\ \quad \textbf{do } \pi_1[i] = q_i \text{ where } Q[i] = \{q_i\} \\ Res \leftarrow \text{COMPARE}(\mathcal{G}, \pi_1, \ell) \end{cases}$

if Q has n blocks

then $\begin{cases} \textbf{if not } BestExist \\ \textbf{then} \begin{cases} \textbf{for } i \leftarrow 0 \textbf{ to } n - 1 \\ \quad \textbf{do } Best[i] = q_i \text{ where } Q[i] = \{q_i\} \\ BestExist \leftarrow \textbf{true} \end{cases} \\ \textbf{else} \begin{cases} \textbf{if } Res = Better \\ \textbf{then } Best \leftarrow \pi_1 \\ \textbf{else if } Res = Equal \\ \textbf{then} \begin{cases} \textbf{for } i \leftarrow 0 \textbf{ to } n - 1 \\ \quad \textbf{do } \pi_2[\pi_1[i]] \leftarrow Best[i] \\ \text{ENTER2}(\pi_2, \vec{G}) \end{cases} \end{cases} \end{cases}$

else $\begin{cases} \textbf{if } Res \neq Worse \end{cases} \begin{cases} C \leftarrow Q[\ell] \\ D \leftarrow Q[\ell] \\ \textbf{for } j \leftarrow 0 \textbf{ to } j < \ell \\ \quad \textbf{do } R[j] \leftarrow Q[j] \\ \textbf{for } j \leftarrow \ell + 1 \textbf{ to } size(Q) \\ \quad \textbf{do } R[j + 1] \leftarrow Q[j] \\ \textbf{while } C \neq \emptyset \\ \textbf{do} \begin{cases} u \leftarrow \text{any element of } C \\ R[\ell] \leftarrow \{u\} \\ R[\ell + 1] \leftarrow D \setminus \{u\} \\ \text{CANON2}(\mathcal{G}, \vec{G}, R) \\ \textbf{for } j \leftarrow 0 \textbf{ to } \ell \\ \quad \textbf{do } \beta'[j] \leftarrow r \text{ where } R[j] = \{r\} \\ \textbf{for each } y \notin \{\beta'[0], \beta'[1], \ldots, \beta'[\ell]\} \\ \quad \textbf{do } \begin{cases} j \leftarrow j + 1 \\ \beta'[j] \leftarrow y \end{cases} \\ \text{CHANGEBASE}(n, \vec{G}, \beta') \\ \textbf{for each } g \in \mathcal{U}_\ell \\ \quad \textbf{do } C \leftarrow C \setminus \{g(u)\} \end{cases} \end{cases}$

Algorithm 7.10: CERT2 $(\mathcal{G}, \vec{G})$

external CANON2()

comment: Set $\vec{G}$ to the identity group with base **I**
for $j \leftarrow 0$ **to** $n - 1$ **do** $\mathcal{U}_J \leftarrow \{\mathbf{I}\}$
$\vec{G} \leftarrow (\mathbf{I}; [\mathcal{U}_0, \mathcal{U}_1, \ldots, \mathcal{U}_{n-1}])$
$P \leftarrow [\{0, 1, \ldots, n\}]$
CANON2$(\mathcal{G}, \vec{G}, P)$
$k \leftarrow 0$
$C \leftarrow 0$
for $j \leftarrow n - 1$ **downto** 1
$\quad$ **do** $\begin{cases} \textbf{for } i \leftarrow j - 1 \textbf{ downto } 0 \\ \quad \textbf{do} \begin{cases} \textbf{if } \{\mu[i], \mu[j]\} \in \mathcal{E}(\mathcal{G}) & \textbf{then } C \leftarrow x + 2^k \\ k = k + 1 \end{cases} \end{cases}$
return (C)

7.4   Isomorphism of other structures

7.4.1   Using known automorphisms

If we know some or all of the automorphisms of the graphs we wish to compute the certificates of, then we can first use Algorithm 6.9 to find the Schreier-Sims representation $\vec{G}$ of the group generated by these automorphisms, and then input $\vec{G}$ to Algorithm 7.10. This will speed up the search for the certificate, because these automorphisms can be used to prune the tree before they are discovered in the search. This is particularly useful when trying to find all graphs with a given automorphism group, say, for example, all graphs on n vertices, that have

$$(0, 1, 2, 3, \ldots, n - 1)$$

as an automorphism. In Figure 7.4, we show the state space tree for the certificate of the graph in Example 7.7 that results when we include as input the automorphism group of this graph. Even in this small example, a savings from searching 16 nodes to 10 nodes results.

7.4.2   Set systems

One common method of determining isomorphism of set systems is to use the following trick. Given a set system $(\mathcal{X}, \mathcal{B})$, a graph $\mathcal{G} = (\mathcal{V}, \mathcal{E})$ is defined as follows:

$\mathcal{V} = \mathcal{X} \cup \mathcal{B}$; and
$\{x, B\} \in \mathcal{E}$ if and only if B, where $x \in \mathcal{X}$ and $B \in \mathcal{B}$.

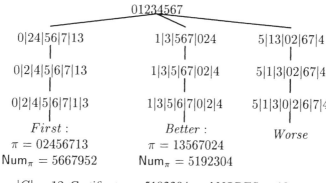

$$01234567$$

0|24|56|7|13	1|3|567|024	5|13|02|67|4
0|2|4|5|6|7|13	1|3|5|67|02|4	5|1|3|02|67|4
0|2|4|5|6|7|1|3	1|3|5|6|7|0|2|4	5|1|3|0|2|6|7|4
First :	*Better* :	*Worse*
$\pi = 02456713$	$\pi = 13567024$	
$\text{Num}_\pi = 5667952$	$\text{Num}_\pi = 5192304$	

$|G| = 12$, Certificate $= 5192304$, and NODES $= 10$.

FIGURE 7.4
State space tree that results from running Algorithm 7.9 on the Graph in Example 7.7.

If the search is initialized with the partition of the vertices into the two blocks $\mathcal{X}$ and $\mathcal{B}$, then Algorithm 7.10 will return a group whose restriction to X will be the automorphism group of the set system. Furthermore it is not difficult to see that two set systems are isomorphic if and only if their corresponding graphs have the same certificates.

Example 7.9 *A set system and its corresponding graph*

The set system.

$B_0 = \{0, 1, 2\}$
$B_1 = \{0, 1, 3\}$
$B_2 = \{0, 2, 4\}$
$B_3 = \{0, 3, 5\}$
$B_4 = \{0, 4, 5\}$
$B_5 = \{1, 2, 5\}$
$B_6 = \{1, 3, 4\}$
$B_7 = \{1, 4, 5\}$
$B_8 = \{2, 3, 4\}$
$B_9 = \{2, 3, 5\}$

The graph.

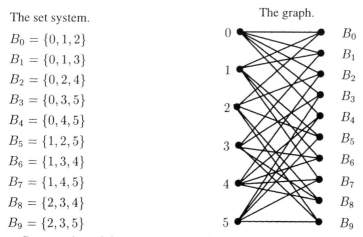

Computation of the automorphism group and certificate for this set system is given in Figure 7.5. Notice that $(2, 3)(4, 5)$, $(1, 2)(3, 4)$, $(0, 1)(2, 3)$ generate the full automorphism group of this set system. $\square$

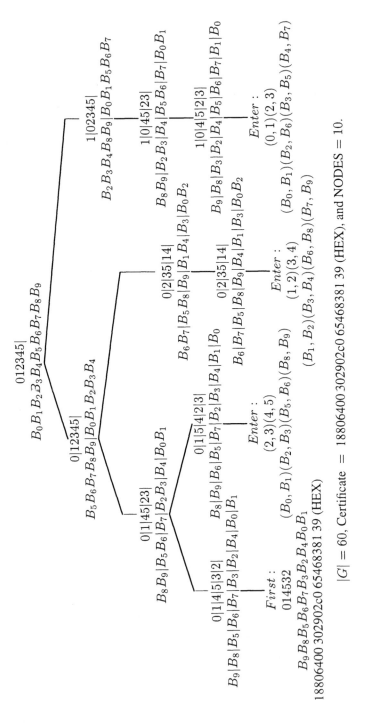

FIGURE 7.5
The state space tree that results from running Algorithm 7.9 on the Graph in Example 7.7.

7.5   Notes

Section 7.1

In [68] it was shown that the isomorphism of graphs whose maximum degree is bounded by a given constant can be determined in polynomial time.

Section 7.2

The technique of using invariants is among the earliest methods used for determining isomorphism; see the description in Kučera [61].

Section 7.3

The method of using balanced binary strings of length $2n$ as a certificate for trees on n vertices is due to Ronald Read [88].

An excellent discussion on writing graph isomorphism programs can be found in Kocay [55]. A somewhat different algorithm, which is very popular, was developed by McKay [70, 71].

Exercises

7.1 Find an isomorphism between the following two graphs. (These are two different representations of the Petersen graph.)

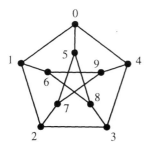
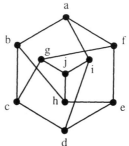

7.2 Use the algorithm described in Section 7.3.1 to compute the certificate for the tree given below

7.3 Use the algorithm described in Section 7.3.1 to compute the tree whose certificate is   0000101110010101100111000011110001111.

7.4 Provide a detailed proof of Theorem 7.1.

7.5 Use Algorithm 7.10 to compute the certificate and automorphism group of the following graph.

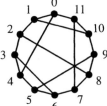

7.6 Use Algorithm 7.10 to compute the automorphism group of the Steiner triple system given below. Compare your result with Exercise 6.4.

$$
\left\{
\begin{array}{lllll}
\{0,1,2\}, & \{3,4,5\}, & \{6,7,8\}, & \{0,3,6\}, \\
\{1,4,7\}, & \{2,5,8\}, & \{0,4,8\}, & \{1,5,6\}, \\
\{2,3,7\}, & \{0,5,7\}, & \{1,3,8\}, & \{2,4,6\}
\end{array}
\right\} .
$$

7.7 Compute the number of non-isomorphic Steiner triple systems of order 13. (See Exercise 4.11.)

7.8 Give a reasonable definition for the automorphism group of a Latin square and use Algorithm 7.10 to compute the automorphism group of the Latin square given in Example 1.1.2.

7.9 Use your solution to Exercise 4.13 and Algorithm 7.10 to determine the number of non-isomorphic Latin squares of order 4.

8

Basis Reduction

8.1 Introduction

Many combinatorial search problems can be reduced to solving a matrix equation of the form

$$AU = B \tag{8.1}$$

for a $(0, 1)$-valued column vector U. For example let $\mathcal{E} = \{e_1, e_2, \ldots, e_m\}$ be the set of edges of a graph $\mathcal{G}$ with vertex set $\mathcal{V} = \{1, 2, \ldots, m\}$. If $\mathcal{T} = \{T_1, T_2, \ldots, T_N\}$ is the set of all triangles in the graph $\mathcal{G}$, then we may define the m by N matrix A by

$$A[i, j] = \begin{cases} 1 & \text{if } e_i \text{ is an edge of } T_j; \\ 0 & \text{if not.} \end{cases}$$

A $(0, 1)$-valued solution U to the matrix equation $AU = B$, where $B = [1, 1, \ldots, 1]^T$, is an edge decomposition of $\mathcal{G}$ into triangles. If a decomposition into other subgraphs is desired, then $\mathcal{T}$ can be chosen to contain these types of subgraphs. In Section 4.5, the problem of choosing members from a given collection $\mathcal{S} = \{S_1, S_2, \ldots, S_n\}$ of subsets of a set $\mathcal{R} = \{1, 2, \ldots, m\}$ such that they partition $\mathcal{R}$ is discussed. If the matrix A is defined by

$$A[i, j] = \begin{cases} 1 & \text{if } i \in S_j; \\ 0 & \text{if not,} \end{cases}$$

and $B = [1, 1, \ldots, 1]^T$, then this problem asks for the $(0, 1)$-valued solutions U to Equation 8.1. Recall that matrix A is called an incidence matrix. When the collection of graphs or sets in the decomposition are assumed to have certain symmetries, the resulting incidence matrices are discussed in Section 6.6.1. These incidence matrices need not be $(0, 1)$-valued matrices; see Example 6.12.

When the size of the incidence matrix A is large, backtracking approaches often fail to find solutions in reasonable time. To get around this problem, the heuristic search techniques of Chapter 5 can be employed. Another successful technique for solving Equation 8.1 has been the method of *basis reduction*. It begins by first transforming Equation 8.1 to the optimization problem of finding a shortest vector in a lattice. Consider the matrix equation

$$\begin{bmatrix} I & \vec{0} \\ A & -B \end{bmatrix} \begin{bmatrix} U \\ 1 \end{bmatrix} = \begin{bmatrix} U \\ \vec{0} \end{bmatrix}, \tag{8.2}$$

where $\vec{0}$ denotes the 0 vector and I the identity matrix. A vector $\vec{b} \in \mathbb{R}^q$ is represented in the computer as a q-dimensional array, whose entries are the components of $\vec{b}$. The following proposition is evident:

LEMMA 8.1   *The $(0, 1)$-valued column vector U solves Equation 8.1 if and only if it solves Equation 8.2.*

If $\vec{b}_1, \vec{b}_2, \ldots, \vec{b}_p \in \mathbb{R}^q$ are real-valued vectors, and $\alpha_1, \alpha_2, \ldots, \alpha_p \in \mathbb{R}$, then

$$\vec{b} = \alpha_1 \vec{b}_1 + \alpha_2 \vec{b}_2 + \cdots + \alpha_p \vec{b}_p$$

is a *linear combination* of $\vec{b}_1, \vec{b}_2, \ldots, \vec{b}_p$ over the real numbers. The set of all linear combinations of $\vec{b}_1, \vec{b}_2, \ldots, \vec{b}_p$ is the vector space

$$\mathsf{Span}_{\mathbb{R}}(\vec{b}_1, \vec{b}_2, \ldots, \vec{b}_p) = \{\alpha_1 \vec{b}_1 + \alpha_2 \vec{b}_2 + \cdots + \alpha_p \vec{b}_p : \alpha_i \in \mathbb{R}, \ 1 \le i \le p\}.$$

If the coefficients $\alpha_1, \alpha_2, \ldots, \alpha_p$ are restricted to be integers, then

$$\mathsf{Span}_{\mathbb{Z}}(\vec{b}_1, \vec{b}_2, \ldots, \vec{b}_p) = \{\alpha_1 \vec{b}_1 + \alpha_2 \vec{b}_2 + \cdots + \alpha_p \vec{b}_p : \alpha_i \in \mathbb{Z}, \ 1 \le i \le p\}$$

is said to be the *lattice* spanned by $\vec{b}_1, \vec{b}_2, \ldots, \vec{b}_p$.

The set of vectors $\vec{b}_1, \vec{b}_2, \ldots, \vec{b}_p$ is said to be *linearly independent* if the only linear combination over the real numbers that is the zero vector is the one in which $\alpha_i = 0$ for all i, $1 \le i \le p$.

Let B be a q by p matrix with linearly independent columns: $\vec{b}_1, \vec{b}_2, \ldots, \vec{b}_p$. Then

$$\mathcal{L} = \mathsf{Span}_{\mathbb{Z}}(\vec{b}_1, \vec{b}_2, \ldots, \vec{b}_p) = \{B\vec{x} : \vec{x} \in \mathbb{Z}^p\}$$

is the lattice with basis (the columns of) B. Not every lattice has a basis of linearly independent vectors. One example is the lattice spanned by the vectors

$$\begin{bmatrix} 2 \\ 1 \end{bmatrix}, \begin{bmatrix} 1 \\ 2 \end{bmatrix}, \text{and} \begin{bmatrix} 1 \\ 1 \end{bmatrix}.$$

None of these vectors can be written as an integer linear combination of the other two. Thus each is required in the span of the lattice. On the other hand,

$$\begin{bmatrix} 2 \\ 1 \end{bmatrix} + \begin{bmatrix} 1 \\ 2 \end{bmatrix} - 3 \begin{bmatrix} 1 \\ 1 \end{bmatrix} = \begin{bmatrix} 0 \\ 0 \end{bmatrix},$$

and so they are not linearly independent. We will only be interested in lattices that have a basis. If $p = q$ and the columns of B are linearly independent, then the lattice $\mathcal{L}$ with basis B is said to be a *full dimensional lattice*.

Example 8.1

Consider the matrix

$$B = \begin{bmatrix} 1 & -2 \\ 3 & 1 \end{bmatrix}.$$

The columns of B are the vectors $\vec{b}_1 = [1, 3]$, and $\vec{b}_2 = [-2, 1]$. It is easy to see that $\vec{b}_1$ and $\vec{b}_2$ are linearly independent. Thus the subspace $\mathsf{Span}_{\mathbb{R}}(\vec{b}_1, \vec{b}_2)$ is the entire xy-plane, but the lattice $\mathcal{L}$ with basis B is a discrete set of points. This is depicted by the following diagram.

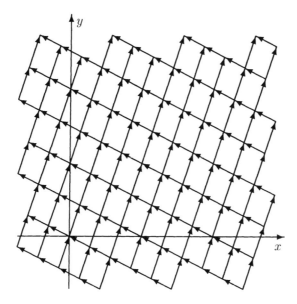

Returning to Equation 8.2, we let $\mathcal{L}$ be the lattice whose basis is the $n + m$ by $n + 1$ matrix M defined by

$$M = \begin{bmatrix} I & \vec{0} \\ A & -B \end{bmatrix}.$$

Then Lemma 8.1 shows that if U is a $(0, 1)$-valued solution to Equation 8.1, then

$$\widehat{U} = \begin{bmatrix} U \\ \vec{0} \end{bmatrix} \in \mathcal{L}.$$

Conversely, if

$$\vec{y} = [u_1, u_2, \ldots, u_n, \underbrace{0, 0, \ldots 0}_{m}]^T \in \mathcal{L},$$

then $\vec{y} = M\vec{x}$ for some integer valued column vector $\vec{x}$. Consequently, from Equation 8.2, it follows that

$$\vec{x} = [u_1, u_2, \ldots, u_n, d]$$

for some integer d. We record this result as Lemma 8.2.

LEMMA 8.2   *Let $\mathcal{L}$ be the lattice with basis*

$$M = \begin{bmatrix} I & \vec{0} \\ A & -B \end{bmatrix}.$$

Then there is a $(0, 1)$-valued column vector

$$\vec{y} = [u_1, u_2, \ldots, u_n, \underbrace{0, 0, \ldots 0}_{m}]^T \in \mathcal{L} \tag{8.3}$$

if and only if there is a $(0, 1)$-valued solution $U = [u_1, u_2, \ldots, u_n]^T$ to the matrix equation $AU = dB$, for some integer d.

Recall that the *Euclidean length* of a vector $\vec{y} \in \mathbb{R}^k$ is

$$||\vec{y}|| = \sqrt{y_1^2 + y_2^2 + \cdots + y_k^2}.$$

Thus, if $\vec{y} \in \mathcal{L}$ has the form of Equation 8.3, then the Euclidean length of $\vec{y}$ is

$$||\vec{y}|| = \sqrt{u_1^2 + u_2^2 + \cdots + u_n^2} \leq \sqrt{n},$$

because $u_i \in \{0, 1\}$ for $i = 1, 2, \ldots, n$. Thus $\vec{y}$ is a vector whose Euclidean length is small when compared to most of the vectors in the lattice $\mathcal{L}$.

This leads us to consider the following optimization problem, Problem 8.1.

Problem 8.1:   Shortest Vector

Instance: a matrix M with integer entries.

Find: the minimum (non-zero) value of $||\vec{y}||$

subject to $\vec{y} \in \mathcal{L}$, the lattice with basis M.

8.2   Theoretical development

Throughout this section, let

$$M = [\vec{b}_1, \vec{b}_2, \ldots, \vec{b}_n]$$

be an m by n matrix considered as a set of column vectors in $\mathbb{Z}^m$, and let

$$\mathcal{L} = \mathsf{Span}_{\mathbb{Z}}(\vec{b}_1, \vec{b}_2, \ldots, \vec{b}_n) = \{y : y = Mx, x \in \mathbb{Z}^n\}$$

be the lattice with basis M. We will describe an algorithm that will find a new basis M' for $\mathcal{L}$ containing vectors that have Euclidean length smaller than the vectors in those in M. The goal is to produce such a basis containing a vector $\vec{y}$ that solves Problem 8.1. First, in order to facilitate the analysis and development of the algorithm, we review some concepts from linear algebra.

Recall that essentially the only way we can alter the appearance of a basis, without changing the vector subspace that it spans, is to either

(i) reorder the vectors in the basis, or

(ii) perform an operation of the form:

$$\text{replace } \vec{b}_j \text{ with } \alpha_1 \vec{b}_1 + \alpha_2 \vec{b}_2 + \cdots + \alpha_n \vec{b}_n$$

where $\alpha_i \in \mathbb{R}$ for all i and $\alpha_j \neq 0$. However, we wish to have the new basis still be a basis for the lattice $\mathcal{L}$. Thus, we also require that $\alpha_i \in \mathbb{Z}$ for all i and that $\alpha_j = \pm 1$. Consequently, we can reduce any operation of type (ii) to performing a sequence of the following three operations:

1. Replace $\vec{b}_i$ with $\vec{b}_i + \vec{b}_j$
2. Replace $\vec{b}_i$ with $-\vec{b}_i + \vec{b}_j$
3. Replace $\vec{b}_i$ with $\vec{b}_i - \vec{b}_j$

For example, to obtain the operation,

$$\text{replace } \vec{b}_2 \text{ by } 2\vec{b}_1 + \vec{b}_2 - 3\vec{b}_3,$$

the following sequence of operations can be performed:

$$\text{replace } \vec{b}_2 \text{ with } \vec{b}_2 - \vec{b}_3$$
$$\text{replace } \vec{b}_2 \text{ with } \vec{b}_2 - \vec{b}_3$$
$$\text{replace } \vec{b}_2 \text{ with } \vec{b}_2 - \vec{b}_3$$
$$\text{replace } \vec{b}_2 \text{ with } \vec{b}_1 + \vec{b}_2$$
$$\text{replace } \vec{b}_2 \text{ with } \vec{b}_1 + \vec{b}_2.$$

If $\vec{x} = [x_1, x_2, \ldots, x_m]$ and $\vec{y} = [y_1, y_2, \ldots, y_m]$ are vectors in $\mathbb{R}^m$, then we denote by $\vec{x} \cdot \vec{y}$ the *dot product* of $\vec{x}$ and $\vec{y}$, defined as

$$\vec{x} \cdot \vec{y} = x_1 y_1 + x_2 y_2 + \cdots + x_m y_m.$$

We say that $\vec{x}$ and $\vec{y}$ are *orthogonal* if $\vec{x} \cdot \vec{y} = 0$. If the vectors in a basis are pairwise orthogonal, we say that it is an *orthogonal basis*. Notice that $||\vec{x}||$, the *Euclidean length* of $\vec{x}$, is $\sqrt{\vec{x} \cdot \vec{x}}$. The *triangle inequality* states that the sum of the lengths of two sides of a triangle is always greater than the length of the third. In terms of vectors $\vec{x}$ and $\vec{y}$, this says that

$$||\vec{x}|| + ||\vec{y}|| \geq ||\vec{x} + \vec{y}||.$$

Observe that, if the basis vectors $\vec{b}_i$ and $\vec{b}_j$ are orthogonal, then by the triangle inequality, we have

$$||\vec{b}_i + \vec{b}_j|| \geq \max\{||\vec{b}_i||, ||\vec{b}_j||\},$$
$$|| - \vec{b}_i + \vec{b}_j|| \geq \max\{||\vec{b}_i||, ||\vec{b}_j||\},$$

and

$$||\vec{b}_i - \vec{b}_j|| \geq \max\{||\vec{b}_i||, ||\vec{b}_j||\}.$$

Hence, none of the three replacement operations can reduce the length of $\vec{b}_i$. Moreover, if all the vectors in the basis are pairwise orthogonal, then there is no way to further reduce the size of the vectors in the basis. Thus, in order to achieve a basis with vectors of minimal length, we will try to make a new basis for the lattice $\mathcal{L}$ in which the vectors in the basis are as pairwise orthogonal as possible. In general, we will not be able to fully orthogonalize M because we can only do the above type (ii) operations with integer coefficients. If real coefficients were allowed, then Algorithm 8.1 would produce a basis in which the vectors are pairwise orthogonal. Algorithm 8.1, the standard *Gram-Schmidt process* of orthogonalization from linear algebra, uses $O(n^3)$ arithmetic operations.

Algorithm 8.1: GRAM-SCHMIDT $(\vec{b}_1, \vec{b}_2, \ldots, \vec{b}_n)$

$\vec{b}_1^* \leftarrow \vec{b}_1$
for $j \leftarrow 2$ **to** n

$\quad$ **do** $\begin{cases} \vec{b}_j^* \leftarrow \vec{b}_j \\ \textbf{for } i \leftarrow 1 \textbf{ to } j - 1 \\ \quad \textbf{do } \begin{cases} \alpha_{ij} \leftarrow (\vec{b}_i^* \cdot \vec{b}_j)/||\vec{b}_i^*||^2 \\ \vec{b}_j^* \leftarrow \vec{b}_j^* - \alpha_{ij}\vec{b}_i^* \end{cases} \end{cases}$

return $([\vec{b}_1^*, \vec{b}_2^*, \ldots, \vec{b}_n^*], \{\alpha_{ij}\}_{i<j})$

A useful tool for measuring the volume of the parallelepiped determined by a set of vectors is the determinant.

Definition 8.1: The *sign* of a permutation on a set $\mathcal{X} = \{1, 2, \ldots, n\}$ is given by

$$\text{sign}(\sigma) = \begin{cases} +1 \text{ if } \sigma \text{ is an even permutation}; \\ -1 \text{ if } \sigma \text{ is an odd permutation}. \end{cases}$$

The *determinant* of the n by n matrix M is

$$\det M = \sum_{\sigma \in \text{Sym}(\mathcal{X})} \text{sign}(\sigma) \prod_{i=1}^{n} M[i, \sigma(i)].$$

Recall that Algorithm 2.19 can be used to compute the sign of a permutation. From Definition 8.1, it is straightforward to see that the determinant of the 2 by 2 matrix M is

$$\det M = M[1, 1]M[2, 2] - M[1, 2]M[2, 1]$$

because the only permutations of $\{1, 2\}$ are $\mathbf{I}$ (the identity permutation) and $(1, 2)$. Similarly, the determinant of the 3 by 3 matrix M is

$$\det M = M[1, 1]M[2, 2]M[3, 3] - M[1, 1]M[2, 3]M[3, 2]$$
$$- M[1, 2]M[2, 1]M[3, 3] + M[1, 2]M[2, 3]M[3, 1]$$
$$+ M[1, 3]M[2, 1]M[3, 2] - M[1, 3]M[2, 2]M[3, 1],$$

because the permutations of $\{1, 2, 3\}$ are $\mathbf{I}$, $(2, 3)$, $(1, 2)$, $(1, 2, 3)$, $(1, 3, 2)$ and $(1, 3)$. Furthermore, if we denote by $M_{\langle ij \rangle}$ the matrix obtained by removing row i and column j, then it is easily seen that we can recursively compute the determinant of M with the formula:

$$\det M = \sum_{j=1}^{n} (-1)^{i+j} M[i, j] \det M_{\langle ij \rangle},$$

where i is any fixed row. Similarly, the determinant can be computed with the formula

$$\det M = \sum_{i=1}^{n} (-1)^{i+j} M[i, j] \det M_{\langle ij \rangle}$$

where j is any fixed column. This method of computing the determinant is called *cofactor expansion* or *Laplace expansion* and will take $O(n!)$ operations. A method that uses only $O(n^3)$ operations is to use elementary row operations to reduce the matrix to upper triangular form; see any text on linear algebra.

Elementary linear algebra establishes Lemma 8.3.

LEMMA 8.3   *Let* $\vec{b}_1, \vec{b}_2, \ldots, \vec{b}_n \in \mathbb{R}^m$ *be linearly independent.*

(i)   *On input* $M = [\vec{b}_1, \vec{b}_2, \ldots, \vec{b}_n]$, *Algorithm 8.1 computes an orthogonal basis*

$$M^* = [\vec{b}_1^*, \vec{b}_2^*, \ldots, \vec{b}_n^*]$$

for the vector space $\mathsf{Span}_{\mathbb{R}}(\vec{b}_1, \vec{b}_2, \ldots, \vec{b}_n)$.

(ii) $\vec{b}_k^*$ *is orthogonal to each vector in*

$$\mathsf{Span}_{\mathbb{R}}(\vec{b}_1^*, \ldots, \vec{b}_{k-1}^*).$$

(iii)   *If* M *is a square matrix, then*

$$|\det M| = |\det M^*| = \prod_{j=1}^{n} ||\vec{b}_j^*||.$$

In view of Lemma 8.3, we define, for any lattice $\mathcal{L}$, with (possibly non-square) linearly independent basis matrix $M = [\vec{b}_1, \vec{b}_2, \ldots, \vec{b}_n]$, the quantity

$$\mathsf{vol}(\mathcal{L}) = \prod_{j=1}^{n} ||\vec{b}_j^*||.$$

$\mathsf{vol}(\mathcal{L})$ is the n-dimensional *volume* of the parallelepiped with vertices

$$\left\{ \sum_{j=1}^{n} \vec{b}_j^* x_j \; : \quad x_j \in \{0, 1\} \quad \forall j \right\},$$

where $[\vec{b}_1^*, \ldots, \vec{b}_n^*]$ is the orthogonal basis obtained by applying Algorithm 8.1. This volume is independent of the choice of linearly independent basis for $\mathcal{L}$.

If M is a basis for a lattice $\mathcal{L}$, we define the *weight* of M to be

$$\mathsf{wt}(M) = \prod_{j=1}^{n} ||\vec{b}_i||.$$

The relationship between $\mathsf{wt}(M)$ and $\mathsf{vol}(\mathcal{L})$ is provided by *Hadamard's inequality*, which we state now.

LEMMA 8.4   *(Hadamard's inequality) For all bases* M *of* $\mathcal{L}$, *we have*

$$\mathsf{wt}(M) \geq \mathsf{vol}(\mathcal{L}).$$

PROOF   Suppose $M = [\vec{b}_1, \vec{b}_2, \ldots, \vec{b}_n]$ is a basis for the lattice $\mathcal{L}$. Then on input M, Algorithm 8.1 constructs, for each $j = 1, 2, \ldots, n$, the vectors

$$\vec{b}_j^* = \vec{b}_j - \sum_{i=1}^{j-1} \alpha_{ij} \vec{b}_i^*,$$

where

$$\alpha_{ij} = \frac{(\vec{b}_i^* \cdot \vec{b}_j)}{||\vec{b}_i^*||^2}.$$

Thus,

$$\vec{b}_j^* \cdot \vec{b}_j^* = \vec{b}_j^* \cdot \left(\vec{b}_j - \sum_{i=1}^{j-1} \alpha_{ij} \vec{b}_i^* \right) = \vec{b}_j^* \cdot \vec{b}_j,$$

because $\{\vec{b}_1^*, \vec{b}_2^*, \ldots, \vec{b}_n^*\}$ is an orthogonal basis. If θ is the angle between $\vec{b}_j$ and $\vec{b}_j^*$, then

$$||\vec{b}_j^*||^2 = \vec{b}_j^* \cdot \vec{b}_j^*$$

$$= \vec{b}_j^* \cdot \vec{b}_j$$

$$= ||\vec{b}_j|| \cdot ||\vec{b}_j^*|| \cos(\theta).$$

Thus $0 \leq \cos(\theta) \leq 1$, and so for all j we have $||\vec{b}_j^*|| \leq ||\vec{b}_j||$. The result now follows. ∎

Hadamard's inequality is illustrated in Example 8.2.

Example 8.2
Let

$$M = \begin{bmatrix} 1 & 1 & 0 & 1 & 3 \\ 1 & 0 & 2 & 1 & 0 \\ 1 & 0 & 1 & 0 & 0 \\ 0 & -1 & 1 & 0 & 3 \\ 1 & 1 & -1 & 0 & 0 \end{bmatrix}.$$

To obtain the determinant of M, we use a cofactor expansion along column 5.

$$\det M = 3 \cdot \det \begin{bmatrix} 1 & 0 & 2 & 1 \\ 1 & 0 & 1 & 0 \\ 0 & -1 & 1 & 0 \\ 1 & 1 & -1 & 0 \end{bmatrix} + (-3) \det \begin{bmatrix} 1 & 1 & 0 & 1 \\ 1 & 0 & 2 & 1 \\ 1 & 0 & 1 & 0 \\ 1 & 1 & -1 & 0 \end{bmatrix}$$

$$= 3(-1) \begin{bmatrix} 1 & 0 & 1 \\ 0 & -1 & 1 \\ 1 & 1 & -1 \end{bmatrix} - 3 \left(-1 \cdot \det \begin{bmatrix} 1 & 0 & 2 \\ 1 & 0 & 1 \\ 1 & 1 & -1 \end{bmatrix} + 1 \cdot \det \begin{bmatrix} 1 & 1 & 0 \\ 1 & 0 & 1 \\ 1 & 1 & -1 \end{bmatrix} \right)$$

$$= 3(-1) \cdot (1) - 3(-1 \cdot 1 + 1 \cdot (-1 + 2))$$

$$= -3.$$

The weight of M is

$$\mathrm{wt}(M) = \sqrt{4} \cdot \sqrt{3} \cdot \sqrt{7} \cdot \sqrt{2} \cdot \sqrt{18} = 12\sqrt{21} \approx 54.990908.$$

Now applying Algorithm 8.1 we obtain:

$$\alpha = \begin{bmatrix} * & \frac{1}{2} & \frac{1}{2} & \frac{1}{2} & \frac{3}{4} \\ * & * & -\frac{3}{2} & 0 & -\frac{3}{4} \\ * & * & * & \frac{2}{3} & -\frac{1}{2} \\ * & * & * & * & 6 \\ * & * & * & * & * \end{bmatrix}$$

and

$$M^* = \begin{bmatrix} 1 & \frac{1}{2} & \frac{1}{4} & \frac{1}{3} & \frac{3}{4} \\ 1 & -\frac{1}{2} & \frac{3}{4} & 0 & -\frac{3}{4} \\ 1 & -\frac{1}{2} & -\frac{1}{4} & -\frac{1}{3} & \frac{3}{4} \\ 0 & -1 & -\frac{1}{2} & \frac{1}{3} & 0 \\ 1 & \frac{1}{2} & -\frac{3}{4} & 0 & -\frac{3}{4} \end{bmatrix}.$$

(Recall that we have defined α_{ij} only for $i < j$.) Observe that

$$\text{vol}(\mathcal{L}) = ||\vec{b}_1^*|| \cdot ||\vec{b}_2^*|| \cdot ||\vec{b}_3^*|| \cdot ||\vec{b}_4^*|| \cdot ||\vec{b}_5^*||$$

$$= \sqrt{4} \cdot \sqrt{2} \cdot \sqrt{\frac{6}{4}} \cdot \sqrt{\frac{1}{3}} \cdot \sqrt{\frac{9}{4}}$$

$$= 3$$

$$= |\det M|.$$

$\Box$

The key to developing an algorithm to find the shortest vector in a lattice is revealed in Lemma 8.6. First, recall Cramer's rule, which is easily established using elementary linear algebra.

LEMMA 8.5 *(Cramer's rule) Let $M = [\vec{b}_1, \vec{b}_2, \ldots, \vec{b}_n]$ be an invertible n by n matrix with entries in $\mathbb{R}$. Then, for any vector $\vec{y} \in \mathbb{R}^n$, the j-th component of the solution $\vec{x}$ to the matrix equation $\vec{y} = M\vec{x}$ is*

$$x_j = \frac{\det M_j}{\det M},$$

where $M_j = [\vec{b}_1, \vec{b}_2, \ldots, \vec{b}_{j-1}, \vec{y}, \vec{b}_{j+1}, \ldots, \vec{b}_n]$.

LEMMA 8.6 *Given a full dimensional lattice $\mathcal{L}$, with basis M, let $\vec{y} = M\vec{x}$ be a shortest vector in $\mathcal{L}$. Then for all $j = 1, 2, \ldots, n$, the j-th component of $\vec{x}$ satisfies*

$$|x_j| \leq \frac{\text{wt}(M)}{\text{vol}(\mathcal{L})}.$$

PROOF Using Lemma 8.5, we have

$$|x_j| = \frac{|\det M_j|}{|\det M|},$$

where $M_j = [\vec{b}_1, \vec{b}_2, \ldots, \vec{b}_{j-1}, \vec{y}, \vec{b}_{j+1}, \ldots, \vec{b}_n]$. By Lemma 8.4, we obtain

$$|\det M_j| \leq ||\vec{b}_1|| \cdot ||\vec{b}_2|| \cdots ||\vec{b}_{j-1}|| \cdot ||\vec{y}|| \cdot ||\vec{b}_{j+1}|| \cdots ||\vec{b}_n||.$$

Also $\vec{b}_j \in \mathcal{L}$, and so $||\vec{y}|| \leq ||\vec{b}_j||$. Thus

$$|\det M_j| \leq ||\vec{b}_1|| \cdot ||\vec{b}_2|| \cdots ||\vec{b}_{j-1}|| \cdot ||\vec{b}_j|| \cdot ||\vec{b}_{j+1}|| \cdots ||\vec{b}_n|| = \mathsf{wt}(M).$$

Therefore $|x_j| \leq \mathsf{wt}(M)/\mathsf{vol}(\mathcal{L})$ as claimed, because $\mathsf{vol}(\mathcal{L}) = |\det M|$ when $\mathcal{L}$ is full dimensional. ∎

As a consequence of Lemma 8.6, we are motivated to find a basis M' such that the ratio

$$\frac{\mathsf{wt}(M')}{\mathsf{vol}(\mathcal{L})}$$

is small. Thus, because $\mathsf{vol}(\mathcal{L})$ depends only on the lattice $\mathcal{L}$ and not on the basis, we search for a basis M' with minimal weight $\mathsf{wt}(M')$. We have the following lower bound on the shortest vector in $\mathcal{L}$.

LEMMA 8.7 *Let* $M = [\vec{b}_1, \vec{b}_2, \ldots, \vec{b}_n]$ *be a basis for the lattice* $\mathcal{L}$. *If* $\vec{b} \in \mathcal{L}$, $b \neq 0$, *then*

$$||\vec{b}|| \geq \min_j ||\vec{b}_j^*||,$$

where $[\vec{b}_1^*, \vec{b}_2^*, \ldots, \vec{b}_n^*]$ *is the result of running Algorithm 8.1 on input* M.

PROOF Let $M = [\vec{b}_1, \vec{b}_2, \ldots, \vec{b}_n]$ be a basis for the lattice $\mathcal{L}$. Then $b \in \mathcal{L}$ implies that there is a $k \leq n$ such that

$$\vec{b} = \sum_{j=1}^{k} \vec{b}_j z_j$$

with $z_j \in \mathbb{Z}, z_k \neq 0$. Algorithm 8.1 sets

$$\vec{b}_j^* = \vec{b}_j - \sum_{i=1}^{j-1} \alpha_{ij} \vec{b}_i^*,$$

where

$$\alpha_{ij} = \frac{(\vec{b}_i^* \cdot \vec{b}_j)}{||\vec{b}_i^*||^2}.$$

Thus, we see that

$$\vec{b} = \sum_{j=1}^{k} \vec{b}_j^* z_j^*,$$

where $z_j^* \in \mathbb{R}$ for $j = 1, 2, \ldots, k - 1$ and $z_k^* = z_k \in \mathbb{Z}$. Now, because $\{\vec{b}_1^*, \vec{b}_2^*, \cdots, \vec{b}_n^*\}$ is an orthogonal basis, we have

$$||\vec{b}|| = \left(\sum_{j=1}^{k} (\vec{b}_j^* \cdot \vec{b}_j^*)(z_j^*)^2 \right)^{1/2}$$

$$\geq |z_k| \cdot ||\vec{b}_k^*||$$

$$\geq \min_{j} ||\vec{b}_j^*||.$$

∎

The basis $M^* = [\vec{b}_1^*, \vec{b}_2^*, \ldots, \vec{b}_n^*]$ computed by Algorithm 8.1 on input M is typically not a basis for the lattice with basis M. This is because the coefficients $\{\alpha_{i,j}\}_{i<j}$ are not necessarily integers. Lemma 8.7 suggests that we should look for a basis that is as close to orthogonal as we can find.

Definition 8.2:   Let $\mathcal{L}$ be a lattice with basis $M = [\vec{b}_1, \vec{b}_2, \ldots, \vec{b}_n]$ and let $M^* = [\vec{b}_1^*, \vec{b}_2^*, \ldots, \vec{b}_n^*]$ be the basis for $\mathsf{Span}_{\mathbb{R}}(M)$ obtained by Algorithm 8.1 on input M. We say that M is a *reduced basis* if:

(a) $|\alpha_{i,j}| \leq \frac{1}{2}$ for all $i < j$; and

(b) $||\vec{b}_{j+1}^* + \alpha_{j,j+1}\vec{b}_j^*||^2 \geq \frac{3}{4}||\vec{b}_j^*||^2$ for all $j = 1, 2, \ldots, n - 1$.

We are interested in this definition because we can establish the following result and obtain an upper bound on the length of the shortest vector in a lattice $\mathcal{L}$.

THEOREM 8.8   *Let $M = [\vec{b}_1, \vec{b}_2, \ldots, \vec{b}_n]$ be a reduced basis for the lattice $\mathcal{L}$. Then the following hold:*

1. $||\vec{b}_1|| \leq 2^{(n-1)/4}\mathsf{vol}(\mathcal{L})^{1/n}$; *and*

2. $\mathsf{wt}(M) \leq 2^{n(n-1)/4}\mathsf{vol}(\mathcal{L})$.

PROOF   Suppose $[\vec{b}_1, \vec{b}_2, \ldots, \vec{b}_n]$ is a reduced basis for the lattice $\mathcal{L}$. Then the orthogonality of $\vec{b}_j^*$ and $\vec{b}_{j+1}^*$ implies, by part (b) of Definition 8.2, that

$$\frac{3}{4}||\vec{b}_j^*||^2 \leq ||\vec{b}_{j+1}^* + \alpha_{j,j+1}\vec{b}_j^*||^2 = ||\vec{b}_{j+1}^*||^2 + \alpha_{j,j+1}^2 ||\vec{b}_j^*||^2.$$

Now, applying part (a), we have $||\vec{b}_{j+1}^*||^2 \geq \frac{1}{2}||\vec{b}_j^*||^2$. Iterating this inequality, it follows that

$$||\vec{b}_j^*||^2 \geq \left(\frac{1}{2}\right)^{j-1}||\vec{b}_1^*||^2. \tag{8.4}$$

Since $\vec{b}_1^* = \vec{b}_1$ we obtain

$$(\text{vol}(\mathcal{L}))^2 = \prod_{j=1}^{n} ||\vec{b}_j^*||^2$$

$$\geq \left(\frac{1}{2}\right)^{\sum_{j=1}^{n}(j-1)} ||\vec{b}_1||^{2n}$$

$$= \left(\frac{1}{2}\right)^{n(n-1)/2} ||\vec{b}_1||^{2n}.$$

This yields part 1.

For part 2, first recall that Algorithm 8.1 sets

$$\vec{b}_j^* = \vec{b}_j - \sum_{i=1}^{j-1}\alpha_{ij}\vec{b}_i^*,$$

where

$$\alpha_{ij} = \frac{(\vec{b}_i^* \cdot \vec{b}_j)}{||\vec{b}_i^*||^2}.$$

Thus,

$$\vec{b}_j = \sum_{i=1}^{j}\alpha_{ij}\vec{b}_i^*,$$

with $\alpha_{jj} = 1$. Furthermore, because the vectors $\{\vec{b}_i^*\}$ are orthogonal it follows that

$$||\vec{b}_j||^2 = \sum_{i=1}^{j}\alpha_{ij}^2||\vec{b}_i^*||^2 \leq ||\vec{b}_j^*||^2 + \frac{1}{4}\sum_{i=1}^{j-1}||\vec{b}_i^*||^2$$

by part (a) of Definition 8.2. Now, using Equation 8.4, we have

$$||\vec{b}_i^*||^2 \leq 2^{j-i}||\vec{b}_j^*||^2,$$

and hence

$$||\vec{b}_j||^2 \leq ||\vec{b}_j^*||^2 \left(1 + \frac{1}{4}\sum_{i=1}^{j-1}2^{j-i}\right) \leq 2^{j-1}||\vec{b}_j^*||^2.$$

Therefore,

$$(\text{wt}(M))^2 = \prod_{j=1}^{n} ||\vec{b}_j||^2$$

$$\leq 2^{n(n-1)/2} \prod_{j=1}^{n} ||\vec{b}_j^*||^2$$

$$= 2^{n(n-1)/2} (\text{vol}(\mathcal{L}))^2,$$

as claimed.   ∎

Example 8.3   *Continuation of Example 8.2*
The following computations show that

$$M' = \begin{bmatrix} 0 & 0 & 1 & 0 & 0 \\ 0 & -1 & 0 & 0 & -1 \\ 0 & -1 & 0 & -1 & 1 \\ -1 & 0 & 0 & 0 & 0 \\ 0 & 0 & 1 & -1 & -1 \end{bmatrix} = \begin{bmatrix} -1 & 0 & -1 & 0 & 0 \\ 0 & 0 & 1 & 0 & 0 \\ 0 & 0 & 0 & 1 & 0 \\ 1 & 1 & 0 & 0 & 0 \\ 0 & 0 & 0 & 0 & 1 \end{bmatrix} M$$

where M is the matrix given in Example 8.2, and that M' is a reduced basis for a
lattice $\mathcal{L}$. From Algorithm 8.1 we have:

$$\alpha' = \begin{bmatrix} * & 0 & 0 & 0 & 0 \\ * & * & 0 & \frac{1}{2} & 0 \\ * & * & * & -\frac{1}{2} & -\frac{1}{2} \\ * & * & * & * & -\frac{1}{2} \\ * & * & * & * & * \end{bmatrix}$$

and

$$M'^* = \begin{bmatrix} 0 & 0 & 1 & \frac{1}{2} & \frac{3}{4} \\ 0 & -1 & 0 & \frac{1}{2} & -\frac{3}{4} \\ 0 & -1 & 0 & -\frac{1}{2} & \frac{3}{4} \\ -1 & 0 & 0 & 0 & 0 \\ 0 & 0 & 1 & -\frac{1}{2} & -\frac{3}{4} \end{bmatrix}$$

Hence part (a) of Definition 8.2 is satisfied. To check part (b) we must show that

$$||\vec{b}_{j+1}'^* + \alpha_{j,j+1}' \vec{b}_j'^*||^2 \geq \frac{3}{4} ||\vec{b}_j^*||^2$$

for each $j = 1, 2, 3, 4$. To see this, observe that for $j = 1, 2, 3, 4$ we have

$$||\vec{b}_{j+1}'^* + \alpha_{j,j+1}' \vec{b}_j'^*||^2 = 2, 2, \frac{3}{2}, \frac{5}{2},$$

and

$$\frac{3}{4}||\vec{b}_j^*||^2 = \frac{3}{4}, \frac{3}{2}, \frac{3}{2}, \frac{3}{4},$$

respectively. Thus, the basis satisfies part (b) and consequently M' is a reduced basis for $\mathcal{L} = \mathcal{L}(M)$. Checking the bounds in Theorem 8.8, we observe that

$$||\vec{b}_1'|| \leq 2^1(\text{vol}(\mathcal{L}))^{1/5}$$

$$= 2 \cdot 3^{\frac{1}{5}}$$

$$\approx 2.4914618;$$

and

$$\text{wt}(M') = \prod_{j=1}^{5} ||\vec{b}_j'|| = 2(6)^{\frac{1}{2}} \approx 4.898979$$

$$\leq 2^5\text{vol}(\mathcal{L}) = 64 \cdot 3^{\frac{1}{5}} \approx 79.7267801.$$

$\square$

8.3   A reduced basis algorithm

Theorem 8.8 shows that, if we have a reduced basis for a lattice $\mathcal{L}$, then it contains a vector that has length less than or equal to $2^{(n-1)/4}\text{vol}(\mathcal{L})^{1/n}$. This is a good start towards finding a shortest vector in $\mathcal{L}$. In this section we present Algorithm 8.2 which computes a reduced basis for $\mathcal{L}$.

Algorithm 8.2: LLL $(\vec{b}_1, \vec{b}_2, \ldots, \vec{b}_n)$

external GRAM-SCHMIDT()
$done \leftarrow$ **false**
$([\vec{b}_1^*, \vec{b}_2^*, \ldots, \vec{b}_n^*], \{\alpha_{ij}\}_{i<j}) \leftarrow$ GRAM-SCHMIDT$(\vec{b}_1, \vec{b}_2, \ldots, \vec{b}_n)$
while **not** $done$

do $\begin{cases} \textbf{for } j \leftarrow 2 \textbf{ to } n \textbf{ do} \begin{cases} \textbf{for } i \leftarrow j - 1 \textbf{ downto } 1 \\ \quad \textbf{do if } |\alpha_{ij}| > \frac{1}{2} \\ \quad \quad \textbf{then } \vec{b}_j \leftarrow \vec{b}_j - \lfloor \alpha_{ij} + \frac{1}{2} \rfloor \vec{b}_i \end{cases} \\ ([\vec{b}_1^*, \vec{b}_2^*, \ldots, \vec{b}_n^*], \{\alpha_{ij}\}_{i<j}) \leftarrow \text{GRAM-SCHMIDT}(\vec{b}_1, \vec{b}_2, \ldots, \vec{b}_n) \\ \textbf{if } ||\vec{b}_{j+1}^* + \alpha_{j,j+1}\vec{b}_j^*||^2 < \frac{3}{4}||\vec{b}_j^*||^2 \quad \text{for some } j \\ \quad \textbf{then } \text{interchange } \vec{b}_j \text{ and } \vec{b}_{j+1} \\ \quad \textbf{else } done \leftarrow \textbf{true} \\ ([\vec{b}_1^*, \vec{b}_2^*, \ldots, \vec{b}_n^*], \{\alpha_{ij}\}_{i<j}) \leftarrow \text{GRAM-SCHMIDT}(\vec{b}_1, \vec{b}_2, \ldots, \vec{b}_n) \end{cases}$

return $([\vec{b}_1, \vec{b}_2, \ldots, \vec{b}_n])$

THEOREM 8.9   *Let $\mathcal{L}$ be a lattice with basis $[\vec{b}_1, \vec{b}_2, \ldots, \vec{b}_n]$, and let*

$$Max = \max\{||b_i||^2 : i = 1, 2, \ldots, n\} \geq 2.$$

Then Algorithm 8.2 finds a reduced basis for $\mathcal{L}$ using at most $O(n^5 \log(Max))$ arithmetic operations.

PROOF   When $|\alpha_{ij}| > \frac{1}{2}$, Algorithm 8.2 replaces $\vec{b}_j$ with

$$\vec{x} = \vec{b}_j - \left\lfloor \alpha_{ij} + \frac{1}{2} \right\rfloor \vec{b}_i.$$

Now

$$
\left| \frac{(\vec{b}_i^* \cdot \vec{x})}{||b_i^*||^2} \right| = \left| \frac{\vec{b}_i^* \cdot \left(\vec{b}_j - \lfloor \alpha_{ij} + \frac{1}{2} \rfloor \vec{b}_i \right)}{||\vec{b}_i^*||^2} \right|
$$

$$
= \left| \frac{(\vec{b}_i^* \cdot \vec{b}_j)}{||\vec{b}_i^*||^2} - \left\lfloor \alpha_{ij} + \frac{1}{2} \right\rfloor \frac{(\vec{b}_i^* \cdot \vec{b}_i)}{||\vec{b}_i^*||^2} \right|
$$

$$
= \left| \alpha_{ij} - \left\lfloor \alpha_{ij} + \frac{1}{2} \right\rfloor \right|
$$

$$
\leq \frac{1}{2}.
$$

Thus when Algorithm 8.1 is called at the end of the loop, the recalculated α_{ij}s will satisfy part (a) of Definition 8.2. It is also easy to see that, if the algorithm terminates, then condition (b) of Definition 8.2 will be met. To see that the algorithm terminates, we define for $\ell = 0, 1, 2, \ldots, n$, the quantity

$$v_\ell = \prod_{j=1}^{\ell} ||\vec{b}_j^*||^2.$$

Notice that each $v_i > 0$. In particular $v_0 = 1$ and $v_n = \text{vol}(\mathcal{L})^2$. Furthermore, elementary linear algebra shows that v_ℓ is the determinant of the ℓ by ℓ matrix whose $[i, j]$ entry is $\vec{b}_i \cdot \vec{b}_j$. In particular, because the $\vec{b}_j$s are linearly independent vectors with integer entries, we have that $v_\ell > 1$. Thus

$$V = \prod_{\ell=1}^{n-1} v_\ell > 1.$$

Furthermore, the value of V changes only when some $\vec{b}_j$ is changed, and this occurs only if

$$||\vec{b}_{j+1}^* + \alpha_{j,j+1} \vec{b}_j^*||^2 < \frac{3}{4} ||\vec{b}_j^*||^2,$$

for some j. In this case, $\vec{b}_j$ and $\vec{b}_{j+1}$ are interchanged, and Algorithm 8.1 recalculates the corresponding orthogonal basis and α_{ij}s. Let $\widehat{b}_1, \widehat{b}_2, \ldots, \widehat{b}_n$ be the new basis after the interchange, and let $\widehat{b}_1^*, \widehat{b}_2^*, \ldots, \widehat{b}_n^*$ be the recomputed orthogonal basis. Then

$$
\begin{aligned}
\widehat{b}_i &= \vec{b}_i \text{ when } i \neq j \text{ or } j+1; \\
\widehat{b}_i^* &= \vec{b}_i^* \text{ when } i \neq j \text{ or } j+1; \\
\widehat{b}_j &= \vec{b}_{j+1}; \text{ and} \\
\widehat{b}_{j+1} &= \vec{b}_j.
\end{aligned}
$$

After the call to Algorithm 8.1, we have

$$
\widehat{b}_j^* = \widehat{b}_j - \sum_{i=1}^{j-1} \widehat{\alpha}_{ij} \widehat{b}_i^*, \tag{8.5}
$$

where

$$
\widehat{\alpha}_{ij} = \frac{\widehat{b}_i^* \cdot \widehat{b}_j}{||\widehat{b}_i^*||^2}
$$

and $|\widehat{\alpha}_{ij}| \leq \frac{1}{2}$. On the other hand, we have

$$
b_{j+1}^* = b_{j+1} - \sum_{i=1}^{j} \alpha_{ij+1} b_i^*, \tag{8.6}
$$

$\widehat{b}_j = \vec{b}_{j+1}$, and $\alpha_{i,j+1} = \widehat{\alpha}_{i,j}$ for $i = 1, 2, \ldots, j-1$. Thus, substituting into Equation 8.5, we see that

$$
\widehat{b}_j^* = \alpha_{j,j+1} \vec{b}_j^* + \vec{b}_{j+1}^*.
$$

Therefore,

$$
||\widehat{b}_j^*||^2 = ||\alpha_{j,j+1} \vec{b}_j^* + \vec{b}_{j+1}^*||^2 < \frac{3}{4} ||\vec{b}_j^*||^2.
$$

Thus when $\vec{b}_j$ and $\vec{b}_{j+1}$ are interchanged, the number v_j is reduced by a factor of $3/4$ (or more). The other v_i are unchanged and so each time there is an interchange, V is reduced by a factor of $3/4$ (or more). Therefore the algorithm terminates, because throughout Algorithm 8.2 we always have $V > 1$. The initial value of v_i is by Lemma 8.4 at most $(Max)^i$. It follows that, prior to the execution of the **while** loop, V is at most $(Max)^{\frac{n(n-1)}{2}}$. Consequently, if t is the number of interchanges performed, then

$$
t \leq \log_{\frac{3}{4}}((Max)^{n(n-1)/2}) = \frac{n(n-1)}{2} \log_{\frac{3}{4}}(Max).
$$

Thus t is $O(n^2 \log(Max))$, and because Algorithm 8.1 is an $O(n^3)$ algorithm, it follows that Algorithm 8.2 uses $O(n^5 \log(Max))$ arithmetic operations. ∎

8.4   Solving systems of integer equations

In this section, we return to the problem of solving a matrix equation of the form $AU = B$ for a $(0, 1)$-valued vector U, where A is an m by n integer valued matrix and $B \in \mathbb{Z}^m$. As outlined in Section 8.1, we set M equal to the following $m + n$ by $n + 1$ matrix:

$$M = \begin{bmatrix} I & \vec{0} \\ A & -B \end{bmatrix},$$

and consider the lattice $\mathcal{L}$ with basis M.

The basic idea of the algorithm is that if U is very short, then there is a good chance that it will appear in a reduced basis M' for $\mathcal{L}$. Thus we check whether the reduced basis contains a vector of the form $[U, \vec{0}]$ with $U \in \{0, 1\}^n$. If it does, then a solution to $AU = dB$ is found for some integer d; otherwise another approach must be taken. We can also check if the reduced basis contains a vector $[U, \vec{0}]$ with $U \in \{0, -1\}^n$, for then $-U \in \{0, 1\}^n$ and $A(-U) = dB$.

Example 8.4   *Using Algorithm 8.2 to find a solution.*

Consider the incidence matrix

$$A = \begin{bmatrix} 1 & 1 & 2 & 0 & 0 & 0 & 1 & 0 & 0 & 0 \\ 0 & 0 & 2 & 2 & 0 & 0 & 0 & 1 & 0 & 0 \\ 0 & 1 & 1 & 1 & 1 & 0 & 0 & 0 & 1 & 0 \\ 0 & 0 & 0 & 2 & 1 & 1 & 0 & 0 & 0 & 1 \\ 0 & 0 & 0 & 0 & 0 & 0 & 2 & 1 & 2 & 0 \\ 0 & 0 & 0 & 0 & 0 & 0 & 0 & 1 & 2 & 2 \\ 1 & 3 & 6 & 6 & 3 & 1 & 3 & 3 & 6 & 3 \end{bmatrix}$$

of Example 6.12.1, in which we have added the orbit length equation in the last row, as suggested by Exercise 6.17. For this matrix, we wish to solve $AU = B$ where

$$B = [1, 1, 1, 1, 1, 1, 7]^{\mathsf{T}}.$$

The basis M is as follows:

$$M = \begin{bmatrix} 1 & 0 & 0 & 0 & 0 & 0 & 0 & 0 & 0 & 0 & 0 \\ 0 & 1 & 0 & 0 & 0 & 0 & 0 & 0 & 0 & 0 & 0 \\ 0 & 0 & 1 & 0 & 0 & 0 & 0 & 0 & 0 & 0 & 0 \\ 0 & 0 & 0 & 1 & 0 & 0 & 0 & 0 & 0 & 0 & 0 \\ 0 & 0 & 0 & 0 & 1 & 0 & 0 & 0 & 0 & 0 & 0 \\ 0 & 0 & 0 & 0 & 0 & 1 & 0 & 0 & 0 & 0 & 0 \\ 0 & 0 & 0 & 0 & 0 & 0 & 1 & 0 & 0 & 0 & 0 \\ 0 & 0 & 0 & 0 & 0 & 0 & 0 & 1 & 0 & 0 & 0 \\ 0 & 0 & 0 & 0 & 0 & 0 & 0 & 0 & 1 & 0 & 0 \\ 0 & 0 & 0 & 0 & 0 & 0 & 0 & 0 & 0 & 1 & 0 \\ 1 & 1 & 2 & 0 & 0 & 0 & 1 & 0 & 0 & 0 & -1 \\ 0 & 0 & 2 & 2 & 0 & 0 & 0 & 1 & 0 & 0 & -1 \\ 0 & 1 & 1 & 1 & 1 & 0 & 0 & 0 & 1 & 0 & -1 \\ 0 & 0 & 0 & 2 & 1 & 1 & 0 & 0 & 0 & 1 & -1 \\ 0 & 0 & 0 & 0 & 0 & 0 & 2 & 1 & 2 & 0 & -1 \\ 0 & 0 & 0 & 0 & 0 & 0 & 0 & 1 & 2 & 2 & -1 \\ 1 & 3 & 6 & 6 & 3 & 1 & 3 & 3 & 6 & 3 & -7 \end{bmatrix}$$

The weight of this basis is $\mathrm{wt}(M) \approx 4504883.126564$. Applying Algorithm 8.2 to M, we obtain the reduced basis

$$M' = \begin{bmatrix} 1 & 0 & 0 & 1 & -1 & 1 & 0 & 0 & 0 & -1 & -1 \\ 0 & 0 & -1 & 0 & 1 & 0 & 1 & -1 & -1 & 0 & 1 \\ 0 & 0 & 0 & 0 & 0 & 0 & -1 & 0 & 0 & 0 & 0 \\ 0 & 0 & 0 & 0 & 0 & 1 & 0 & 0 & 0 & 1 & 0 \\ 0 & 0 & 1 & 1 & 0 & 0 & -1 & 1 & 0 & 0 & 0 \\ 0 & 1 & 0 & 0 & -1 & 0 & -1 & 0 & 0 & -1 & -1 \\ 0 & 0 & 0 & 0 & 0 & 0 & 0 & 0 & 0 & 1 & 1 \\ 0 & 0 & 0 & 1 & 0 & 0 & 0 & 0 & 1 & 0 & 0 \\ 0 & 0 & 0 & 0 & 0 & 1 & -1 & 1 & 0 & 0 & -1 \\ 0 & 0 & 0 & 0 & 0 & 0 & 0 & 0 & 0 & 0 & 1 \\ 1 & 0 & -1 & 0 & 0 & -1 & 1 & -2 & -1 & -1 & 1 \\ 0 & 0 & 0 & 0 & 0 & 0 & 0 & -1 & 1 & 1 & 0 \\ 0 & 0 & 0 & 0 & 1 & 0 & 0 & 0 & -1 & 0 & 0 \\ 0 & 1 & 1 & 0 & -1 & 0 & 0 & 0 & 0 & 0 & 0 \\ 0 & 0 & 0 & 0 & 0 & 0 & 0 & 1 & 1 & 1 & 0 \\ 0 & 0 & 0 & 0 & 0 & 0 & 0 & 1 & 1 & -1 & 0 \\ 1 & 1 & 0 & 0 & 1 & -1 & 1 & -1 & 0 & 0 & 1 \end{bmatrix}$$

The weight of the new basis is $\mathrm{wt}(M') \approx 10571.993190$, which is, as expected, less than $\mathrm{wt}(M)$. Also, column 4 of M' is of the form

$$[u_1, u_2, \ldots, u_{10}, 0, 0, 0, 0, 0, 0, 0]^\mathsf{T}$$

with each $u_i \in \{0, 1\}$. This gives the solution

$$U = [u_1, u_2, \ldots, u_{10}]^\mathsf{T} = [1, 0, 0, 0, 1, 0, 0, 1, 0, 0]^\mathsf{T}$$

to $AU = dB$ with $d = 1$.

$\square$

In Example 8.4, Algorithm 8.2 succeeds in finding a solution to $AU = dB$. However, there are many situations in which it will not succeed. This is because the weight of reduced basis obtained by Algorithm 8.2 may be still too large to guarantee that the basis contains the shortest vectors in the lattice (see Example 8.5). Therefore, we will now consider another method to reduce the weight of the basis. Suppose the current basis is $M = [\vec{b}_1, \vec{b}_2, \ldots, \vec{b}_n]$ and consider combinations of the form

$$\vec{v} = \vec{b}_i + \epsilon\, \vec{b}_j$$

where $\epsilon = \pm 1$. If $||v|| < \max\{||\vec{b}_i||, ||\vec{b}_j||\}$, then one of $||\vec{b}_i||$ and $||\vec{b}_j||$ can be replaced by v and the weight of the basis will be reduced. In order to facilitate the implementation of this idea, we will use the array Δ defined by

$$\Delta_{i,j} = \vec{b}_i \cdot \vec{b}_j.$$

Let $k \in \{i, j\}$ be such that $||\vec{b}_k|| = \max\{||\vec{b}_i||, ||\vec{b}_j||\}$. When $\vec{b}_k$ is replaced by $\vec{v}$, the only entries of Δ that need to be recomputed are $\Delta_{h,k}$ and $\Delta_{k,h}$ for each h. Observe, for $h \neq k$, that

$$\Delta_{k,h} = \vec{v} \cdot \vec{b}_h$$
$$= (\vec{b}_i + \epsilon\, \vec{b}_j) \cdot \vec{b}_h$$
$$= (\vec{b}_i \cdot \vec{b}_h) + \epsilon\, (\vec{b}_j \cdot \vec{b}_h)$$
$$= \Delta_{i,h} + \epsilon\, \Delta_{j,h},$$

and

$$\Delta_{k,k} = \vec{v} \cdot \vec{v}$$
$$= (\vec{b}_i + \epsilon\, \vec{b}_j) \cdot (\vec{b}_i + \epsilon\, \vec{b}_j)$$
$$= \Delta_{i,i} + \Delta_{j,j} + 2\epsilon\, \Delta_{i,j}.$$

Thus, Δ can be updated in $O(n)$ operations. This method of reducing the weight of the basis is simple to implement, and we present it as Algorithm 8.3. There are $\binom{n}{2}$ pairs of vectors to check and so this algorithm runs in $O(n^3)$ time.

Algorithm 8.3: WEIGHTREDUCTION $(\vec{b}_1, \vec{b}_2, \ldots, \vec{b}_n)$

global $\Delta_{ij}, 1 \leq i, j \leq n$

for $i \leftarrow 1$ **to** n

do $\begin{cases} \textbf{for } j \leftarrow i+1 \textbf{ to } n \\ \textbf{do} \begin{cases} \textbf{for each } \epsilon \in \{-1, 1\} \\ \textbf{do} \begin{cases} \textbf{if } \Delta_{i,i} < \Delta_{j,j} \textbf{ then } k \leftarrow j \textbf{ else } k \leftarrow i \\ \vec{v} \leftarrow \vec{b}_i + \epsilon \vec{b}_j \\ \textbf{if } ||\vec{v}||^2 < \Delta_{k,k} \\ \textbf{then} \begin{cases} \Delta_{k,k} \leftarrow \Delta_{i,i} + \Delta_{j,j} + 2\epsilon \Delta_{i,j} \\ \textbf{for } h \leftarrow 1 \textbf{ to } n \\ \quad \textbf{do if } h \neq i \textbf{ and } h \neq j \\ \quad\quad \textbf{then } \begin{cases} \Delta_{k,h} \leftarrow \Delta_{i,h} + \epsilon \Delta_{j,h} \\ \Delta_{h,k} \leftarrow \Delta_{k,h} \end{cases} \\ \textbf{if } k \neq i \\ \quad \textbf{then } \begin{cases} \Delta_{k,i} \leftarrow \Delta_{i,i} + \epsilon \Delta_{j,i} \\ \Delta_{i,k} \leftarrow \Delta_{k,i} \end{cases} \\ \quad \textbf{else } \begin{cases} \Delta_{k,j} \leftarrow \Delta_{i,j} + \epsilon \Delta_{j,j} \\ \Delta_{j,k} \leftarrow \Delta_{k,j} \end{cases} \\ \vec{b}_k \leftarrow \vec{v} \end{cases} \end{cases} \end{cases} \end{cases}$

return $([\vec{b}_1, \vec{b}_2, \ldots, b_n])$

We see that progress can be made, using these algorithms, to reduce the weight of the basis and possibly converge to a solution. It turns out that using a combination of Algorithms 8.2 and 8.3 is often superior for reducing the weight than using either of them alone. Among the many possible ways to do this, a successful and simple method is given in Algorithm 8.4.

Algorithm 8.2 and Algorithm 8.1 work through the basis $[\vec{b}_1, \vec{b}_2, \ldots, \vec{b}_n]$ from left to right. Thus the order in which the basis vectors appear has an effect on the outcome of Algorithm 8.2 and thus also on the outcome of Algorithm 8.4. For a given input basis, it may be prudent to order the basis vectors in increasing order of length, while for another basis we may prefer a decreasing order, and still another could do well with random ordering. Since the algorithms run relatively quickly, many such approaches can be tried. One can even try sorting the basis before each call to Algorithm 8.3 and also to include in the **while** loop additional calls to Algorithm 8.2. The optimal approach depends on the input basis and is best determined by experimentation.

Algorithm 8.4: KR $(M = [\vec{b}_1, \vec{b}_2, \ldots, \vec{b}_n])$

external LLL(), WEIGHTREDUCTION()

Sort the basis vectors so that $||\vec{b}_1|| < ||\vec{b}_2|| < \cdots < ||\vec{b}_n||$

$M \leftarrow$ LLL(M)

for $i \leftarrow 1$ **to** n

$\quad$ **do** $\begin{cases} \textbf{for } j \leftarrow 1 \textbf{ to } n \\ \quad \textbf{do } \Delta_{i,j} \leftarrow \vec{b}_i \cdot \vec{b}_j \end{cases}$

$weight \leftarrow \prod_{i=1}^{n} \sqrt{\Delta_{i,i}}$

$done \leftarrow$ **false**

while not $done$ $\begin{cases} M \leftarrow \text{WEIGHTREDUCTION}(M) \\ new \leftarrow \prod_{i=1}^{n} \sqrt{\Delta_{i,i}} \\ \textbf{if } new < weight \\ \quad \textbf{then } weight \leftarrow new \\ \quad \textbf{else } done \leftarrow \textbf{true} \end{cases}$

return (M)

Example 8.5

Consider the incidence matrix

$$A = \begin{bmatrix} 2 & 1 & 0 & 0 & 0 & 2 & 2 & 0 & 0 & 1 & 2 & 0 & 0 & 0 & 2 & 0 & 0 & 0 & 1 \\ 1 & 2 & 1 & 1 & 0 & 0 & 0 & 0 & 0 & 0 & 0 & 0 & 2 & 2 & 0 & 2 & 0 & 2 & 0 \\ 0 & 0 & 1 & 1 & 2 & 1 & 1 & 0 & 2 & 0 & 1 & 1 & 0 & 1 & 1 & 1 & 0 & 0 & 0 \\ 0 & 0 & 0 & 0 & 2 & 2 & 2 & 1 & 0 & 0 & 0 & 2 & 0 & 0 & 0 & 2 & 2 & 0 \\ 0 & 0 & 0 & 0 & 0 & 0 & 0 & 1 & 1 & 1 & 2 & 1 & 1 & 1 & 1 & 2 & 1 & 1 \\ 15 & 15 & 15 & 15 & 30 & 30 & 30 & 5 & 30 & 15 & 30 & 30 & 30 & 30 & 30 & 30 & 30 & 30 & 15 \end{bmatrix}$$

of Example 6.12.2, in which we have added the orbit length equation in the last row, as suggested by Exercise 6.17. For this matrix, we wish to solve $AU = B$, where

$$B = [1, 1, 1, 1, 1, 35]^{\mathsf{T}}.$$

Applying Algorithm 8.2 to the basis

$$M = \begin{bmatrix} I & \vec{0} \\ A & -B \end{bmatrix},$$

we obtain the basis M' given below

$$\left[\begin{array}{ccccccccccccccccccc|c}
0 & 0 & -1 & -1 & 0 & -1 & -1 & -1 & 0 & 1 & 0 & 1 & 0 & 0 & 0 & 0 & -1 & -3 & 3 & -2 \\
0 & 0 & 1 & 0 & -1 & 1 & 0 & 1 & 0 & -1 & 0 & -1 & 0 & 1 & 0 & 0 & 1 & 3 & -3 & 2 \\
-1 & 0 & 0 & 1 & 1 & -1 & 1 & 0 & 0 & 0 & 0 & 0 & 0 & 0 & 0 & 0 & 0 & 0 & 0 & 0 \\
1 & 0 & 0 & 0 & 0 & 0 & 0 & -1 & 0 & -1 & 0 & -1 & 0 & 0 & 0 & 0 & 0 & 0 & 0 & 0 \\
0 & 0 & 0 & -1 & 0 & 0 & 0 & 0 & 0 & 0 & 0 & 0 & 0 & 0 & 0 & 0 & 0 & -1 & 1 & -1 \\
0 & -1 & 0 & 1 & 0 & 0 & 0 & 0 & 0 & 0 & 0 & 0 & -1 & 0 & 0 & 0 & 0 & 2 & -3 & 2 \\
0 & 1 & 0 & 0 & 0 & 0 & 0 & 0 & 0 & 0 & 0 & 0 & 0 & 0 & 0 & 0 & 0 & 0 & 0 & 0 \\
0 & 0 & 0 & 0 & 0 & 0 & 0 & 0 & 0 & 0 & 0 & 0 & 0 & 0 & 0 & 0 & 1 & 0 & 0 & 0 \\
0 & 0 & 0 & 0 & 0 & 0 & -1 & 0 & -1 & 0 & 0 & 0 & 0 & 0 & 0 & 0 & 0 & 1 & -1 & 1 \\
0 & 0 & 0 & 0 & 0 & 1 & 0 & -1 & 0 & -1 & 0 & -1 & 0 & -1 & 0 & 1 & 0 & 1 & -1 & 1 \\
0 & 0 & 0 & 0 & 0 & 0 & 1 & 1 & 0 & 0 & -1 & 0 & 1 & 0 & 0 & 0 & 1 & 0 & 1 & -1 \\
0 & 0 & 0 & 0 & 0 & 0 & 0 & 0 & 1 & 0 & 0 & 0 & 0 & 0 & 0 & 0 & 0 & 0 & 0 & 0 \\
0 & 0 & 0 & 0 & 0 & 0 & 0 & 0 & 0 & 0 & 0 & 0 & 1 & -1 & -1 & 0 & 0 & 0 & 0 & 0 \\
0 & 0 & 0 & 0 & 0 & 0 & 0 & 0 & 0 & 1 & 0 & 0 & -1 & 0 & 0 & 0 & 0 & 0 & 0 & 0 \\
0 & 0 & 0 & 0 & 0 & 0 & 0 & 0 & 0 & 0 & 1 & 0 & 0 & 0 & 0 & 0 & 0 & 0 & 0 & 0 \\
0 & 0 & 0 & 0 & 0 & 0 & 0 & 0 & 0 & 0 & 0 & 1 & 0 & 0 & 0 & 0 & 0 & 1 & -1 & 1 \\
0 & 0 & 0 & 0 & 0 & 0 & 0 & 0 & 0 & 0 & 0 & 0 & 0 & 1 & 0 & 0 & 0 & 0 & 0 & 0 \\
0 & 0 & 0 & 0 & 0 & 0 & 0 & 0 & 0 & 0 & 0 & 0 & 0 & 0 & 1 & 0 & 0 & 0 & 0 & 0 \\
0 & 0 & 0 & 0 & 0 & 0 & 0 & 0 & 0 & 0 & 0 & 0 & 0 & 0 & 0 & 1 & 0 & 0 & 0 & 0 \\ \hline
0 & 0 & -1 & 0 & -1 & 0 & 0 & 0 & 0 & 0 & 0 & 0 & 0 & 0 & 0 & 0 & 0 & -1 & 1 & -1 \\
0 & 0 & 1 & 0 & -1 & 0 & 0 & 0 & 0 & 0 & 0 & 0 & 0 & 0 & 0 & 0 & 0 & 2 & -2 & 2 \\
0 & 0 & 0 & 0 & 1 & -1 & 0 & 0 & -1 & 0 & 0 & 0 & -1 & 0 & 0 & 0 & 0 & 0 & 0 & 0 \\
0 & 0 & 0 & 0 & 0 & 0 & 0 & 0 & 0 & 0 & 0 & 0 & 0 & 0 & 0 & 0 & 0 & -1 & -1 & 0 \\
0 & 0 & 0 & 0 & 0 & 1 & 0 & 0 & 1 & 0 & 0 & 0 & 1 & 0 & 0 & 0 & 0 & 0 & 1 & 0 \\
0 & 0 & 0 & 0 & 0 & 0 & 0 & 0 & 0 & 0 & 0 & 0 & 0 & 0 & 0 & 0 & 0 & 0 & 0 & 5
\end{array}\right]$$

The weight of this new basis is $\text{wt}(M') \approx 13152322.331817$. Unfortunately there is no column of the form

$$[u_1, u_2, \ldots, u_{19}, 0, 0, 0, 0, 0, 0]^{\mathsf{T}}$$

with each $u_i \in \{0, 1\}$ or with each $u_i \in \{0, -1\}$. Applying Algorithm 8.4 to the basis M', we obtain the basis M'', which is as follows:

$$
\left[\begin{array}{ccccccccccccccccccc|c}
0 & 0 & 0 & 1 & -1 & 1 & 0 & 0 & 0 & 0 & 0 & -1 & 0 & 0 & 0 & 0 & 0 & 0 & 0 & 0 \\
0 & 0 & 0 & 0 & 1 & 0 & 0 & 0 & 0 & 0 & 0 & 0 & 0 & 0 & -1 & 0 & -1 & -1 & 0 & 0 \\
-1 & 0 & -1 & 0 & 0 & 0 & 0 & 0 & 0 & 0 & 0 & 0 & 0 & 0 & 0 & 0 & 0 & 0 & 0 & 0 \\
1 & 0 & -1 & 0 & 0 & 0 & 0 & 0 & 0 & 0 & -1 & 1 & 0 & 0 & 0 & 0 & 1 & 0 & 0 & 0 \\
0 & 0 & 0 & 0 & 0 & 0 & 0 & 0 & 0 & 0 & 0 & -1 & 0 & 0 & 0 & 0 & 0 & 0 & -1 & 0 \\
0 & 0 & 0 & 0 & 0 & 0 & 0 & 0 & 0 & 0 & 0 & 1 & -1 & 1 & 0 & 0 & 0 & -1 & -1 & 0 \\
0 & 0 & 0 & 0 & 0 & 0 & 0 & 0 & 0 & 0 & 0 & 0 & 1 & 0 & 0 & 0 & 0 & 0 & 0 & 0 \\
0 & 0 & 0 & 0 & 0 & 1 & 0 & 0 & 0 & 1 & 0 & 0 & 0 & 0 & 0 & 0 & 1 & 1 & 0 & 1 \\
0 & 0 & 0 & 0 & 0 & 0 & 1 & 0 & 0 & 0 & 0 & 0 & 0 & 0 & 0 & 0 & 0 & 0 & 0 & 0 \\
0 & -1 & 0 & -1 & 0 & -1 & 0 & 0 & 0 & -1 & 0 & 0 & 0 & 0 & 1 & 0 & 0 & -1 & 0 & 0 \\
0 & 0 & 0 & 0 & 0 & 0 & 0 & 1 & 0 & 0 & 0 & 0 & -1 & 0 & 0 & 0 & 0 & 1 & 0 & 0 \\
0 & 0 & 0 & 0 & 0 & 1 & -1 & 0 & 0 & 0 & 0 & 0 & 1 & 0 & 0 & 0 & -1 & -1 & 0 & 0 \\
0 & 0 & 0 & 0 & 0 & 0 & 0 & 0 & 0 & 0 & 0 & 0 & 0 & 1 & 0 & 0 & 0 & 0 & 0 & 0 \\
0 & 0 & 1 & 0 & 0 & 0 & 0 & 0 & 1 & 0 & 0 & 0 & 0 & 0 & 0 & 0 & 0 & 0 & 0 & 0 \\
0 & 0 & 0 & 0 & 0 & 0 & 0 & -1 & 0 & 1 & 0 & 0 & 0 & 0 & 0 & 0 & 0 & 0 & 0 & 0 \\
0 & 0 & 0 & 0 & 0 & 0 & 0 & 0 & -1 & 0 & 1 & 0 & 0 & 0 & 0 & 0 & 0 & 0 & 0 & 0 \\
0 & 0 & 0 & 0 & 0 & 0 & 0 & 0 & 0 & 0 & 0 & 0 & -1 & 0 & -1 & 0 & 0 & 0 & 0 & 0 \\
0 & 0 & 0 & 0 & 0 & 0 & 0 & 0 & 0 & 0 & 0 & 0 & 0 & -1 & 1 & 0 & 0 & 0 & 0 & 0 \\
0 & 1 & 0 & 0 & 0 & 0 & 0 & 0 & 0 & 0 & 0 & 0 & 0 & 0 & 0 & 1 & 0 & -1 & 0 & 0 \\
\hline
0 & 0 & 0 & 1 & -1 & 0 & 0 & 0 & 0 & 1 & -1 & 0 & 0 & 0 & 0 & 0 & -1 & 0 & 0 & 0 \\
0 & 0 & 0 & 1 & 1 & 0 & 0 & 0 & 0 & -1 & 1 & 0 & 0 & 0 & 0 & 0 & 0 & 1 & 0 & 0 \\
0 & 0 & -1 & 0 & 0 & 0 & 1 & 0 & 0 & 0 & 0 & 0 & 1 & 0 & 0 & 0 & 0 & 0 & 0 & 0 \\
0 & 0 & 0 & 0 & 0 & 0 & 0 & 0 & 0 & 0 & 0 & 0 & 0 & 0 & 0 & 0 & 1 & -1 & 0 & 1 \\
0 & 0 & 1 & -1 & 0 & -1 & 0 & 0 & 0 & 0 & 0 & -1 & 0 & 0 & 0 & 0 & 0 & 0 & 0 & 0 \\
0 & 0 & 0 & 0 & 0 & 0 & 0 & 0 & 0 & 0 & 0 & 0 & 0 & 0 & 0 & 0 & 0 & 0 & 0 & 5
\end{array}\right]
$$

The weight of this new basis is $\mathrm{wt}(M'') \approx 927342.111629$, and this is less than $\mathrm{wt}(M')$ for the previous basis. Furthermore, column 17 has the form

$$[u_1, u_2, \ldots, u_{19}, 0, 0, 0, 0, 0, 0]^\mathsf{T}$$

with each $u_i \in \{0, 1\}$. This gives the solution

$$
\begin{aligned}
U &= [u_1, u_2, \ldots, u_{19}]^\mathsf{T} \\
&= [0, 0, 0, 1, 0, 0, 0, 1, 0, 0, 0, 0, 0, 0, 0, 0, 0, 0, 1]^\mathsf{T}
\end{aligned}
$$

to $AU = dB$ with $d = 1$.   ⬜

8.5 The Merkle-Hellman knapsack system

A *public key cryptosystem* is a method of secure transmission of messages in which the sender looks up the receiver's key K from a publicly available list of keys $\mathcal{K}$. From the key, an encryption rule is determined, which is used to encode *plaintext* messages $\mathcal{P}$ into *ciphertext* messages $\mathcal{C}$. A ciphertext message is then transmitted to the receiver. Only the receiver knows the decryption rule, corresponding to the *key* K, that will decode messages sent to him. Anyone else will have great difficulty in decrypting the ciphertext, even though they know the public key K and the encryption rule used.

The well-known Merkle-Hellman knapsack cryptosystem was first described by Merkle and Hellman in 1978. This public key cryptosystem, and several variants of it, were broken in the early 1980s. In this section we will show how basis reduction can be used to break the Merkle-Hellman knapsack cryptosystem.

The term "knapsack" is actually a misnomer. The Knapsack problem, as it is usually defined, is a problem involving selecting objects with given weights and profits in such a way that a specified capacity is not exceeded and a maximum profit is attained (see Problem 1.4). The Merkle-Hellman knapsack cryptosystem is instead based on Problem 8.2.

Problem 8.2:   Subset Sum

Instance: positive integers $a_1, \ldots a_n$ and z. The a_is are called *sizes* and z is called the *target sum*.

Find: a 0-1 vector $U = [u_1, \ldots, u_n]$ such that

$$\sum_{i=1}^{n} u_i a_i = z.$$

Problem 8.2 is a search problem which is known to be NP-hard. Among other things, this means that there is no known polynomial-time algorithm that solves it. But even if a problem has no polynomial-time algorithm to solve it in general, this does not rule out the possibility that certain special cases can be solved in polynomial time. This is indeed the situation with the Subset Sum problem.

we define a list of sizes, $[a_1, \ldots, a_n]$, to be *superincreasing* if

$$a_j > \sum_{i=1}^{j-1} a_i$$

for $2 \leq j \leq n$. If the list of sizes is superincreasing, then the Subset Sum problem can be solved very easily in time $O(n)$ by a greedy algorithm, and a solution U (if it exists) must be unique. The algorithm to do this is presented in Algorithm 8.5.

Algorithm 8.5: SUPERINCREASINGSOLVER $(a_1, a_2, \ldots, a_n, z)$

for $i \leftarrow n$ **downto** 1

$$\mathbf{do} \begin{cases} \mathbf{if}\ z \geq a_i \\ \qquad \mathbf{then} \begin{cases} z \leftarrow z - a_i \\ u_i \leftarrow 1 \end{cases} \\ \qquad \mathbf{else}\ u_i \leftarrow 0 \end{cases}$$

if $z = 0$
   then $U \leftarrow [u_1, \ldots, u_n]$ is the solution
   else there is no solution.

Suppose $A = [a_1, \ldots, a_n]$ is superincreasing, and consider the function

$$e_A : \{0, 1\}^n \rightarrow \left\{ 0, \ldots, \sum_{i=1}^{n} a_i \right\}$$

defined by the rule

$$e_A(u_1, \ldots, u_n) = \sum_{i=1}^{n} u_i a_i.$$

Is e_A a possible candidate for an encryption rule? Since A is superincreasing, e_A is an injection, and the algorithm presented in Algorithm 8.5 would be the corresponding decryption algorithm. However, such a system would be completely insecure because anyone can decrypt a message that is encrypted in this way.

The strategy therefore is to transform the list of sizes in such a way that it is no longer superincreasing. The receiver will be able to apply an inverse transformation to restore the superincreasing list of sizes. On the other hand, an observer who does not know the transformation that was applied is faced with what looks like a general, apparently difficult, instance of the Subset Sum problem when he tries to decrypt a ciphertext.

One suitable type of transformation is a *modular transformation*. That is, a prime modulus p is chosen such that

$$p > \sum_{i=1}^{n} a_i,$$

as well as a multiplier m, where $1 \leq m \leq p - 1$. Then we define

$$t_i = m a_i \bmod p,$$

$1 \leq i \leq n$. The list of sizes $T = [t_1, \ldots, t_n]$ will be the public key used for encryption. The values m, p used to define the modular transformation are secret. The complete description of the Merkle-Hellman knapsack cryptosystem is given in Definition 8.3.

Definition 8.3: *Merkle-Hellman knapsack cryptosystem.* Let $A = [a_1, \ldots, a_n]$ be a superincreasing list of integers, let $p > \sum_{i=1}^{n} a_i$ be prime, and let $1 \leq m \leq p - 1$. For $1 \leq i \leq n$, define

$$t_i = ma_i \bmod p,$$

and denote $T = [t_1, \ldots, t_n]$. Let the set of plaintext messages be $\mathcal{P} = \{0, 1\}^n$, let the set of ciphertext messages be $\mathcal{C} = \{0, \ldots, n(p-1)\}$, and let the set of keys be

$$\mathcal{K} = \{(A, p, m, T)\},$$

where A, p, m, and T are constructed as described above. T is public, and p, m and A are secret.

For $K = (A, p, m, T)$, and $[u_1, \ldots, u_n] \in \mathcal{P}$, define

$$e_K(u_1, \ldots, u_n) = \sum_{i=1}^{n} u_i t_i \in \mathcal{C}.$$

For $y \in \mathcal{C}$, define $z = m^{-1}y \bmod p$ and solve Problem 8.2 with sizes $a_1, \ldots, a_n$ and target sum z, obtaining $d_K(y) = [u_1, \ldots, u_n] \in \mathcal{P}$.

The following small example illustrates the encryption and decryption operations in the Merkle-Hellman knapsack cryptosystem.

Example 8.6

Suppose

$$A = [2, 5, 9, 21, 45, 103, 215, 450, 946]$$

is the secret superincreasing list of sizes. Suppose $p = 2003$ and $m = 1289$. Then the public list of sizes is

$$T = [575, 436, 1586, 1030, 1921, 569, 721, 1183, 1570].$$

Now, if the sender wants to encrypt the plaintext $U = [1, 0, 1, 1, 0, 0, 1, 1, 1]$, she computes

$$y = 575 + 1586 + 1030 + 721 + 1183 + 1570 = 6665.$$

When the ciphertext y is received, we first compute

$$z = a^{-1}y \bmod p$$
$$= 317 \times 6665 \bmod 2003$$
$$= 1643.$$

Then we solve the instance of Problem 8.2, with sizes $s_1, s_2, \ldots, s_n$ and target sum z, using Algorithm 8.5. The plaintext $(1, 0, 1, 1, 0, 0, 1, 1, 1)$ is obtained.

$\Box$

By the early 1980s, the Merkle-Hellman knapsack cryptosystem had been broken by Shamir. Shamir was able to use an integer programming algorithm of Lenstra to break the system. This allowed the receivers's trapdoor (or an equivalent trapdoor) to be discovered by a cryptanalyst. This cryptanalyst can decrypt messages exactly as the intended receiver does. To circumvent these attacks on the Merkle-Hellman knapsack cryptosystem other variants were introduced. These variants had the effect of increasing the density of the Subset Sum problem. The *density* of an instance of Problem 8.2, with sizes $A = [a_1, \ldots, a_n]$, is defined by

$$\partial(A) = \frac{n}{\log_2(\max_j a_j)}.$$

If $\partial(A) > 1$, then there will be in general many subsets of the a_is with the same sum. These instances of Problem 8.2 could not be used in a cryptosystem. Consequently the interesting case is when $\partial(A) \leq 1$. Basis reduction can be used to solve almost all instances of Problem 8.2 with $\partial(A)$ sufficiently small.

We begin by reducing Problem 8.2 to Problem 8.1. The method and reasoning is exactly the same as in Section 8.4. However this time the incidence matrix $[a_1, a_2, \ldots, a_n]$ has only one row. The basis is thus

$$M = \begin{bmatrix} 1 & 0 & \cdots & 0 & 0 \\ 0 & 1 & \cdots & 0 & 0 \\ & & \vdots & & \vdots \\ 0 & 0 & \cdots & 1 & 0 \\ a_1 & a_2 & \cdots & a_n & -z \end{bmatrix}. \tag{8.7}$$

Example 8.7

In Example 8.6, the public list of sizes is

$$T = [575, 436, 1586, 1030, 1921, 569, 721, 1183, 1570]$$

and the received ciphertext is $y = 6665$. Thus the lattice we wish to reduce is

$$M = \begin{bmatrix} I & \vec{0} \\ T & -y \end{bmatrix}.$$

Applying Algorithm 8.2, we obtain the basis

$$
M' = \begin{bmatrix}
-2 & 0 & 1 & 0 & 1 & 0 & 1 & -1 & -2 & 1 \\
-1 & 0 & -1 & 1 & -1 & 0 & 0 & -1 & 0 & 1 \\
1 & 0 & 0 & -1 & -1 & -1 & 1 & 0 & -1 & 0 \\
0 & -1 & -2 & -1 & 1 & 0 & 1 & 1 & 0 & -1 \\
0 & 0 & 1 & 1 & 0 & 1 & 0 & -1 & -1 & -2 \\
0 & 1 & 0 & 0 & 2 & 0 & 0 & 0 & 0 & 0 \\
0 & -1 & 0 & 2 & -1 & -1 & 1 & 1 & 1 & 1 \\
0 & 1 & 0 & -1 & 0 & -1 & 1 & 1 & 2 & 0 \\
0 & 0 & 0 & 0 & 0 & 1 & 1 & 0 & 1 & 2 \\
0 & 1 & 0 & 0 & 0 & 1 & 0 & 2 & 0 & 0
\end{bmatrix},
$$

which has the solution

$$
U = [1, 0, 1, 1, 0, 0, 1, 1, 1]
$$

in column 7. In fact U is the original plaintext. ▯

It is interesting that the method of basis reduction breaks the Merkle-Hellman knapsack cryptosystem without determining the multiplier or modulus used.

If the reduced basis obtained by Algorithm 8.2 on the input basis given in Equation 8.7 fails to contain a solution, then further reduction methods such as Algorithm 8.4 can be applied. Alternatively, we can change the form of the basis. A different basis that has been studied is

$$
M = \begin{bmatrix}
1 & 0 & \cdots & 0 & \frac{1}{2} \\
0 & 1 & \cdots & 0 & \frac{1}{2} \\
& & \vdots & & \vdots \\
0 & 0 & \cdots & 1 & \frac{1}{2} \\
\hline
a_1 N & a_2 N & \cdots & a_n N & -zN
\end{bmatrix}, \tag{8.8}
$$

where $N = \lceil \frac{1}{2}\sqrt{n} \,\rceil$. Consider the lattice $\mathcal{L}$ with the basis $M = [\vec{b}_1, \vec{b}_2, \ldots, \vec{b}_n]$ given in Equation 8.8. If $U = [u_1, u_2, \ldots, u_n]$ is a solution to the Subset Sum problem

$$
\sum_{i=1}^{n} u_i a_i = z,
$$

then

$$
\vec{y} = [\vec{y}_1, \ldots, \vec{y}_n, 0] = \sum_{i=1}^{n} u_i \vec{b}_i - \vec{b}_{n+1}
$$

is in $\mathcal{L}$, and $y_i \in \{-\frac{1}{2}, \frac{1}{2}\}$, for $i = 1, 2, \ldots, n$. To recover the solution U from $\vec{y}$ we simply set $u_i = y_i + \frac{1}{2}$ for $i = 1, 2, \ldots, n$. Observe that $\|\vec{y}\| = \frac{1}{2}\sqrt{n}$, and so $\vec{y}$ is a vector of short length in $\mathcal{L}(M)$.

TABLE 8.1
Subset Sum data.

	Basis 8.7		Basis 8.8	
∂	LLL	KR	LLL	KR
0.650	55	64	99	100
0.700	38	53	98	98
0.800	24	31	90	87
0.900	19	15	82	79
0.930	13	20	76	83
0.960	12	16	73	78
0.990	10	17	68	78

If the density of the Subset Sum problem is small, then the size of the a_is are large and consequently most of the vectors in the lattice (with basis (8.7) or (8.8)) will have relatively large length. Therefore, an algorithm that reduces the weight of the basis will have a good possibility of finding a solution. In this chapter we have presented two such algorithms, Algorithm 8.2 and Algorithm 8.4. The success in finding the shortest vector in the lattice with either of these algorithms is not guaranteed.

A *lattice oracle* is an algorithm that is guaranteed to return the shortest vector in the lattice in polynomial time. No such algorithm is known. If such an oracle exists, then it has been shown that the shortest vector in the lattice with basis 8.7 corresponds with high probability to the solution of the Subset Sum problem, whenever the density is less than 0.6463. If we instead use basis 8.8, then the shortest vector corresponds to a solution with high probability when the density is less than 0.9408. Therefore, theoretically, basis (8.8) is superior to basis (8.7). To support this, we give experimental evidence in Table 8.1. For each density listed in Table 8.1 we generated 100 random Subset Sum problems of size 20. These were constructed by choosing 20 random non-negative integers less than $\lfloor 2^{20/\partial} \rfloor$. The target sum was created by choosing a random subset of these 20 integers and then summing the entries. In the table we report for each basis and each reduction algorithm the number of successes in solving the 100 random Subset Sum problems.

8.6 Notes

Section 8.2

Analysis similar to that given in this section can be found in [79, 64, 96, 19].

Section 8.3

Algorithm 8.2 appears in [64] where it was introduced as a method for factoring polynomials with rational coefficients. It is often also called the L^3, or the Lovasz algorithm, and it is a crucial component in many number theoretic algorithms [19].

Section 8.4

The multi-row situation when A is the orbit incidence matrix for constructing designs (see Section 6.6.1) was first investigated by Kreher and Radziszowski [58, 60]. In particular Algorithms 8.3 and 8.4 were first described in [58, 59, 60]. To the best of our knowledge no theoretical or experimental analysis has been obtained in this situation. On the other hand, several thousand new combinatorial designs were discovered using the basis reduction algorithm of Kreher and Radziszowski.

Section 8.5

The Merkle-Hellman knapsack cryptosystem was presented in [74]. This system was broken by Shamir [98], and the "iterated" version of the system was broken by Brickell [11]. The analysis using a lattice oracle and giving the bounds on the density appears in the article [23]. For more information, see the survey article by Brickell and Odlyzko [12] and the *Handbook of Applied Cryptography* [73]. Algorithm 8.3 which first appeared in [58, 59] was also used by Schnorr and Euchner to solve Subset Sum problems in [95].

Exercises

8.1 Give a complete proof of Lemma 8.1.

8.2 Which of the following vectors are in the lattice given in Example 8.1?
  - (a) $[-1, 18]$
  - (b) $[4, 12]$
  - (c) $[1, 6]$
  - (d) $[1, 10]$
  - (e) $[1, -11]$

8.3 Consider the lattice with basis
$$M = \begin{bmatrix} 1 & -2 \\ 3 & 1 \end{bmatrix}$$
  that is displayed in Example 8.1.
  - (a) Compute $\mathsf{wt}(M)$ and $\mathsf{vol}(\mathcal{L})$ and verify Hadamard's inequality for this lattice.
  - (b) Show geometrically (i.e., draw a picture) that for this lattice
$$\mathsf{wt}(M) \geq \mathsf{vol}(\mathcal{L}).$$

8.4 Let $[\vec{b}_1, \vec{b}_2, \ldots, \vec{b}_n]$ be a basis. Give a formal proof that an operation of the form

$$\text{replace } \vec{b}_j \text{ with } \alpha_1 \vec{b}_1 + \alpha_2 \vec{b}_2 + \cdots + \vec{b}_j + \cdots + \alpha_n \vec{b}_n$$

can be obtained by performing a sequence of the following three operations:
   (a) Replace $\vec{b}_i$ with $\vec{b}_i + \vec{b}_j$,
   (b) Replace $\vec{b}_i$ with $-\vec{b}_i + \vec{b}_j$, and
   (c) Replace $\vec{b}_i$ with $\vec{b}_i - \vec{b}_j$, where $i \neq j$.

8.5 Using Algorithm 8.1, work out by hand an orthogonal basis for the lattice spanned by

$$M = \begin{bmatrix} 1 & 1 & 2 \\ 1 & 0 & 1 \\ 1 & 1 & 0 \end{bmatrix}.$$

Check your results by computer. What is the volume of the lattice spanned by the columns of M? Verify Lemma 8.4 for the matrix M.

8.6 Consider the matrix

$$M = \begin{bmatrix} 0 & 2 & 3 & 1 \\ 1 & 0 & -1 & 5 \\ 2 & -2 & 2 & 0 \\ 2 & 2 & 0 & -2 \end{bmatrix}.$$

   (a) Show that M is a reduced basis.
   (b) Verify the inequalities of Theorem 8.8 for the matrix M.

8.7 Use the algorithms in Section 8.4 to construct a Steiner triple system of order 9 that has

$$g = (0, 1, 2)(3, 4, 5)(6, 7, 8)$$

as an automorphism.

8.8 Suppose the Merkle-Hellman knapsack cryptosystem has as its public list of sizes the vector

$$T = [1394, 1256, 1508, 1987, 439, 650, 724, 339, 2303, 810].$$

Suppose an observer discovers that $p = 2503$.
   (a) By trial and error, determine the value a such that the list $a^{-1}T \bmod p$ is a permutation of a superincreasing list.
   (b) Show how the ciphertext 5746 would be decrypted.
   (c) Use basis reduction to decrypt the ciphertext 5746.

8.9 Develop an algorithm similar to Algorithm 8.3 that reduces the weight of the basis by considering combinations of the form

$$\vec{v} = \epsilon_1 \vec{b}_{i_1} + \epsilon_2 \vec{b}_{i_2} + \epsilon_3 \vec{b}_{i_3},$$

where $\epsilon_i = \pm 1$, for $i = 1, 2, 3$.

8.10 For each of the following Subset Sum problems compute their density and use basis reduction to find a solution.
   (a)

$$A = [283615655564068, 796478694573302, 600340146256703,$$

$$732983327534134, 786266787523357, 105515816928335,$$

$$112897627203131, 330057122934813, 1089988300272331,$$

$$1051338601577848, 1109763392717310, 145117009247205,$$

283635625683684, 6217169571139, 909231046365184,

740552083084632, 767717555811633, 222691570662389,

287870530458475, 250604219988445],

$z = 5055299030829558$

(b)

$A = [7960137240, 7503674315, 8975593017, 6982240834, 750319933,$

2263778309, 5779454351, 2189761281, 6377653436, 1899000113,

560590007, 6148611908, 5254132888, 4377585063, 1837007135,

8439676091, 4254195333, 5970662702, 1507562435, 1826255982],

$z = 37987557118$

(c)

$A = [806109, 408997, 1169428, 1011478, 1150062, 1182254, 658173,$

1198146, 1199680, 430790, 774558, 850850, 916096, 1085626,

164865, 288661, 260406, 619265, 1030628, 946958],

$z = 8440889$

Bibliography

[1] E. AARTS AND J.K. LENSTRA. *Local Search in Combinatorial Optimization*, John Wiley & Sons, 1997.

[2] G.E. ANDREWS. *The Theory of Partitions*, Addison-Wesley, 1976.

[3] S. BAASE. *Computer Algorithms, Introduction to Designs and Analysis (Second Edition)*, Addison-Wesley, 1988.

[4] L. BABEL. Finding maximum cliques in arbitrary and in special graphs. *Computing* **15** (1991), 321–341.

[5] E. BALAS AND C.S. YU. Finding a maximum clique in an arbitrary graph. *SIAM Journal of Computing* **5** (1986), 1054–1068.

[6] K.P. BOGART. *Introductory Combinatorics (Second Edition)*, Harcourt, Brace, Jovanovich, 1990.

[7] J.A. BONDY AND U.S.R. MURTY. *Graph Theory with Applications*, American Elsevier, 1976.

[8] R.M. BRADY. Optimization strategies gleaned from biological evolution, *Nature* **317** (1985), 804–806.

[9] G. BRASSARD AND P. BRATLEY. *Algorithmics Theory and Practice*, Prentice-Hall, 1988.

[10] D. BRELAZ. New methods to color vertices of a graph. *Communications of the ACM* **22** (1979), 251–256.

[11] E.F. BRICKELL. Breaking iterated knapsacks. *Lecture Notes in Computer Science* **218** (1986), 342–358. (Advances in Cryptology – CRYPTO '85.)

[12] E.F. BRICKELL AND A.M. ODLYZKO. Cryptanalysis, a survey of recent results. In *Contemporary Cryptology, The Science of Information Integrity*, G.J. Simmons, ed., IEEE Press, 1992, pp. 501–540.

[13] C. BRON AND J. KERBOSCH. Algorithm 457: finding all cliques of an undirected graph H. *Communications of the ACM* **16** (1973), 375–577.

[14] R.A. BRUALDI. *Introductory Combinatorics (Second Edition)*, Prentice-Hall, 1992.

[15] G. BUTLER. *Fundamental Algorithms for Permutation Groups, (Lecture*

Notes in Computer Science, vol. 559), Springer-Verlag, 1991.

[16] P.J. CAMERON. *Combinatorics: Topics, Techniques, Algorithms*, Cambridge University Press, 1994.

[17] V. ČERNY. A thermodynamical approach to the traveling salesman problem. *Journal of Optimization Theory and Applications* **45** (1985), 41–55.

[18] N. CHRISTOFIDES. *Graph Theory: An Algorithmic Approach*, Academic Press, 1975.

[19] N. COHEN. *A Course in Computational Algebraic Number Theory*, Springer-Verlag, 1993.

[20] C.J. COLBOURN AND J.H. DINITZ, EDS. *The CRC Handbook of Combinatorial Designs*, CRC Press, 1996.

[21] J.H. CONWAY AND R.K. GUY. *The Book of Numbers*, Springer-Verlag, 1996.

[22] T.H. CORMEN, C.E. LEISERSON AND R.L. RIVEST. *Introduction to Algorithms*, MIT Press, McGraw-Hill, 1990,

[23] M.J. COSTER, A. JOUX, B.A. LAMACCHIA, A.M. ODLYZKO, C.P. SCHNORR AND J. STERN. Improved low-density subset algorithms, *Computational Complexity* **2** (1992), 111–128.

[24] G.A. CROES. A method for solving traveling salesman problems. *Operations Research* **6** (1958), 791–812.

[25] J.H. DINITZ AND D.R. STINSON. A fast algorithm for finding strong starters. *SIAM Journal on Algebraic and Discrete Methods* **2** (1981), 50–56.

[26] J.D. DIXON AND B. MORTIMER. *Permutation Groups*, Springer-Verlag, 1996.

[27] A.A. EL GAMAL, L.A. HEMACHANDRA, I. SHPERLING AND V.K. WEI. Using simulated annealing to design good codes. *IEEE Transactions on Information Theory* **33** (1987), 116–123.

[28] S. EVEN. *Algorithmic Combinatorics*, MacMillan, 1973.

[29] S. EVEN. *Graph Algorithms*, Computer Science Press, 1979.

[30] T.C. FRENZ AND D.L. KREHER. Enumerating cyclic Steiner systems, *Journal of Computational Mathematics and Computational Computing* **11** (1992), 23–32.

[31] M.R. GAREY AND D.S. JOHNSON. *Computers and Intractibilty: A Guide to the Theory of NP-Completeness*, Freeman, 1979.

[32] I.M. GESSEL AND R.P. STANLEY. Algebraic enumeration. In *Handbook of Combinatorics*, R.L. Graham, M. Grötschel and L. Lovász, eds., Elsevier Science, 1995, pp. 1021–1061.

[33] P.B. GIBBONS. Computational methods in design theory. In *The CRC*

Handbook of Combinatorial Designs, C.J. Colbourn and J.H. Dinitz, eds., CRC Press, 1996, pp. 718–753.

[34] F. GLOVER. Future paths for integer programing and links to artificial intelligence. *Computers and Operations Research* **5** (1986), 533–549.

[35] F. GLOVER AND M. LAGUNA. Tabu Search. In *Modern Heuristic Techniques for Combinatorial Problems*, C.R. Reeves, ed., John Wiley & Sons, 1993.

[36] L.A. GOLDBERG. *Efficient Algorithms for Listing Combinatorial Structures*, Cambridge University Press, 1993.

[37] I.P. GOULDEN AND D.M. JACKSON. *Combinatorial Enumeration*, John Wiley & Sons, 1983.

[38] R.L. GRAHAM, M. GRÖTSCHEL AND L. LOVÁSZ, EDS. *Handbook of Combinatorics*, Elsevier Science B.V., 1995.

[39] P. HANSEN. The steepest ascent mildest descent heuristic for combinatorial programing. *Congress on Numerical Methods in Combinatorial Optimization*, 1986, Capri, Italy.

[40] A. HERTZ AND D. DE WERRA. The tabu search metaheuristic: how we used it. *Annals of Mathematics and Artificial Intelligence* **1** (1991), 111–121.

[41] J.H. HOLLAND. *Adaptation in Natural and Artificial Systems*, University of Michigan Press, 1975.

[42] I.S. HONKALA AND P.R.J. ÖSTERGÅRD. Code Design. In *Local Search in Combinatorial Optimization*, E. Aarts and J.K. Lenstra, eds., John Wiley & Sons, 1997.

[43] E. HOROWITZ AND S. SAHNI. *Fundamentals of Computer Algorithms*, Computer Science Press, 1978.

[44] T.C. HU. *Combinatorial Algorithms*, Addison-Wesley, 1982.

[45] T.R. JENSEN AND B. TOFT. *Graph Coloring Problems*, John Wiley & Sons, 1995.

[46] M. JERRUM. A compact representation for permutation groups. *Journal of Algorithms* **7** (1986), 60–78.

[47] P. JOG, J.Y. SUH AND D.V. GUCHT. The effects of population size, heuristic crossover and local improvement on a genetic algorithm for the travelling salesman problem. In *Proccedings of Third International Conference on Genetic Algorithms*, J.D. Schaffer, ed., Morgan Kaufman, 1989.

[48] D.S. JOHNSON, C.R. ARAGON, L.A. MCGEOGH AND C. SCHEVON. Optimization by simulated annealing: an experimental evaluation. Part I, graph partitioning. *Operations Research* **37** (1989), 865-892.

[49] D.S. JOHNSON AND M.A. TRICK, EDS. *Cliques, Coloring and Satisfiability: Second DIMACS Implementation Challenge*, American Mathemati-

cal Society, 1996.

[50] S.M. JOHNSON. Generation of permutations by adjacent transpositions. *Mathematics of Computation* **17** (1963), 282–285.

[51] R.M. KARP. Reducibility among combinatorial problems. In *Complexity of Computer Computations*, R.E. Miller and J.W. Thatcher, eds., Plenum Press, 1972, pp. 85-103.

[52] B.W. KERNIGHAN AND S. LIN. An efficient heuristic procedure for partitioning graphs. *Bell Systems Technical Journal* **49** (1970), 291–307.

[53] S. KIRKPATRICK, C.D. GELLAT AND M.P. VECCHI. Optimization by simulated annealing. *Science* **220** (1983), 671–680.

[54] D.E. KNUTH. Efficient representation of perm groups. *Combinatorica* **11** (1991), 33–44.

[55] W.L. KOCAY. On writing isomorphism programs. In *Computational and Constructive Design Theory*, W.D. Wallis, ed., Kluwer, 1996, pp. 135–175.

[56] D.C. KOZEN. *The Designs and Analysis of Algorithms*, Springer-Verlag, 1992.

[57] E.S. KRAMER AND D.M. MESNER. *t*-Designs on hypergraphs. *Discrete Mathematics* **15** (1976), 263–296.

[58] D.L. KREHER AND S.P. RADZISZOWSKI. Finding simple *t*-designs by using basis reduction. *Congressus Numerantium* **55** (1986), 235–244.

[59] D.L. KREHER AND S.P. RADZISZOWSKI. Solving subset-sum problems with the L^3 algorithm. *Journal of Combinatorial Mathematics and Combinatorial Computing* **3** (1988), 49–63.

[60] D.L. KREHER AND S.P. RADZISZOWSKI. Constructing 6-$(14, 7, 4)$ designs. *Contemporary Mathematics* **111** (1990), 137–151.

[61] L. KUČERA. *Combinatorial Algorithms*, Adam Hilger, 1990.

[62] J. LAGARIAS AND A. ODLYZKO. Solving low-density subset sum problems. *Journal of the ACM* **32** (1985), 229–246.

[63] E.L. LAWLER, J.K. LENSTRA, A.H.G. RINNOOY KAN AND D.B. SHMOYS. *The Traveling Salesman Problem: A Guided Tour of Combinatorial Optimization*, John Wiley & Sons, 1985.

[64] A.K. LENSTRA, H.W. LENSTRA, JR., AND L. LOVASZ. Factoring polynomials with rational coefficients. *Mathematics Annals* **261** (1982), 515–534.

[65] S. LIN. Computer solutions of the traveling salesman problem. *Bell Systems Technical Journal* **44** (1965), 224–226.

[66] C.C. LINDNER AND C.A. RODGER. *Design Theory*, CRC Press, 1997.

[67] J.H. VAN LINT AND R.M. WILSON. *A Course in Combinatorics*, Cambridge University Press, 1992.

[68] E.M. LUKS. Isomorphism of graphs of bounded valence can be tested in polynomial time. In *Proceedings of the 21st IEEE Symposium on the Foundations of Computer Science,* 1980, pp. 42–49.

[69] H.B. MANN. On orthogonal Latin squares. *Bulletin of the American Mathematical Society* **50** (1950), 418–423.

[70] B.D. McKAY. *Naughty User's Guide (version 1.5),* Computer Science Department, Australian National University.

[71] B.D. McKAY. Practical graph isomorphism. *Congressus Numerantium* **30** (1981), 45–87.

[72] K. MEHLHORN. *Data Structures and Algorithms 2: Graph Algorithms and NP-Completeness,* Springer-Verlag, 1984.

[73] A.J. MENEZES, P.C. VAN OORSCHOT AND S.A. VANSTONE. *Handbook of Applied Cryptography,* CRC Press, 1996.

[74] R.C. MERKLE AND M.E. HELLMAN. Hiding information and signatures in trapdoor knapsacks. *IEEE Transactions on Information Theory,* **24** (1978), 525–530.

[75] N. METROPOLIS, A.W. ROSENBLUTH, A.H. TELLER AND E. TELLER. Equation of state calculation by fast computing machines. *Journal of Chemical Physics* **21** (1953), 1087–1091.

[76] H. MÜHLENBEIN, M. GORGES-SHCLEUTER AND O. KRÄMER. New solutions to the mapping problem of parallel systems — the evolution approach. *Parallel Computing* **4** (1987), 269–279.

[77] H. MÜHLENBEIN, M. GORGES-SHCLEUTER AND O. KRÄMER. Evolution algorithms in combinatorial optimization. *Parallel Computing* **7** (1988), 65-85.

[78] W. MYRVOLD, T. PRSA AND N. WALKER. A dynamic programming approach for testing clique algorithms, preprint (1997).

[79] G.L. NEMHAUSER AND L.A. WOLSEY. *Integer and Combinatorial Optimization,* John Wiley & Sons, 1988.

[80] A. NIJENHUIS AND H.S. WILF. *Combinatorial Algorithms (Second Edition),* Academic Press, 1978.

[81] T.A.J. NICHOLSON. A sequential method for discrete optimization problems and its application to the assignment, traveling salesman and three scheduling problems. *Journal of the Institute of Mathematics and its Applications* **13** (1965), 362–375.

[82] C.H. PAPADIMITRIOU. *Computational Complexity,* Addison-Wesley, 1994.

[83] C.H. PAPADIMITRIOU AND K. STEIGLITZ. *Combinatorial Optimization: Algorithms and Complexity,* Prentice-Hall, 1982.

[84] P.W. PURDOM, JR. AND C.A. BROWN. *The Analysis of Algorithms,* Holt,

Reinhart and Winston, 1985.

[85] N.J. RADCLIFFE. Equivalence class analysis of genetic algorithms. *Complex Systems* 5 (1991), 183–205.

[86] N.J. RADCLIFFE AND P. SURRY. Formae and the variance of fitness. In *Foundations of Genetic Algorithms 3*, D. Whitley and M. Vose, eds., Morgan Kaufman, 1995.

[87] V.J. RAYWARD-SMITH, I.H. OSMAN, C.R. REEVES AND G.D. SMITH, EDS. *Modern Heuristic Search Methods*, John Wiley & Sons, 1996.

[88] R.C. READ. The coding of various kinds of unlabeled trees. In *Graph Theory and Computing*, R.C. Read, ed., Academic Press, 1972, pp. 153–182.

[89] C.R. REEVES, ED. *Modern Heuristic Techniques for Combinatorial Problems*, John Wiley & Sons, 1993.

[90] E.M. REINGOLD, J. NIEVERGELT AND N. DEO. *Combinatorial Algorithms*, Prentice-Hall, 1977.

[91] S. REITER AND G. SHERMAN. Discrete Optimizing. *Journal of the Society for Industrial and Applied Mathematics* 13 (1965), 864–889.

[92] F.S. ROBERTS. *Applied Combinatorics*, Prentice-Hall, 1984.

[93] J.J. ROTMAN. *An Introduction to the Theory of Groups*, Springer Verlag, 1995.

[94] C. SAVAGE. A survey of combinatorial Gray codes. *SIAM Review*, 39 (1997), 605–629.

[95] C.P. SCHNORR AND M. EUCHNER. Lattice basis reduction: improved practical algorithms and solving subset sum problems. *Lecture Notes in Computer Science* 529 (1991), 68–85. (Fundamentals of Computation Theory – FCT '91.)

[96] A. SCHRIJVER. *Theory of Linear and Integer Programming*, John Wiley & Sons, 1986.

[97] R. SEDGEWICK. *Algorithms (Second Edition)*, Addison-Wesley, 1988.

[98] A. SHAMIR. A polynomial-time algorithm for breaking the basic Merkle-Hellman cryptosystem. *IEEE Transactions on Information Theory* 30 (1984), 699–704.

[99] C.C. SIMS. Computational methods for permutation groups. In *Computational Problems in Abstract Algebra*, Pergamon Press, 1970, pp. 169–184.

[100] R.P. STANLEY. *Enumerative Combinatorics, Volume 2*, Cambridge University Press, 1998.

[101] D. STANTON AND D. WHITE. *Constructive Combinatorics*, Springer-Verlag, 1986.

[102] D.R. STINSON. Hill-climbing algorithms for the construction of combinatorial designs. *Annals of Discrete Mathematics* 26 (1985), 321–334.

[103] D.R. STINSON. *An Introduction to the Designs and Analysis of Algorithms (Second Edition)*, Charles Babbage Research Centre, 1987.

[104] D.R. STINSON. *Cryptography Theory and Practice*, CRC Press, 1995.

[105] H.J. STRAIGHT. *Combinatorics: An Invitation*, Brooks/Cole, 1993.

[106] H.F. TROTTER. PERM (Algorithm 115). *Communications of the ACM* **5** (1962), 434–435.

[107] A. TUCKER. *Applied Combinatorics (Third Edition)*, John Wiley & Sons, 1995.

[108] P. VAN LAARHOVEN AND E.K.L. AARTS. *Simulated Annealing: Theory and Applications*. Kluwer Academic Publishers, 1988.

[109] W.D. WALLIS. *Combinatorial Designs*, Marcel Dekker, 1988.

[110] W.D. WALLIS, ED. *Computational and Constructive Design Theory*, Kluwer Academic Publishers, 1996.

[111] M.B. WELLS. *Elements of Combinatorial Computing*, Pergamon Press, 1971.

[112] D.B. WEST. *Introduction to Graph Theory*, Prentice-Hall, 1996.

[113] H.S. WILF. *Algorithms and Complexity*, Prentice-Hall, 1986.

[114] H.S. WILF. *Combinatorial Algorithms: An Update*, SIAM, 1989.

[115] S. GILL WILLIAMSON. *Combinatorics for Computer Science*, Computer Science Press, 1985.

Algorithm Index

319

Problem Index

Index